INSTITUTE OF HUMAN RELATIONS

YALE UNIVERSITY

MATHEMATICO-DEDUCTIVE THEORY OF ROTE LEARNING

MATHEMATICO-DEDUCTIVE THEORY OF ROTE LEARNING

A STUDY IN SCIENTIFIC METHODOLOGY

BY

CLARK L. HULL
CARL I. HOVLAND
ROBERT T. ROSS
MARSHALL HALL
DONALD T. PERKINS
FREDERIC B. FITCH

GREENWOOD PRESS, PUBLISHERS
WESTPORT, CONNECTICUT

Reprinted by permission
of Yale University Press

First Greenwood Reprinting 1970

Library of Congress Catalogue Card Number 74-95126

SBN 8371-3126-X

Printed in the United States of America

CONTENTS

FOREWORD

The Institute of Human Relations has supported the preparation of this volume mainly because it furnishes a demonstration of the use of the logico-empirical method of science in psychology. The method has been developed over a period of years by logicians, mathematicians, physicists, and others but has not until recently been successfully employed in the biological or in the psychological sciences. In a volume entitled *The Axiomatic Method in Biology*, J.H. Woodger has shown how it may be applied to certain problems in biology. In the present volume by Professor Hull and his associates its possible uses in psychology are clearly demonstrated. Other psychologists, notably Thurstone and Lewin, have also used this method.

This volume also illustrates one of the ways in which coöperative research may be effectively carried on. The process that was employed is quite analogous to that used in the planning and construction of a building. The senior author has acted as the planning and supervising architect while his collaborators have checked the plans and supplied much of the technical skill necessary for their execution. The work has been carefully coördinated throughout with the result that the end product is a unified whole.

The volume illustrates still a third principle which is important in the work of the Institute of Human Relations. It is that two or more scientific systems may be put together into a larger and more comprehensive system. The authors have shown that some of the postulates in this system of *rote learning* are identical with some of the principles of *conditioned responses*. In another publication Professor Hull has demonstrated further that at least some of the empirical facts about *trial and error learning* can be deductively predicted from a set of postulates that overlap those of *conditioned responses* and of *rote learning*. It thus appears that three kinds of learning - rote, conditioned response, and trial and error - which have been regarded by some psychologists as fundamentally different will probably turn out to be only different aspects or manifestations of the same underlying principles. By extending this procedure, other forms of learning which are characterized in the textbooks of psychology as "insight," "suggestion," "imitation," and so on, may eventually be encompassed in a system which will cover a wide range of social behavior.

Mark A. May, *Director,*
Institute of Human Relations

December, 1939

PREFACE

The present monograph originated during the summer of 1931 in a small class exercise prepared for the senior author by William M. Lepley, then a graduate student at Pennsylvania State College. The exercise suggested the presumptive identity of the remote excitatory tendencies in rote learning reported by Ebbinghaus and the delayed or trace conditioned reflexes reported by Pavlov. In due course Dr. Lepley published his ingenious idea, with suitable elaboration.

Taking Dr. Lepley's hypothesis as a point of departure, the senior author of the present monograph published in 1935, as an illustration of formal theoretical methodology believed suitable for psychology, a small rote-learning system consisting of eleven theorems. The method of formally deriving these theorems from the definitions and postulates of the system was analogous to that used in geometry. This attempt at rigorous theory revealed serious defects both as to the postulates employed and as to the versatility of the geometrical methodology for mediating quantitative deductions in the field of behavior. These evident inadequacies gave rise (1) to an effort so to recast the postulates and concepts as to bring them into line with the known facts of rote learning, and (2) to an attempt to utilize the great resources of ordinary mathematics in the derivation of the theorems of a revised system. After working out the mathematical proofs of the first five theorems, it was necessary to seek expert mathematical assistance, first from Robert T. Ross, later from Marshall Hall, and finally from Donald T. Perkins, as each in turn was called from the Yale campus. Meanwhile, numerous conversations with Professor J.H. Woodger, of the University of London, had convinced us of the indispensability of the methods and symbolism of symbolic logic for the rigorous formulation of definitions and postulates in any genuinely scientific system. Accordingly, Frederic B. Fitch was induced to participate in the enterprise by supplementing with symbolic logic the work of the mathematicians. Finally, after a considerable number of mathematically proved but as yet unverified theorems had become available, Carl I. Hovland became interested in experimentally checking their empirical soundness. This latter work has since proceeded with great ingenuity and facility, and is still under way.

While it is believed by the authors that the present monograph contributes to an understanding of the learning processes, it is judged that its chief value consists in the large-scale pioneering demonstration of the logico-empirical methodology in the field of behavior. Moreover, the monograph is also illustrative of a quite different type of methodology in science - that of the coordination of the efforts of a number of specialists in a scientific enterprise which would have been quite impossible of execution by any one of them alone.

This latter methodology may be illustrated by the procedure employed in deriving the proofs of the theorems and corollaries: The senior author, on the basis of a careful preliminary behavioral analysis, would hand the mathematician a tentative set of theorem propositions, with the request that they be derived mathematically from the set of behavior postulates and definitions previously formulated. In due course the detailed proofs would be turned in. Occasionally, however, one of the proposed propositions would be rejected by the mathematician as inconsistent with the postulates. In such cases the defective proposition would be recast to agree with the postulates, after which the proof would be worked out for the revised theorem proposition the same as for the others.

While there has been much work done in conferences and otherwise which cannot be adequately allocated, it is possible to state approximately the portions of the present monograph for which the several authors are responsible. Dr. Ross rendered invaluable aid in the early attempts to give the postulates a mathematical formulation, though he cannot be held responsible for any defects in their final state. Dr. Ross and Dr. Hall each independently reworked the proofs of Theorems I to V inclusive. Dr. Hall derived the proofs of Theorems XXI to LIV inclusive, together with those of a number of other theorems, the proofs of which are not included in the present monograph. Dr. Perkins derived the proofs of Theorems VI to XX inclusive, together with the proofs of practically all of the corollaries in the monograph. Dr. Fitch derived the symbolic logic formulations of the definitions and postulates, and the illustrative proof for one corollary. The senior author, with much assistance from the other authors, wrote the introductory sections of the monograph, the concluding section, the verbal formulations of the definitional, postulational, theorem, and corollary propositions, and the B sections of all the postulates, theorems, and corollaries. Dr. Hovland wrote the C sections attached to most of the corollaries. Professor Mark A. May, in his capacity as Director of the Institute of Human Relations, aided greatly by material and moral support. Mr. Robert A. Lorenzini drew the numerous diagrams of the monograph. Miss Ruth Hays prepared the immensely difficult Vari-typer manuscript of which the present text is a photo-offset reproduction.

Scientific theory proceeds by trial-and-error, by a series of successive approximations. The present monograph is the second systematic trial made in this field. It has already revealed to the authors both defects in the postulate set here employed and promising new vistas as yet unworked; even now we can see rather clearly the nature of the next approximation. That so many specialists working on this complexly integrated enterprise should always have understood one another is too much to hope. Under these circumstances it is inevitable that errors have occurred, though we have striven constantly to avoid them. Perhaps such errors as may be found will be forgiven us in view of the frankly pioneering nature of the work.

C.L.H.

December, 1939

CONCERNING SCIENTIFIC THEORETICAL METHODOLOGY

Introduction

'It is quite clear that the beginnings of science evolve from the activities of everyday life. In this humble setting there gradually accumulate a body of observations, on the one hand, and simultaneously a parallel body of ideas or interpretations of these observations, on the other. Thus, even from the very first, scientific development has involved an intimate interaction between observations and ideas. Actually, observations give rise to ideas, and ideas lead to further observations. The orderly arrangement of the observations constitutes the empirical component of science, and the logical systematization of the ideas concerning these observations constitutes the theoretical component. Thus comes about the logico-empirical nature of well-developed sciences.

As observations and related ideas multiply, certain observations demanded by ideas can not be made except under special conditions. The activities involved in the creation of these conditions constitute the substance of experimentation. It is true that occasionally, even in advanced stages of scientific development, an experiment may be performed through mere curiosity as to what will be the outcome. Indeed, when experiments and observations are very simple, ideas may play a minimal rôle in their design and execution. As a general rule, however, theoretical considerations play an increasing rôle in the design of experiments as the complexity of the experiment increases and as the science develops. The rôle of ideas in the conduct of an experiment, as distinguished from its design, is much less prominent.

The more precise scientific experiments are usually carried out in specially constructed laboratories, with the aid of delicate apparatus and elaborate procedures for making exact measurements, for carrying out control experiments, and for the statistical evaluation of results secured. Likewise, the theoretical component of scientific methodology has its own characteristic procedures and problems. Since the present monograph is designed in part as a study in scientific methodology with special emphasis on the side of systematic theory, it has been thought desirable to include a brief summary of some of the more important aspects of the methodology of theoretical systematization.[1]

1. The remarks on scientific methodology which follow are intended to be neither complete nor exhaustive, but rather to emphasize a number of points which have appeared to be of special importance in the light of the experience of constructing the present system. Persons wishing a thorough discussion concerning scientific methodology should consult a modern systematic work on logic and scientific method. The following are recommended: John Dewey, *Logic, The Theory of Inquiry*; M.R. Cohen and Ernest Nagel, *An Introduction to Logic and Scientific Method*.

Systematic natural-science theory properly consists of three distinguishable portions: (1) a set of definitions of the critical (indispensable) terms employed in the system; (2) a set of postulates concerning presumptive relationships among the natural phenomena represented by the terms; and (3) a hierarchy of interlocking theorems ultimately derived from the postulates by a rigorous logical process. The function of the definitions of critical terms of a natural-science system is to make clear the relationship of these terms (and so of the system as a whole) to the relevant part of nature which the system concerns. The postulates of a theoretical system, if valid, are presumptive natural laws. The hierarchy of theorems, corollaries, etc., makes up the bulk of the system. Scientific theorems are "if - then" statements; i.e., they ordinarily state in effect that *if* such and such antecedent conditions exist, *then* such and such consequences will follow. The validity of a scientific theoretical system is dependent upon the extent of the agreement between the theorems, on the one hand, and observations of the natural phenomenon to which they refer, on the other.

Definitions

So far as the mere logic of a scientific system is concerned, the critical terms might perfectly well be replaced by a set of purely arbitrary signs, such as X's, Y's, and Z's. If this were done a person unacquainted with the natural phenomena represented by the signs could deduce the formal theorems of the system but would be completely unable to determine their meaning or whether they agree with known facts. It is only as the critical terms or signs of a system are associated with the phenomena which they represent that it becomes possible to determine either that a given postulate is supposed to be operative in a given natural situation, or that a theorem which results from the operation of certain postulates in a given situation agrees with empirically observed fact. In scientific theory both these conditions are absolutely necessary.

The definitions which are given for purposes of ordinary exposition do not satisfy the more exacting requirements of rigorous systematization. In this connection Dewey remarks,

"While all language or symbol-meanings are what they are as parts of a system, it does not follow that they have been determined on the basis of their fitness to be such members of a system; much less on the basis of their membership in a comprehensive system. The system may be simply the language in common use. Its meanings hang together not in virtue of their examined relationship to one another, but because they are current in the same set of group habits and expectations. They hang together because of group activities, group interests, customs and institutions. Scientific language, on the other hand, is subject to a test over and above this criterion. Each meaning that enters into the language is expressly determined in its relation to other members of the language system. In all reasoning or ordered discourse this criterion takes precedence over that instituted by connection with cultural habits. The resulting difference in the two types of language-meanings fundamentally fixes the difference between what is called common sense and what is called science." (*5*, 49-50.)

In precise definition it is necessary to divide the terms into two groups. Since definition of a term always requires the use of one or more other terms, it follows that in any formal system there must be a number of terms which are not really defined at all. Such concepts are called *undefined notions*; their meanings are merely explained or elucidated in ordinary language. Once an adequate set of such primitive concepts or undefined notions is available, the remaining critical terms employed in the system may presumably be defined with precision.

In the past, ordinary word languages were employed in the definitions associated with deductive systems; this was true of the Greek formulations of geometrical postulates, and of the eight definitions of Newton's *Principia* (*45*,1). With care, ordinary words may serve fairly well, and it is probable that in the early stages of the formulation of any natural-science theory it is economical to use this medium. Moreover, there seems no other way in which to elucidate the meaning of the undefined notions wherever the thing represented by the term can not be brought into the physical presence of the reader and directly pointed out. But once the undefined terms or signs have been elucidated, the elaborate methodology of symbolic logic (*66*,26) is available for effecting elegant and unambiguous definitions.

At this point we meet the interesting problem of the definition of terms representing unobservables; familiar physical concepts of this sort are: *gravity, energy, molecule, atom,* and *electron*. In the present system we have the parallel problem of defining such terms (p. 22 ff.) as *stimulus trace, excitatory potential* (E), *inhibitory potential* (I), *reaction threshold* (*ℓ*), and so on. Since in any truly scientific system, all unobservables must be linked to one or more observables by unambiguous logical relationships, it would seem that they should be definable by means of such relationships. An attempt to do this was made in the construction of the present system, but it was abandoned when it was found to involve a restatement of a good share of the postulates of the system in the definition of each term. Possibly it is for this reason that signs representing unobservables are usually placed among the undefined notions. This practice has been followed in the present system.

In this connection it is to be noted that unobservables fall into two distinguishable classes: (1) those which, if conditions were as favorable as they might conceivably be, could really be observed, and (2) those which are inherently unobservable. An example of the former is the class of blood vessels known as capillaries. At the time when Harvey had to postulate such entities in order to complete his theory of the circulation of the blood (*15*,81) the "anastamosis of veins and arteries" (*5*) was unobservable because sufficiently powerful microscopes were not available. Here we meet the puzzling problem presented by the hypothetical case in which a theoretically useful postulated entity of the first class turns out to be non-existent in the observational sense when conditions chance to become suitable for such observation. In such an event it would seem reasonable to retain the concept as long as it continues to mediate all other empirically valid theorems; in that case it would pass substantially into the status of the second group of concepts, i.e., it would be regarded as a purely symbolic construct.

It is important to note that a meticulous definition, however perfect from a logical point of view, does not guarantee the scientific validity of the concept defined; this can only be determined by empirical trial. A term to be at all useful ordinarily represents a class or group of phenomena all of which under specified conditions behave according to the same natural principles or laws. If it is discovered empirically that some of the individual phenomena included within a definition behave according to a different set of principles or laws, the concept must either be abandoned or redefined. It thus appears that even definitions must ultimately be validated by observation; they are not subject to the mere whim or arbitrary will of the theorist.

Postulates

At the outset of the consideration of the postulates of a scientific system, it should be observed that there are two distinguishable classes of such assumptions, though as a rule only one of these is discussed as such. The postulates which are usually, and properly, under active discussion in a scientific system are those which purport to be natural laws. Thus Newton, in his *Principia*, presents only three postulates, his three laws of motion (*45*, 13). In the derivation of his theorems, however, he utilizes without specifically listing them a very complex set of mathematical (logical) assumptions. These latter assumptions are examples of the second type of postulates necessarily assumed in the construction of any scientific system. Such logical assumptions are usually not mentioned in the evolution of scientific systems because, for one thing, the sophisticated reader is supposed to be familiar with them.

In works on logic, three criteria are ordinarily given for the postulates of any logical system. They are that the postulates shall be (1) as *few* as possible, (2) *consistent* with each other, and (3) *sufficient* to mediate the deduction, as theorems, of all relevant facts (*66*). This formulation in relation to a growing science represents an ideal, a goal or state of things to be striven for and to be approximated as closely as possible; it emphatically is not the state of things fully attained during the growth of a system, particularly during the early stages of such growth.[2] It is only in theology and in metaphysics generally that finalistic or closed systems are put forward. Scientific systems develop

2. See Dewey (*5*). In this work Dewey goes into these problems in convincing detail.

by a complex trial-and-error process, by an elaborate series of successive approximations.

Because of the growth character of scientific systems, it is impossible to state in advance how many postulates will be required. In case it is later found possible to derive one postulate from a combination of others, the situation is remedied by simply inserting the derived postulate among the theorems, thereby reducing the number of postulates by one. Possibly because it is still in process of rapid development, the number of scientific postulates at the base of even our greatest scientific systematization, mathematical physics, is not precisely known. This has been estimated by W.M. Malisoff as 39.[3] The theoretical system concerning rote learning presented in the present monograph has 18 stated postulates, but further developments now clearly in view will certainly increase this number.

The great reason why inconsistent postulates cannot be tolerated in a scientific theoretical system is that they would lead to different theorems concerning the outcome of the same dynamic situation. Unfortunately, it is not always possible by a mere inspection of a set of postulates to tell in advance whether or not they will permit the generation of conflicting theorems. About the only thing to do in the actual construction of a scientific system is to be always on the alert for such an eventuality. In case such evidence of inconsistency should actually occur, the remedy would be to set up the conditions presupposed in common by the inconsistent theorems, and determine which (if either) of the two alternative theoretical outcomes turns out to agree with observation. The postulate which mediated the theorem shown by this procedure to be false should be discarded, or at least be recast in such a way as to eliminate the false implication.

The principle of sufficiency is the critical consideration in evaluating the postulates of a scientific system. In effect this principle means that the postulates must be sufficient to permit the derivation of theorems which will state the nature of the outcome of all the dynamic situations possible of combination from the conditions implicit in the several postulates as those under which each is operative.

It is even more difficult in the case of the principle of sufficiency than in that of consistency to determine by mere inspection whether or not a postulate set is adequate. Ordinarily this can only be determined by patiently deriving the theorems one by one and observing whether agreement does or does not exist. When a theorem is found to disagree with fact (assuming the logic to be sound) it inevitably means that the postulate set actually involved in its derivation must be revised in some sense until agreement is reached. Among the causes of disagreement between a theorem and observation are: the invalidity of at least one postulate; the action of some principle or principles not yet known and so not included in the postulate set; and the entrance of a factor into the empirical situation which, while definitely postulated, was not recognized as active

3. In a private communication, cited with permission of the author.

in the particular situation. In case postulational error is suspected, the suspicion falls more or less over the entire group of postulates involved in the derivation of the "sour" theorem. There seems to be no simple formula for determining which, or how many, postulates may be in error, though the intimate knowledge both of the facts and of his system is likely to suggest to the investigator where the trouble lies. The same general rule seems to apply in the case of a suspected new principle. The question of the surreptitious entrance of a principle, recognized by the system, into the empirical dynamic situation (or the failure to be active of a principle which was supposed to be so) while in part a matter to be decided by the sagacity of the experimentalist may itself also turn out to be a matter of the faulty formulation of the conditions under which the relevant postulates operate. In the first case the conditions laid down may be too wide; in the second, too narrow. It is evident, however, that the establishment of the validity of postulates, as sets and individually, is a gradual matter - one of successive approximations and usually one of slow growth.[4]

It is relatively easy, then, to show that *something* about a scientific situation involving theory is invalid. This requires the disagreement of only one of the theorems of a system with an observation, though such disagreement may represent defect in the experiment which led to the observation rather than defect in the system. The establishment of the *validity* of a scientific system is, on the other hand, exceedingly difficult. Indeed, it seems to be quite generally agreed that the establishment of the absolute validity of a postulate set or even of an individual postulate is an impossibility.[5] Though principles are the most important products of scientific effort, apparently the most that can be attained in determining their validity is to build up for them a favorable presumption, or probability, of impressive magnitude.

What is meant by "favorable presumption or probability" may perhaps best be explained by assuming a definitely artificial situation: suppose that by some miracle a scientist should come into the possession of a set of postulates none of which had ever been employed, but which was known to satisfy the logical criterion of yielding large numbers of empirically testable theorems. Suppose, further, that a very large number of such theorems should be deducible by special automatic logical calculation machines, all theorems to be turned over to the scientist at once. Then these theorems would be placed in a box, thoroughly mixed, drawn out one at a time, and compared with empirical fact. Assuming that no failures of agreement occurred for a long time, it would be proper to say that each succeeding agreement would in-

4. The gradual growth of physical theory, as well as its present inadequacies, is shown by Einstein and Infeld (*9*), and by Reiche (*51*). It is evident that as yet no theoretical system has been able to reach perfection, despite a wide range of most impressive achievements.

5. In this connection consult Dewey (*5*). Nagel has a very concise treatment of the subject (*44*, especially p. 60 *ff.*).

crease the probability that the next drawing from the box would also result in an agreement, just as each successive uninterrupted drawing of white marbles from a large box suspected of containing some black marbles would increase progressively the probability that the next drawing would also yield a white marble. But, just as the probability of drawing a white marble will always lack something of certainty even with the best conceivable score, so the validation of scientific principles must always lack something of being complete. Theoretical "truth" appears in the last analysis to be a matter of greater or less probability. It is consoling to know that this probability frequently becomes very high indeed.

Despite much belief to the contrary, it seems likely that logical (mathematical) principles are essentially the same in their mode of validation; they appear to be merely rules of symbolic manipulation which have been found by trial in a great variety of situations to mediate the deduction of existential sequels verified by observation. Thus logic in science is conceived to be primarily a tool or instrument useful for the derivation of dependable expectations regarding the outcome of dynamic situations. Except for occasional chance successes, it requires sound rules of deduction, as well as sound empirical postulates, to produce sound theorems. By the same token, each observationally confirmed theorem increases the justified confidence in the logical rules which mediated the deduction, as well as in the "empirical" postulates themselves. The rules of logic are more dependable, and consequently less subject to debate, presumably because they have survived a much longer and more exacting period of trial than is the case with most scientific postulates. Probably it is because of the widespread and relatively unquestioned acceptance of the ordinary logical assumptions, and because they come to each individual investigator ready-made, and usually without any appended history, that logical principles are so frequently regarded with a kind of religious awe as a subtle distillation of the human spirit; that they are regarded as never having been, and as never to be, subjected to the usual tests of validity applicable to ordinary scientific principles.[6]

It is probable that all scientific principles must find ultimate validation in this somewhat roundabout and indirect manner, though at first sight it might seem that some postulates (natural laws) can be determined completely by direct experiment. Take, for example, Postulate 13 (p. 67) of the present system, which purports to state the principle of the diminution of inhibitory potential. Ellson (*10*) has plotted this law to a first approximation from the behavior of albino rats in a situation where the habit of pressing a simple bar to secure food suffered experimental extinction. There emerged from the experiment the general shape of the gradient involved, and a suggestion as to certain constants. Even so, there still remains the problem of the

6. See Dewey (*5*), especially page 157 *ff*.

generality of the law: does it apply to rats in other situations? Does it apply to higher mammals as well as to rats? Even if it should be found to hold for human subjects in conditioned reflex situations, this does not make it certain that it will hold for the "inhibitory potential" supposed to be generated in rote-learning situations. Ultimately the answer to such inevitable questions as to the *generality* even of carefully determined experimental "laws," must be left to systematic trial and error. Many of these may permit relatively direct determination, as in the case of Ellson's experiment, but frequently the law may be operating in such a complex situation that the shape of the gradient, and even its existence, can only be determined very indirectly.[7]

Since it appears probable that everything which exists at all in nature exists in some amount, it would seem that the ultimate form of all scientific postulates should be quantitative. Nevertheless it is a fact that many scientific principles, at least when first stated, are qualitative or, at most, only quasi-quantitative. For example, Pavlov states (*47*, 58) that with the passage of time following experimental extinction, an extinguished habit will recover its original strength. A step in the direction of quantification would have been to determine experimentally that recovery is more rapid at first than later. Pavlov publishes evidence from which such a statement might have been drawn. Such a formulation of the law of spontaneous recovery would correspond rather accurately in stage of development to the formulation of the law of falling bodies previous to the time of Galileo.

The complete quantification of a natural law ordinarily involves three things. The first is a precise statement of one variable as a function of at least one other variable. This normally involves the use of an equation, whose complexity is dependent upon the relationship involved. For example, Ellson found as a first approximation that the law of spontaneous recovery with rats was expressible as a function of time by means of a relatively simple exponential equation (*10*, 357). Secondly, there is ordinarily involved the empirical determination of the values of at least one constant which, while in part a function of the units employed in the measurement of the two variables involved, is also in part dependent upon, and inherent in, the phenomena concerned. Thus we have the simple equation for accelerated motion due to gravity:

$$s = \frac{1}{2} gt^2$$

7. The situation here sketched appears in the natural sciences exactly as in the so-called social sciences. The principle that gravity operates according to a gradient such that objects attract each other inversely as the square of the distance separating their centers of gravity, was first detected in relatively simple astronomical situations. But then the question arose as to whether this is true of terrestrial objects. Experiment verified this expectation. Still later the question arose as to whether the law is general enough to hold for unobservable entities such as molecules, atoms, electrons, protons, neutrons, *etc.* The indirect evidence, which alone is available in these latter situations, seems to point to a negative (*51*).

in which g is a constant. In case t is taken in seconds and s in feet, then g is approximately 32.16, depending somewhat upon the place where the determination is made. The corresponding equation tentatively being tried out for the law of spontaneous recovery is (see p. 67):

$$I_n(t) = I_n e^{-dt}$$

in which e is the mathematical constant 2.718 used as the basis of the Napierian logarithms;[8] t is the time in hours since the termination of the extinction; and d is a constant of the type here under discussion. Unfortunately the value of d has not yet been determined empirically. Serious attempts have, however, been made to determine empirically certain other constants involved in the present system, such as F, K, L, and σ_L (see Problems I and VIII, pp. 103 ff. and 170 ff.)

It is inevitable that if genuine measurement is to occur, there must be available a unit of measurement. This presents a peculiarly difficult problem where the entity to be measured is an unobservable. In the present system this problem has been solved in about the same manner that similar problems are solved in the physical sciences. In physics, energy (an unobservable) is measured *indirectly* by means of the equation,

$$\text{Energy} = \frac{1}{2} Mv^2$$

where the values in the right-hand member of the equation are all ultimately measurable in terms of ordinary objective scales such as those employed in the measurement of space and time. In the present system the unobservable "excitatory potential" (E) is measured in units of ΔE, where ΔE is defined as the increment of excitatory potential to the evocation of a particular syllable reaction by one syllable-presentation cycle (p. 36). Following this, ΔI, L, and σ_L are all determined indirectly in terms of ΔE (see Problems I and VIII).

The matter of the mode of presentation of scientific postulates is of great importance. It is clear that the choice in this respect is limited to a certain extent by the available knowledge concerning the postulate. If the law is known only qualitatively it is usually stated in ordinary language. Such purely qualitative postulates may, however, be stated with much greater precision by means of the symbolism developed by symbolic logic. Quasi-quantitative postulates may also be stated in ordinary word language and, alternatively, in terms of symbolic logic, though sometimes a mathematical notation, usually involving $<$'s, $>$'s, etc., may advantageously be employed. But in the case where a fully developed natural law is involved, the method always to be preferred is a mathematical equation. Even when the substance of the postulates is available in the form of mathematical equations, symbolic logic may be of great value in supplementing the merely mathematical expressions, mainly, perhaps, by aiding in the precise statement of the exact conditions under which the law stated in the postulate is supposed to operate. The present

8. Any other arbitrary value, such as 10, might here be used quite as well except that mathematicians have a fondness for the value, e (2.718) because of its great advantages in other situations.

system furnishes illustration of such use of symbolic logic. In the physical sciences these conditions, perhaps because they are so well known, are ordinarily not stated even in words.

Theorems

The theorems,[9] together with their proofs, normally make up the main bulk of any theoretical system. In the discourse of informal exposition there is rarely any attempt at a clear statement of definitions, of postulates, or of the logical steps whereby conclusions are reached. Too often, indeed, a writer has given very little attention to these critical matters; such a writer when questioned is usually quite unable to say whether his system is based on two or on twenty postulates. When a writer does not himself know what assumptions he has made, the deductions are open to all sorts of fallacies. And while the first draft of a theory might very well be informal in nature, it should later be reworked in some formal manner to facilitate the removal of presumptive fallacies. Unfortunately, for most readers the presentation of formally logical proofs is the most repellent form of exposition. But even if it be thought desirable to employ an informal mode of presentation, the exposition can, for purposes of publication, be translated back into the usual mode. The logical interlude will almost certainly repay the effort involved through enabling the thinker to detect a number of serious fallacies. Such an excursion into logic is likely to prove a wholesome, even if a somewhat disconcerting, exercise.

Probably the simplest and most natural formalization of theory employing the ordinary word language, and available to everyone without special training, is that traditional in geometry (*30*; *31*). This procedure has the great virtue of leading the theorizer at least to attempt definitions of his terms, explicit statements of his postulates, and coherent sequences in the formal steps according to which his theorems are mediated from logically preceding propositions. While actually an extremely imperfect logical medium and by no means a guarantee against serious fallacies, the method employed in geometry is nevertheless an incomparable advance over informal expository discourse. It is especially useful in revealing to the thinker unstated premises which he is tacitly employing without realization.

The technique of symbolic logic appears to be a singularly elegant and precise tool for deriving the implications of postulate sets of the most diverse sorts, particularly where only qualitative or quasi-quantitative postulates are available. While it is true that surprisingly little use has been made of symbolic logic in the construction of dynamic natural-science systems, it would seem that it has great potentialities. The increasing attention being paid by natural-science

9. The term "theorem" as here used includes all logically proved propositions of whatever nature. It thus includes what are called corollaries in the system later to be presented, as well as what are called theorems proper.

theorists to symbolic logic augurs well for an increase in the quality of theories to be put forward in the future, especially where they are based for the most part on qualitative postulates.

In science, the worker does the best he is able with what is available; if only qualitative postulates are to be had, he does what he can with them. It is always to be hoped, however, that such postulates will be only a temporary expedient. The great reason why qualitative postulates are so unsatisfactory is that they have so little deductive fertility. In a broad scientific sense they fail on the criterion of sufficiency; it is impossible to derive from them, even with the powerful devices of symbolic logic, more than a small portion of the phenomena which the postulates clearly concern. One of the more important reasons for this relative sterility of qualitative postulates in behavior situations is that very commonly action potentials of opposing sign are operative simultaneously, and the theoretical outcome is dependent upon which of the opposing potentials is dominant. This dominance, obviously, cannot be determined until the amount of each separate potential can be represented by exact symbolism. When the postulates can be written out in equations, or in words which readily generate equations, and especially when, in addition, the constants making up important portions of the equations are known from empirical determination, the rich store of powerful devices which mathematicians have invented, at once becomes available. Judging by our experience with the present system, the change from qualitative to quantitative postulates with known constants increases the fertility of the postulate set between ten and fifty times.

Much has been said in the preceding pages about the necessity of an empirical check on the theorems of a scientific system. It is in this connection that theorems primarily concerned with non-observable elements, such as excitatory and inhibitory potentialities, present a serious problem because they can not be subjected to direct empirical test. Indeed, if all of the theorems of a system concerned unobservables, its scientific validity could not be determined at all; such a system would be essentially metaphysical so far as its validation is concerned. However, if theorems concerned primarily with unobservables subsequently aid in the deriving of theorems which concern observables, the difficulty is fully met. Through the observational verification of these latter theorems the empirically untestable theorems, along with the postulates employed in the derivation of both sets of theorems, become subject to indirect verification or refutation and consequently become scientifically legitimate and respectable.[10]

10. The emphasis in the preceding pages on the use of logic in system construction must not be mistaken to imply that this exhausts the role played by logic in science. Actually, hardly a detail even of experimental procedure lacks its logical aspect. An especially illuminating example is seen in the experimental testing of a theorem of a formal scientific system. Such a theorem ordinarily states, at least by implication, that (1) if such and such antecedent conditions occur, (2) then such and such consequences will follow. The experimentalist has the task of bringing about these antecedent conditions and these only. This is rarely or never possible; the best that can be done is to eliminate from the antecedent conditions all those which are presumably active or relevant to the expected outcome. In order to make sure that the supposedly inactive conditions really are inactive, control experiments, often of the most intricate nature, must be performed. And when the experimental results are secured,

Is Rigorous Deductive Systematization Possible In the Behavior Sciences?

In the early days of the psychology laboratory it was not uncommon to meet with the view that experiments on the mind were impossible. For the most part, the objections were made by philosophers, the reasons adduced naturally being mainly metaphysical in nature.[11] That question was settled, like most scientific questions, by trial; it is now firmly established that the empirical component of the logico-empirical methodology is generally applicable to the behavior of humans as well as of the lower animals. There still remains the question of whether the logical component of the logico-empirical methodology is also applicable to the behavioral sciences. It is doubtful whether many scientists at present would categorically deny such a possibility. Actually this question, just as that concerning the empirical component, must await the verdict of trial. The present monograph is, in fact, one such trial.

Many persons, while willing to grant the possibility that the sciences of human behavior may ultimately be susceptible to organization in a strict natural-science theoretical system, are inclined to doubt whether the time is yet ripe for the attempt to be made. This raises the question as to what may be expected to mark the proper time to make such an attempt. This question, just as that concerning the empirical methodology, must in the end be settled by trial. In a certain practical sense, one may say the time is ripe for the attempt whenever someone has the impulse to make it. However, the history of science does offer us certain pertinent suggestions concerning the conditions favorable for such a venture.

While the logical and the empirical naturally develop more or less together, history shows that usually during the early stages of development the empirical phase is dominant. It is much later that the logical, or deductive, aspect of natural science manifests itself in the form of large theoretical systematizations. This, of course, is quite to be expected; since logic cannot work in a vacuum, there must be available a supply of preliminary concepts and tentative principles before it can begin systematization. These initial concepts and principles are yielded automatically (*5*) by familiarity with the concrete phenomena of the science in question. There is no reason to believe that the science of human behavior will prove to be any exception to this rule of development.

complex statistical procedures must be employed to make reasonably sure that the critical values obtained are not due to mere sampling or chance. In all this, logic plays a vital role. Unfortunately the scope of the present methodological remarks precludes any detailed discussion of these important matters.

11. For example, Ebbinghaus remarks in the preface to his classical work:

> "The principal objections which, as a matter of course, rise against the possibility of such a treatment are discussed in detail in the text and in part have been made objects of investigation. I may therefore ask those who are not already convinced *a priori* of the impossibility of such an attempt to postpone their decision about its practicability." (*8*)

It would seem that in the field of individual (as distinguished from social) human behavior the conditions for the development of systematic theory are definitely favorable. Wundt founded the first psychological laboratory at Leipzig in 1879. Following this beginning, laboratories of psychology sprang up with great rapidity in many different parts of the world. During the sixty-year period which has since elapsed these laboratories have produced an immense volume of meticulous experimental observation. This is particularly true in the field of rote learning, which was initiated fifty-four years ago by the publication of Ebbinghaus's epoch-making monograph, *Über das Gedachtnis*. It would seem that this extended engrossment in the empirical could hardly have failed to shape, at least in a preliminary manner, the more obvious concepts, and to give rise to numerous promising, if tentative, formulations of the more important dynamic relationships existing between such concepts. Such general considerations as the above certainly do not suggest any insuperable obstacles in the way of the application of the complete natural-science methodology to some aspects, at least, of human behavior.

TYPICAL ROTE-LEARNING EXPERIMENTAL PROCEDURE

As already pointed out, the precise experimental investigation of the psychology of rote learning was begun by Hermann Ebbinghaus over a half century ago. As the material to be memorized, Ebbinghaus employed lists of meaningless syllables consisting of a vowel placed between two consonants, such as FID, KEX, ZAT, etc. The occasional syllables so constructed which were (or closely resembled) words known to the subjects, were carefully excluded from such lists. Even after such exclusions, many so-called nonsense syllables in experimental use do have a suggestion of meaning (*14*;*28*). Moreover, syllables so constructed and chosen are by no means of equal learning difficulty. Nevertheless their introduction into memory experimentation represented a notable technical advance. According to the theory here to be put forward this is mainly because they present the subject with material to which, as remote stimuli (i.e., stimulus traces) he has relatively few reaction potentials already connected, certainly nothing like so many as is the case with isolated words and, even more, connected discourse. The relative elimination of the effects of this particular type of previous learning from the experimental situation thus greatly facilitates the ultimate analysis of the essential phenomena involved in rote learning.

The technique employed by Ebbinghaus was exceedingly simple; he wrote down the syllables in a vertical list on a piece of paper, and learned them by merely reading them from top to bottom. The progress of learning was shown by the number of syllables which could be written down after each reading. Subsequent investigators such as Müller and Schumann (*43*), Müller and Pilzecker (*42*), gradually eliminated the defects of Ebbinghaus's technique. This was accomplished in part by the presentation of the syllables, one at a time, through the use of specially designed automatic memory machines, and in part by numerous other technical devices which need not concern us here. The laboratory situation of a typical modern memory experiment is shown in Plate 1. The syllables are usually printed in bold-faced type such as,

HAJ

ZOX

KEB

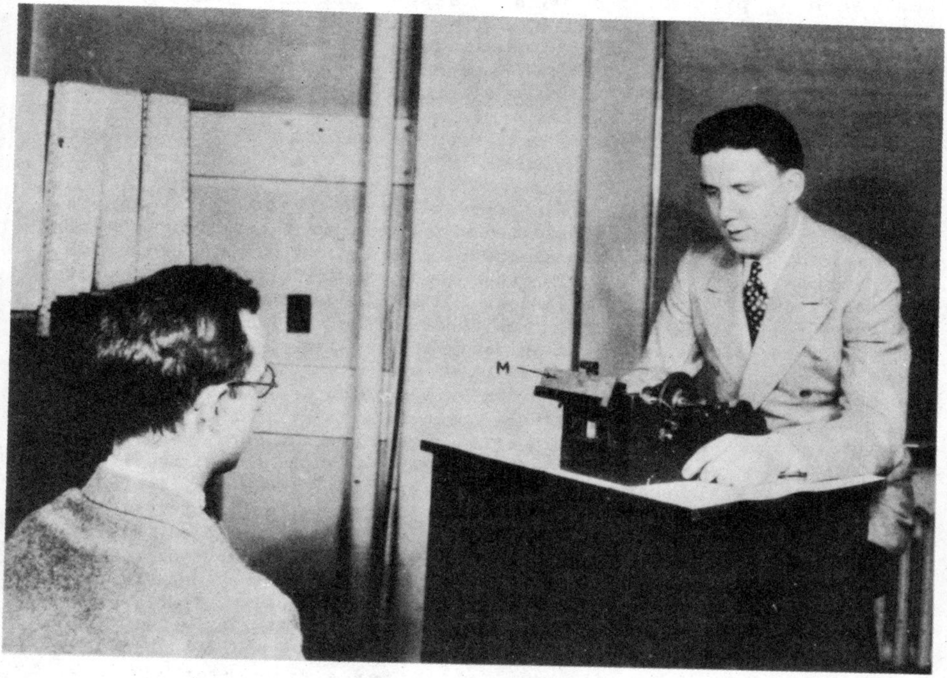

Plate 1. Small motor-driven "memory machine" in use. The young man with spectacles is the subject, and the other is the experimenter. At first the subject reads the nonsense syllables as they appear at the window in the front of the machine. Each syllable is exposed for a constant interval, e.g., two seconds. On the second and all subsequent presentations, the subject tries to anticipate each successive syllable except the first by speaking it just before it is shown at the window. The experimenter observes the syllables through the mirror, *M*, and records the subject's reactions on a specially prepared form which rests on the stand at the right of the apparatus (see p. 17). Normally a low screen shuts off the subject's view of the recording process.

They are attached to the surface of the memory drum which revolves, preferably step-wise, behind a window in the front of the apparatus. This mechanism will hold a syllable before the window without movement a specified period (e.g., two seconds) at the end of which time the drum will make a quick turn of a few degrees, removing the first syllable from view and presenting the second. This step-wise movement of the drum is continued until all of the syllables have been presented. By the learning procedure known as "massed" practice, only a very short period intervenes with no syllable in view, e.g., one equal to that given to the presentation of a single syllable, after which the first syllable again appears, followed as before by the remaining syllables of the series, and so on until the series as a whole is learned.

Before the experiment begins the experimenter instructs the subject somewhat as follows:

"This is an experiment in learning lists of nonsense syllables, and not a psychological test. We are interested in certain complex relationships of the learning process common to all people, and not at all concerned with your personal reactions. Shortly after the apparatus starts you will see a three-letter syllable in the window. You are to pronounce this syllable and those that follow it as you see them. After you have seen the list once, you are to endeavor to anticipate each syllable, except the first which is merely a cue syllable; in other words, as you see one syllable you are to pronounce the syllable that will follow it *before* it appears. If you can not anticipate a syllable, pronounce it when it appears. If you think you know what a syllable will be, but are not sure, guess; because it will not hurt your score any more than to say nothing, and if you get it right it will count as a success. If you anticipate a syllable incorrectly, correct yourself as soon as the syllable itself appears. Try always to speak the syllables as distinctly as possible. Please do not try to think ahead more than one step at a time, or to count, or to make up fanciful connections between the syllables to assist the learning process. Don't try to use any special system in your learning. Simply associate each syllable with the next one as the series moves along."

Under these conditions the subject, after a longer or shorter series of revolutions of the drum (presentations of the series as a whole) learns to pronounce each syllable of the series while its immediate predecessor in the series is in view. This procedure is appropriately known as the anticipation method of rote learning. It has the advantage that the progress of learning is measured in an objective manner while it is taking place, i.e., without interruption; learning is shown by the increasing number of correct anticipations which occur as the successive presentations are made or, what comes to the same thing, by the decreasing number of failures to make such anticipations (see bottom line of sample record, p. 17).

A second advantage of the modern technique of memory experimentation is that it permits keeping a detailed record of the subject's responses. The recording of the subject's behavior is greatly facilitated by the use of specially printed or mimeographed forms. Such a record of a typical rote-learning experiment is shown in detail on page 17. If the reader will study this with a little care it should materially facilitate his understanding of the remainder of the present monograph.

The rote-learning record of subject B.D.S. A minus sign signifies that the subject made no reaction; a plus sign, that he reacted correctly; and a syllable, that the subject responded with this syllable reaction, which was incorrect. Both failures to respond (-'s) and incorrect reactions are counted as failures. Minor changes have been made in this record to enhance its expository value.

Ordinal number of syllables	Syllables in order of presentation	Ordinal number of syllable-cycle presentations or "repetitions"																			Indicators of learning difficulty		
		1	2	3	4	5	6	7	8	9	10	11	12	13	14	15	16	17	18	19	Presentation at first success	Presentation at last failure	Total No. of failures
Cue	MOY																						
1	ZIV	-	+	+	+	+	+	+	+	+	+	+	+	+	+	+	+	+	+	+	2	1	1
2	FAP	-	-	+	+	+	+	+	+	+	+	+	+	+	+	+	+	+	+	+	3	2	2
3	DOB	-	-	-	+	+	+	+	+	+	+	+	-	+	+	+	+	+	+	+	4	12	4
4	NIJ	-	-	-	-	-	-	-	+	YUF	+	+	+	+	+	+	+	+	+	+	8	9	8
5	WOX	-	-	-	-	-	-	-	+	+	+	+	+	+	+	+	+	+	+	+	8	7	7
6	YUF	-	-	-	-	-	-	-	-	-	-	+	+	+	+	+	+	+	+	+	11	10	10
7	HIB	-	-	-	-	-	-	-	-	-	-	+	+	+	WOX	+	+	+	+	+	11	14	11
8	ZET	-	-	-	-	-	-	-	-	-	-	+	+	+	-	+	+	+	+	+	11	14	11
9	YIL	-	-	-	-	-	-	-	-	-	-	-	-	+	+	+	+	-	+	+	13	17	13
10	VAW	-	-	-	-	-	-	-	-	-	-	-	-	+	-	+	+	-	+	+	13	17	14
11	NEF	-	-	-	-	-	-	-	-	-	-	-	-	-	-	-	+	+	+	+	16	15	15
12	PIM	-	-	-	-	-	-	+	+	+	-	-	+	+	+	+	+	+	+	+	7	11	8
13	WUB	-	-	-	-	+	+	+	+	+	-	+	+	-	+	+	+	+	+	+	5	13	6
14	FOV	-	-	-	-	-	+	+	+	+	+	+	+	+	+	+	+	+	+	+	6	5	5
15	JEZ	-	-	-	-	+	-	+	+	+	+	+	+	+	+	+	+	+	+	+	5	6	5
Total No. of failures per presentation		15	14	13	12	10	10	8	6	7	8	4	4	2	4	1	0	2	0	0			

The first thing to be noted in this record is that at the completion of learning the subject pronounces each syllable while its predecessor on the list stands at the window. Thus, upon seeing MOY, ZIV is spoken; shortly after ZIV appears, FAP is spoken; shortly after FAP is seen, DOB is spoken, and so on. In this connection it will be noticed that while there are sixteen syllables in the list, only fifteen are numbered. This is because only fifteen of them are learned. The first syllable (MOY) is not learned; it serves merely as the cue syllable, i.e., the immediate stimulus for the anticipatory recall of ZIV.

A second thing to be noted is that the score at the first presentation, i.e., the first revolution of the drum, is an uninterrupted series of -'s. This is, of course, quite inevitable since the subject has never seen the list before. Possibly as a result of this fact, i.e., because the score at the first presentation can be taken wholly for granted, some investigators do not include the first syllable-presentation cycle in their records. It is feared that this practice sometimes leads certain writers to overlook in their reports this initial revolution of the drum. It is evident, however, that this syllable-presentation cycle contributes to the learning process very much as do the others, as shown by the fact that at the second presentation subjects are frequently able to recall one or more syllables correctly.

In this connection it is important to observe that by the anticipatory method of rote learning here under consideration, it necessarily comes about that the learning which occurs at one presentation can not possibly show itself until the next following presentation (*cf.* the record of syllable 1). Since nothing is known regarding the course of learning during each presentation, it is assumed as a convenient fiction that the increment in the learning occurs abruptly at the termination of each presentation, displaying itself in the form of reaction at the beginning of the next presentation. Figure 22 is a diagrammatic attempt to illustrate this important principle.

There are three distinct measures of the difficulty of learning a given syllable, each with its characteristic theoretical status; the score for each syllable by these methods is shown in the three columns placed at the right-hand margin of the record (p. 17). The measure most commonly used by experimentalists is that of the total number of failures before complete learning is manifested. Thus the record of the first syllable shows one failure before permanent success, the second shows two failures, the third syllable, four failures, the seventh syllable, eleven failures, and so on.

A second measure of syllable-learning difficulty is the ordinal number of the presentation at which the first success occurred. For example, an examination of the record will show that the first success of the first syllable occurred at the second syllable-cycle presentation; that of the third syllable occurred at the fourth presentation; that of the seventh syllable, at the eleventh presentation, and so on.

The third measure of syllable-learning difficulty is the number of the syllable-presentation cycle at the last failure preceding final success. Examination of the record will show, for example, that in the case of the second syllable the last failure occurred at the second presentation, i.e., at the presentation immediately preceding the first success. The third syllable, however, shows quite a different situation in this respect; the last failure occurs at the twelfth presentation, whereas the first success occurs at the fourth presentation. Examination of the record shows that the difference in the two cases is caused by the fact that with the third syllable the subject failed once after having recalled it eight times. Syllables 7, 8, and 9, among others, show single lapses of this kind, and syllables 10 and 13 show two such lapses. In the theory later to be presented, this phenomenon is attributed to oscillation or variability in the reaction threshold (Postulate 15) and, as such, leads to a number of interesting theoretical expectations susceptible of empirical verification.

Failures of recall may take either of two forms: the subject may say nothing, or he may make an incorrect reaction. In the latter case the nature of the reaction is recorded, as shown in the reproduced record. Among the incorrect reactions two important types may be distinguished: *anticipatory errors* and *perseverative errors*. Anticipatory errors are much the more common of the two. An example of anticipatory error is the YUF reaction appearing during the ninth presentation, in place of NIJ. This is called an anticipatory (antedating) reaction because it occurs earlier in the series than, according to the instructions to the subject (p.16), it properly should. The appearance of WOX in the fourteenth presentation, in place of HIB, is an example of a perseverative error; i.e., WOX appears *later* than it properly should. Such errors are counted as failures along with the -'s, without distinction.

A glance at the pattern of minus signs in this record shows, despite a certain amount of irregularity characteristic of all mammalian behavior, that they take the general form of a pyramid with its apex to the right. Further scrutiny will show that the apex of this pyramid is placed somewhat asymmetrically; it appears lower, i.e., closer, to the posterior end of the series than to the anterior end. A glance at the three indices of syllable difficulty shows in clear numerical terms that this is the case; the point of maximum learning difficulty appears to be in the neighborhood of the tenth syllable. There is also a parallel and doubtless related asymmetry in that the final syllable is definitely more difficult to learn than is the first syllable.

In connection with the concept of the "last failure," there enters necessarily the notion of *permanent success*. In the theoretical treatment later to be presented, the final error is that error which would be followed by an infinite number of correct reactions if the experiment were to continue indefinitely. It is evident, however, that the ordinary memory experiments approach this ideal only approximately. Many experimenters terminate the learning when the subject has been able to anticipate all the syllables correctly at a given syllable-presentation cycle. A glance at the record of subject B.D.S. will show that by this criterion learning would have stopped with the sixteenth repetition. However, the criterion employed with this subject was *two* perfect recalls of the series as a whole on successive presentations. It may be seen that this more rigorous criterion changes the score somewhat, since on the seventeenth presentation failures occurred at both syllables 9 and 10, thus necessitating two additional repetitions. According to both theory and empirical observations, if a still more exacting criterion of learning were employed, such as four perfect recalls of the series as a whole on successive presentations, it is highly probable that additional lapses would have occurred. For most purposes, the procedure employing an arbitrarily limited criterion of learning may do no harm provided the criterion adopted is applied consistently. It does mean, however, that the various measures of learning difficulty as a function of syllable position in the series differ somewhat from what they would be if a more exacting criterion of learning were applied.

A simple but important principle related to the foregoing observation is that the number of revolutions of the memory drum, i.e., the number of the presentations and so the aggregate length of time required to memorize a rote series, is dependent upon, indeed is the same as *that required for the single syllable manifesting the greatest learning difficulty* as shown by the criterion of the presentation at the last failure.

THE SYSTEM

The exposition of a mathematical theory in the behavior-science field encounters many obstacles. The major difficulty is caused by the fact that many readers interested in behavior theory do not have sufficient command of advanced mathematics or of symbolic logic readily to follow proofs by either symbolism. This presents a serious dilemma because rigorous proofs can hardly be expressed in any other manner. It is hoped, however, that this expository difficulty has been largely obviated in the present work where, in general, the device employed has been to set forth parallel statements (1) in ordinary language and (2) in the language of mathematics or symbolic logic, or both. For the most part the undefined notions and the definitions are stated in duplicate. The first portion of each consists mainly of a statement in ordinary language which can be understood by anyone. In most cases this is immediately followed by the second portion which makes a parallel statement in symbolic logic. The second portion has the same number as the first in each case, except that it is distinguished by an accent or prime mark.

The postulates are similarly stated in duplicate, the two portions in this case being marked *A* and *A′* respectively. Following the *A′* (symbolic-logic) statement will be found in each case a somewhat elaborate explanatory section marked *B*. This same general method has also been followed in the presentation of the theorems, but in this latter case there are often three parallel sections, labeled *A*, *B*, and *C* respectively. The *A*-section presents the formal statement of the theorem together with the mathematical proof. The *B*-section follows with a common-sense account of the same matter in ordinary word language, usually the application of simple arithmetic to a particular illustrative example. The *C*-section (usually found only in connection with corollaries) illustrates the point by citing relevant experimental evidence bearing on the empirical validity of the proposition as presented in the *A*-section. By this arrangement it is hoped that all readers can secure a clear understanding of the theoretical propositions from the *B*-section, and an objective indication of the scientific validity of the propositions capable of empirical verification from the *C*-section.[1] Meanwhile the non-mathematical reader has the objective evidence before him in the *A*-section that a *bona fide* attempt at rigor of deduction has at least been made public, and that this can be challenged at any time by other individuals who have the special equipment for so doing.

In the construction and presentation of this system we have tried, in so far as circumstances would permit, to conform to the canons of scientific procedure, some of the more important principles of which are stated in a preceding section

1. It happens that in the present system a considerable number of the propositions formally listed as theorems concern unobservables and so are not capable of direct empirical verifications; these propositions naturally have no *C*-section. Usually the propositions formally listed as corollaries need little explanation, and therefore in these cases the *B*-section is omitted.

devoted to methodology. The principles there laid down represent ideals. Despite our best efforts we have probably attained none of these ideals completely, and a number have not been approached even closely. Some of the more obvious defects of the system will be called to the reader's attention as occasion offers.

Undefined Concepts

U1. *Syllable exposure* (slex): A class of events each of which may be described as the stationary presence in the window of a memory machine (Plate 1) of a syllable consisting of a vowel placed between consonants in a combination not used as a word by the subject. The syllable is supposed to be printed in such a way as to reflect clearly a characteristic pattern of light rays. The subject may or may not be present.

U1'. "*a* is a syllable exposure" will be expressed thus in symbolic logic: $a \; \varepsilon$ slex.

U2. *Subject* (sb): A class each member of which is a normal human being who, previous to the rote-learning experiment, has acquired a set of phonetic reading habits, who has received adequate instruction as to his duties as subject (p. 16), and who has consented in good faith to perform these duties.

U2'. "*a* is a subject" will be expressed thus in symbolic logic: $a \; \varepsilon$ sb.

U3. *Conjunction* (cn): The conjunction of a syllable exposure with a subject is the event consisting of the impingement of the light waves reflected from the syllable exposure in the window of the memory machine upon the visual organs of the subject.

U3'. "The event *a* is the conjunction of the event *b* with the event *c*" will be expressed thus in symbolic logic: *a* cn (*b*, *c*).

U4. *Reaction* (rn): The subject's act of speaking a syllable according to his previously acquired habits of speech.

U4'. "*a* is a reaction of *b*" will be expressed thus in symbolic logic: *a* rn *b*.

U5. *Syllable congruence* (slcg): Two syllable exposures are congruent if they are exposures of the same syllable. Two syllable reactions are congruent if they are acts of speaking the same syllable. A syllable reaction and a syllable exposure are congruent if one is an act of speaking the same syllable as the syllable of which the other is an exposure.

U5'. "*a* is congruent with *b*" will be expressed thus in symbolic logic: *a* slcg *b*.

U6. *Stimulus trace* (tr): The stimulus trace of a syllable presentation (see Def. 17) is a progressively changing activity within the subject's body corresponding uniquely to the syllable presentation in question. The beginning of the activity coincides with the beginning of the syllable presentation in question, and the end of the activity coincides with the end of the syllable-presentation cycle (see Def. 21) in which the syllable presentation occurs.

The concept *stimulus trace* has substantially the status of a symbolic or logical construct. While there are physiological indications that the expression represents an entity which may ultimately be observable in some indirect manner, for the present purposes it may be regarded as an unobservable. The existence of this hypothetical entity is explicitly assumed by Postulate 1. A somewhat more detailed indication of its characteristics than is strictly necessary for the derivation of the theorems of the present monograph is given by Figure 1 and several following diagrams.

In order to attain unambiguous statements regarding this concept, a special notation will be employed. In general the left-hand subscript refers to the syllable-of-origin of the stimulus trace, and the right-hand subscript indicates the particular segment of the trace under consideration. For example, in Figure 2 the expression ${}_{2}tr_{5}$ means the segment concurrent with syllable presentation 5 (JUB) of the stimulus trace originating in syllable 2 (ZIT). Frequently the syllable-of-origin of a trace is represented by *s*, and the particular segment of a trace under consideration is represented by *n* thus: ${}_{s}tr_{n}$. Accordingly the first or initial segment of a trace is represented by ${}_{s}tr_{s}$. In a similar manner all the segments of any given stimulus trace are represented by the expression, ${}_{s}tr_{s \ldots N}$ where N is the last syllable of the series, i.e., the complete trace originating at syllable *s* will have segments concurrent with every syllable from *s* to the last syllable of the series, inclusive. By an analogous notation, ${}_{0 \ldots N}tr_{n}$ represents a compound trace at syllable *n*, i.e., the compound trace at *n* includes the *n*-segment of each of the stimulus traces originating in all syllables from the cue syllable (numbered zero) to syllable *n* inclusive. For example, in Figure 2, the compound trace at syllable ZIT represents the ZIT-segment of stimulus traces originating in syllables KEM, FAP, and ZIT itself.

U6'. "*a* is the stimulus trace of *b*" will be expressed thus in symbolic logic: *a* tr *b*.

U7. *Excitatory potential* (E): A cumulative and relatively permanent organization (within the subject's body) which, after the inhibitory potential (see U8) has been deducted from it, must exist in amount greater than the threshold ℓ (see D81) in order that the concurrent syllable presentation shall, by way of the concurrent compound trace segment, evoke a syllable reaction congruent with the following syllable presentation.

The concept *excitatory potential* is a logical construct and clearly is unobservable. The existence of this hypothetical entity is assumed by Postulate 2. In the notation adopted for the representation of these values, e.g., ${}^{2,6,15}E_{3,4}$, the first subscript (counting from the left) represents the syllable-presentation interval at which the particular excitatory potential is active; the second subscript indicates the syllable-reaction potentiality active at the syllable-presentation interval indicated by the first subscript; the first superscript (counting from the left) indicates the ordinal number of the massed practice involved; the second superscript indicates the number of repetitions (syllable-presentation cycles) which have contributed to the building up of the excitatory potential in the massed practice in question; and the third superscript indicates the number of syllables in the series being learned. For example, ${}^{M,R,N}E_{n,r}$ means the amount of potentiality for reaction corresponding to syllable r to occur during the presentation interval of syllable n in a series of N syllables after R repetitions of the Mth massed practice. In this notation, $\Delta\,{}^{N}E_{r-1,r}$ represents the increment (input) of excitatory potential resulting from a single (unspecified) repetition of the series as a whole in an unspecified massed practice for reaction r to occur concurrently with the presentation of syllable r-1 in a rote series of N syllables.

U7'. "x is the (gross) excitatory potential at time t for the nth syllable presentation of syllable-presentation cycle P to evoke a reaction congruent with the syllable exposure involved in the (n+1)st syllable presentation" will be expressed thus in symbolic logic: x E (t, n, P).

U8. *Inhibitory potential* (I): A cumulative and relatively permanent organization (within a subject's body) which comes into action in differential amounts at the several syllable presentations of a syllable-presentation cycle, and which, at any syllable presentation, acts to diminish the functional potentiality of the concurrent excitatory potential.

The concept *inhibitory potential*, like that of excitatory potential, is a logical construct and also clearly represents an unobservable entity. The existence of this hypothetical entity is assumed by Postulate 5. The positional notation devised for the representation of these values resembles in most respects that adopted for the representation of excitatory potentials. For example, the positions of superscripts and subscripts in the expression ${}^{M,R,N}_{\;\;s}I_{n,r}$ have the following meaning: the position of superscript N indicates the number of syllables in the series being learned; that of R represents the number of repetitions which have occurred (by a particular massed practice); that of M is the ordinal number of the massed practice in which the repetitions have occurred; that of the left-hand subscript, s, indicates the number of the syllable-of-origin of the stimulus trace with which the inhibitory potential is associated; that of the right-hand subscript n indicates the particular segment of the said stimulus trace with which the inhibitory potential is associated; and that of the second right-hand subscript, r, indicates the number of the syllable reaction the learning of which was associated with the origin of the inhibitory potential in question.

In addition to the *positional* significance of the superscripts and subscripts appearing in the above expression, the symbols M, R, N, s, n, and r also have specifically the significance there indicated, wherever found throughout the present system.

U8'. "x is part of the total inhibitory potential which operates at time t to diminish the (gross) excitatory potential for the compound stimulus trace concurrent with the nth syllable presentation of a syllable-presentation cycle

P to evoke a reaction congruent with the syllable exposure involved in the $(n+1)$st syllable presentation; where *x* is the part which is a summation of increments each of which, in some syllable-presentation cycle *Q* congruent with *P*, arose concurrently with the setting up of the gross excitatory potential (E) for the syllable exposure involved in the $(r-1)$st syllable presentation of *Q* to evoke a reaction congruent with the *r*th syllable presentation of *Q* and each of which is associated with the *n*th segment of the stimulus trace originating at the sth syllable presentation of *Q*" will be expressed thus in symbolic logic:
x I (t, n, P, s, r).

U9. *Reaction threshold* (rnth): The value of the effective excitatory potential (see D79) which, at a given syllable presentation, separates those values which are great enough to evoke a correct reaction from those values which are not.

U9'. "*x* is the reaction threshold for syllable presentation *a* to evoke a correct reaction" will be expressed thus in symbolic logic: *x* rnth *a* .

U10. *Constant of decay of excitatory potential* (b): This constant appears in the expression (Postulate 12) describing the law of decay of excitatory potential [$E_{n,r}(t) = E_{n,r}\, e^{-bt}$ at any point, *n*, which results from any given massed practice.

U10'. In the symbolic logic this numerical constant appears in Postulate 12 and in the definition of $dsxp_n$.

U11. *Constant of decay of inhibitory potential* (d): This constant appears in the expression (Postulate 13) describing the law of decay of inhibitory potential [$I_n(t) = I_n\, e^{-dt}$ at any point, *n*, which results from any given massed practice.

U11'. In the symbolic logic this numerical constant appears in Postulate 13 and in the definition of $I_n(t)$.

U12. *Minimal reaction latency* (v): This constant appears in the equation (Postulate 18) expressing the reaction latency as a function of effective excitatory potential [$T = v + (h-v)\, e^{-k\bar{E}_{r-1,r}}$]. This equals the value of *T* when $R = \infty$ in the equation, $T = v + (h-v)\, e^{-kR\Delta\bar{E}_{r-1,r}}$. This latter equation has been obtained by a process of curve fitting from an empirical investigation by Simley (*59*) in which reaction latencies were taken during extensive over-learning of paired associates.

U12'. In the symbolic logic this numerical constant appears in Postulate 18.

U13. *Basic reaction latency* (h): This constant appears in the equation (Postulate 18) expressing the reaction latency as a function of effective excitatory potential [$T = v + (h-v)\, e^{-k\bar{E}_{r-1,r}}$]. This equals the value of *T* when $R = 0$ in the equation, $T = v + (h-v)\, e^{-kR\Delta\bar{E}_{r-1,r}}$. This latter equation has been obtained by a process of curve fitting from an empirical investigation by Simley (*59*) in which reaction latencies were taken during extensive over-learning of paired associates.

U13'. In the symbolic logic this numerical constant appears in Postulate 18.

U14. *Exponential constant of reaction latency* (k): This constant appears in the equation (Postulate 18) expressing the reaction latency as a function of effective excitatory potential $[T = \mathrm{v} + (\mathrm{h}-\mathrm{v})\, e^{-\mathrm{k}\bar{\mathrm{E}}_{r-1,r}}]$.

U14'. In the symbolic logic this numerical constant appears in Postulate 18.

U15. *Beginning time* (bg): The time, in seconds, at which an event begins.

U15'. "x is the time at which a begins" will be expressed thus in symbolic logic: x bg a .

U16. *Ending time* (nd): The time, in seconds, at which an event ends.

U16'. "x is the time at which a ends" will be expressed thus in symbolic logic: x nd a .

Definitions

D1. The *duration* (du) of an event is the length of time between its beginning and end.

D1'. $\mathrm{du} = \hat{t}\hat{a}[t = \mathrm{nd}{`}a - \mathrm{bg}{`}a]$

D2. A *syllable-exposure cycle* (slexcy) is a sequence of from 6 to 100 distinct (non-congruent) syllable exposures (regardless of whether a subject is present), the duration of all exposures being the same (1 to 4 seconds) with a constant time interval between exposures equal to, or less than, one-tenth of one syllable exposure, the final syllable exposure of the sequence being followed by a non-exposure interval equal to or greater than one syllable exposure.

D2'. $\mathrm{slexcy} = \hat{P}[P\ \varepsilon\ \mathrm{ser}\ .\ \mathrm{C}{`}P \subset \mathrm{slex}\ .\ P \mathrel{\dot{\subset}} \dot{-}\mathrm{slcg}\ .\ 6$
$\leq \mathrm{Nc}{`}\mathrm{C}{`}P \leq 100 : (\exists k)(a, b) : a\ P\ b\ .\ \sim(a\ P^2\ b)\ .\supset .\ 0$
$< \mathrm{bg}{`}b - \mathrm{nd}{`}\mathrm{a} = k \leq (\mathrm{du}{`}a \div 10)\ .\ 1 \leq \mathrm{du}{`}a = \mathrm{du}{`}b \leq 4]$

D3. The *syllable exposure period* (slexpd) of a given syllable exposure cycle is the duration of any syllable exposure of that syllable exposure cycle.

D3'. $\mathrm{slexpd} = \hat{t}\hat{P}[P\ \varepsilon\ \mathrm{slexcy} : (a) : a\ \varepsilon\ \mathrm{C}{`}P\ .\supset .\ t = \mathrm{du}{`}a]$

D4. The *syllable-inter-exposure interval* (slinexin) of a syllable-exposure cycle is the length of time from the end of one syllable exposure of that syllable-exposure cycle to the beginning of the next succeeding syllable exposure of the same syllable-exposure cycle.

D4'. $\mathrm{slinexin} = \hat{t}\hat{P}[P\ \varepsilon\ \mathrm{slexcy} : (a, b) : a\ P\ b\ .\ \sim(a\ P^2 b)\ .\supset .\ t$
$= \mathrm{bg}{`}b - \mathrm{nd}{`}a]$

D5. The *syllable-exposure interval* (slexin) of a syllable-exposure cycle is the length of time from the beginning of one syllable exposure of that syllable-exposure cycle to the beginning of the next succeeding syllable exposure of the same syllable-exposure cycle.

D5′. slexin = $\hat{t}\hat{P}[P \; \varepsilon$ slexcy : (a, b) : $a \; P \; b \; . \sim(a \; P^2 b)$.

$\supset . \; t =$ bg‘b − bg‘a]

D6. *Syllable exposure number* (slexnm): The syllable exposures of a syllable-exposure cycle are numbered from first to last, the first syllable exposure of the syllable-exposure cycle being numbered zero and the last being numbered *N*

The numbers so arrived at are known as the *numbers of the syllable exposures*, and are usually represented by the symbol *n*. However, in symbolic logic *n* has a wider use; and N‘*P* generally replaces *N*.

D6′. slexnm = $\hat{n}\hat{a}\hat{P}[n =$ Nc‘$\overrightarrow{P}$‘a . $P \; \varepsilon$ slexcy . $a \; \varepsilon$ C‘P]

D6.1′. N = $\hat{n}\hat{P}[n =$ Nc‘C‘P − 1]

D7. The *cue syllable exposure* (cuslex) of a syllable-exposure cycle is the syllable exposure which precedes all others in that syllable-exposure cycle and whose syllable-exposure number is therefore *zero*.

D7′. cuslex = $\hat{a}\hat{P}$[slexnm‘(a, P) = 0]

D8. The *terminal syllable exposure* (tmslex) of any syllable-exposure cycle is the last syllable exposure of that syllable-exposure cycle.

D8′. tmslex = $\hat{a}\hat{P}[P \; \varepsilon$ slexcy . a = B‘$\breve{P}$]

D9. A syllable-exposure cycle has *syllable-exposure-cycle congruence* (slexcycg) to a second syllable-exposure cycle when the total number of syllable exposures, the syllable-exposure periods, and the syllable-exposure intervals of the two syllable-exposure cycles, are equal and when the syllable exposures having the same syllable-exposure numbers in the two syllable-exposure cycles are congruent.

D9′. slexcycg = $\hat{P}\hat{Q}$[slexpd‘P = slexpd‘Q . slexin‘P

= slexin‘Q . N‘P = N‘Q : (a, b) : slexnm‘(a, P)

= slexnm‘(b, Q) . $\supset$. a slcg b :. $P, Q \; \varepsilon$ slexcy]

D10. Congruent syllable exposures are said to be instances of the same *syllable* (sl).

D10′. sl = $\overrightarrow{\text{slcg}}$

D11. Syllable-exposure cycles which have syllable-exposure congruence to one another are said to be instances of the same *rote series* (rtsr).

D11′. rtsr = $\overrightarrow{\text{slexcycg}}$

D12. *Syllable number* (slnm): The syllables of a rote series are numbered from first to last, the first syllable of the rote series being numbered zero and the last being numbered *N*. The numbers so arrived at are known as the *syllable numbers of the respective syllables* and are usually represented by the number n.

D12'. slnm = $\hat{n}\hat{\alpha}\hat{\lambda}[(\exists P, \alpha) : \lambda = \text{rtsr}'P \,.\, n = \text{slexnm}'(a, P) \,.$
$\alpha = \text{sl}'a]$

D13. The *cue syllable* (cusl) of a rote series is the syllable which precedes all others in that rote series and whose syllable number is therefore zero.

D13'. cusl = $\hat{\alpha}\hat{\lambda}[\text{slnm}'(\alpha,\lambda) = 0]$

D14. The *initial syllable* (insl) of a rote series is the syllable whose syllable number is 1 in that series.

D14'. insl = $\hat{\alpha}\hat{\lambda}[\text{slnm}'(\alpha,\lambda) = 1]$

D15. The *terminal syllable* (tmsl) of a rote series is the syllable which comes last in the rote series.

D15'. tmsl = $\hat{\alpha}\hat{\lambda}[(P) : \lambda = \text{rtsr}'P \,.\supset .\, \alpha = \text{sl}'\text{B}'\breve{P}]$

D16. The *inter-cycle interval* (incyin) between two successive congruent syllable-exposure cycles is the length of time from the end of the last syllable exposure of the first of these syllable-exposure cycles to the beginning of the first syllable exposure of the other.

D16'. incyin = $\hat{x}\hat{P}\hat{Q}[P \text{ slexcycg } Q \,.\, x = \text{bg}'\text{B}'Q - \text{nd}'\text{B}'\breve{P}]$

D17. A *syllable presentation* (slpn) is the conjunction of a syllable exposure with a subject.

D17'. slpn = $\hat{a}[(\exists b, c) : b \,\varepsilon\, \text{slex} \,.\, c \,\varepsilon\, \text{sb} \,.\, a \text{ cn } (b, c)]$

D18. A *syllable-presentation subject* (slpnsb) of a syllable presentation is the subject whose conjunction with a syllable exposure constitutes the syllable presentation.

D18'. slpnsb = $\hat{a}\hat{b}[(\exists c) \,.\, a \,\varepsilon\, \text{sb} \,.\, c \,\varepsilon\, \text{slex} \,.\, b \text{ cn } (c, a)]$

D19. One syllable presentation has *syllable-presentation congruence* (slpncg) with a second syllable presentation when the two syllable exposures involved in the two syllable presentations are congruent and the subject involved in the two syllable presentations is the same.

D19'. slpncg = $\hat{a}\hat{b}[(\exists c, d, e) : d \text{ slcg } e \,.\, a \text{ cn } (d, c) \,.$
$b \text{ cn } (e, c)]$

D20. The *syllable-presentation-cycle conjunction* (slpncycn) of a syllable-exposure cycle with a subject is the sequence of syllable presentations each of which is a conjunction of a syllable exposure of the syllable-exposure cycle with the subject in question and each of which begins and ends at the same time as its corresponding syllable exposure.

D20′. $\mathrm{slpncycn} = \hat{P}\hat{Q}\hat{a}\ [\ a\ \varepsilon\ \mathrm{sb}\ .\ Q\ \varepsilon\ \mathrm{slexcy} : (b, c) : b\,(P \dot{-} P^2)\,c$ $\equiv .\ (\exists d, e) .\ d\,(Q \dot{-} Q^2)\,e\ .\ b\ \mathrm{cn}\ (d, a)\ .\ c\ \mathrm{cn}\ (e, a)\ .\ \mathrm{du}'b = \mathrm{du}'c$ $= \mathrm{slexpd}'Q\ .\ \mathrm{bg}'b = \mathrm{bg}'d\ .\ \mathrm{bg}'c = \mathrm{bg}'e]$

D21. A *syllable-presentation cycle* (slpncy) or *repetition* is the (syllable-presentation cycle) conjunction of a syllable-exposure cycle with a subject.

The expression *syllable-presentation cycle*, while clumsy, is a more precise term than the usually employed term *repetition*. It does some violence to the English language to speak of the first syllable-presentation cycle as a repetition because something can hardly be repeated which has not previously occurred. Nevertheless, wherever the term *repetition* occurs in the present monograph, it is used in the strict sense of *syllable-presentation cycle*.

D21′. $\mathrm{slpncy} = \hat{P}\,[(\exists Q, c) : P\ \mathrm{slpncycn}\ (Q, c) : (a, b) : a$ $= \mathrm{slpnsb}'b\ .\ a = \mathrm{slpnsb}'\mathrm{B}'P\ .\ \mathrm{bg}'\mathrm{B}'P \leq \mathrm{bg}'b\ .\ \mathrm{nd}'b$ $\leq \mathrm{nd}'\mathrm{B}'P + \mathrm{slinpnin}'\breve{P}\ .\ \supset .\ b\ \varepsilon\ \mathrm{C}'P]$

D22. The *syllable-presentation period* (slpnpd) of a syllable-presentation cycle is equal in duration to the syllable exposure period of the syllable-exposure cycle whose (syllable-presentation cycle) conjunction with a subject constitutes the syllable-presentation cycle in question.

D22′. $\mathrm{slpnpd} = \hat{t}\hat{P}\,[(\exists Q, a) : P\ \mathrm{slpncycn}\ (Q, a)\ .\ t = \mathrm{slexpd}'Q]$

D23. The *syllable-inter-presentation interval* (slinpnin) of a syllable-presentation cycle is equal in duration to the syllable-inter-exposure interval of the syllable-exposure cycle whose (syllable-presentation cycle) conjunction with a subject constitutes the syllable-presentation cycle in question.

D23′. $\mathrm{slinpnin} = \hat{t}\hat{P}\,[(\exists Q, a) : P\ \mathrm{slpncycn}\ (Q, a)\ .\ t$ $= \mathrm{slinexin}'Q]$

D24. The *syllable-presentation interval* (slpnin) of a syllable-presentation cycle is equal to the syllable-exposure interval of the syllable-exposure cycle whose (syllable-presentation cycle) conjunction with a subject constitutes the syllable-presentation cycle in question.

D24′. $\mathrm{slpnin} = \hat{t}\hat{P}\,[(\exists Q, a) : P\ \mathrm{slpncycn}\ (Q, a)\ .\ t$ $= \mathrm{slexin}'Q]$

D25. *Syllable presentation number*: (slpnnm): The syllable presentations of a syllable-presentation cycle are numbered from first to last, the first syllable presentation of the syllable-presentation cycle being numbered zero and the last being numbered N.

The numbers so arrived at are known as the syllable-presentation numbers of the respective syllable presentations and are usually represented by the number n.

D25'. $\mathrm{slpnnm} = \hat{n}\hat{a}\hat{P}[n = \mathrm{Nc}'\vec{P}'a \,.\, P \;\varepsilon\; \mathrm{slpncy} \,.\, a \;\varepsilon\; \mathrm{C}'P]$

D26. The *cue syllable presentation* (cuslpn) of a syllable-presentation cycle is the syllable presentation which precedes all others in that syllable-presentation cycle and whose syllable-presentation number is therefore *zero*.

D26'. $\mathrm{cuslpn} = \hat{a}\hat{P}[\mathrm{slpnnm}'(a,P) = 0]$

D27. The *terminal syllable presentation* (tmslpn) of a given syllable-presentation cycle is the last syllable presentation of that syllable-presentation cycle.

D27'. $\mathrm{tmslpn} = \hat{a}\hat{P}[P \;\varepsilon\; \mathrm{slpncy} \,.\, a = \mathrm{B}'\breve{P}]$

D28. A syllable-presentation cycle has *syllable-presentation-cycle congruence* (slpncycg) to a second syllable-presentation cycle when the two syllable-exposure cycles involved in the two syllable-presentation cycles have syllable-exposure-cycle congruence to one another and when the subject involved in the two syllable-presentation cycles is the same.

D28'. $\mathrm{slpncycg} = \hat{P}\hat{Q}[(\exists R, a, S) : P \;\mathrm{slpncycn}\; (R,a) \,.\, Q$

$\mathrm{slpncycn}\; (S,a) \,.\, R \;\mathrm{slexcycg}\; S]$

D29. A *rote learning* (rtln) consists of all the syllable-presentation cycles congruent to a given syllable-presentation cycle.

D29'. $\mathrm{rtln} = \overrightarrow{\mathrm{slpncycg}}$

D30. A *rote practice* (rtpr) consists of a finite sequence of congruent syllable-presentation cycles to which belong all syllable presentations (to the subject in question) occurring between the beginning and end of the sequence.

D30'. $\mathrm{rtpr} = \hat{T}[\dot{\exists}!T \,.\, T \;\varepsilon\; \mathrm{refl} \,.\, T \;\varepsilon\; \mathrm{trans} \,.\, T \;\varepsilon\; \mathrm{connex} \,.$

$T \subset \mathrm{slpncycg} \,.\, \mathrm{C}'T \;\varepsilon\; \mathrm{NC\ induct} : (Q,S) : Q\,T\,S \,.\supset .\, \mathrm{bg}'\mathrm{B}'Q$

$\leq \mathrm{bg}'\mathrm{B}'S :. (a,b) : a = \mathrm{slpnsb}'b \,.\, a = \mathrm{slpnsb}'\mathrm{B}'\mathrm{B}'T$

$\mathrm{bg}'\mathrm{B}'\mathrm{B}'T < \mathrm{nd}'b \,.\, \mathrm{bg}'b < \mathrm{nd}'\mathrm{B}'\mathrm{Cnv}'\mathrm{B}'\breve{T} + \mathrm{slpnin}'\mathrm{B}'T$

$+ \mathrm{slinpin}'\mathrm{B}'T \,.\supset .\, (\exists P) \,.\, P \;\varepsilon\; \mathrm{C}'T \,.\, b \;\varepsilon\; \mathrm{C}'P]$

D31. The *ordinal number* (*R*) of a syllable-presentation cycle in a rote practice is the number of syllable-presentation cycles up to and including the syllable-presentation cycle in question, taken in the order in which they occur.

It is to be noted that *R* as used in D31 becomes R in the symbolic logic of D31'.

D31'. $\mathrm{R} = \hat{n}\hat{P}\hat{T}[n = \mathrm{Nc}'\vec{T}'P \,.\, T \;\varepsilon\; \mathrm{rtpr} \,.\, P \;\varepsilon\; \mathrm{C}'T]$

D32. A *massed practice* (mspr) is a rote practice which includes all those successive congruent syllable-presentation cycles, and only those, which are separated by inter-cycle intervals which are equal to the sum of one syllable-presentation interval and one syllable inter-presentation interval, the last syllable presentation of the massed practice being followed by an interval equal to or greater than the said inter-cycle interval.

D32'. $\text{mspr} = \hat{T}[T \in \text{rtpr} : (P, Q) : \text{R}'(P, T) + 1 = \text{R}'(Q, T) .$
$\supset . \text{bg}'\text{B}'Q - \text{nd}'\text{B}'\breve{P} = \text{slpnin}'P + \text{slinpnin}'P :. \sim(\exists S) : S$
$\text{slpncycg B}'T : \text{bg}'\text{B}'S = \text{nd}'\text{B}'\text{Cnv}'\text{B}'\breve{T} + \text{slpnin}'\text{B}'T$
$+ \text{slinpnin}'\text{B}'T . \text{v} . \text{nd}'\text{B}'\breve{S} + \text{slpnin}'\text{B}'T + \text{slinpnin}'\text{B}'T$
$= \text{bg}'\text{B}'\text{B}'T]$

D33. The *termination of massed practice* (prtm) occurs at that instant of time which is later than the end of the last syllable presentation of the last syllable-presentation cycle of the massed practice in question, by an amount equal to the sum of one syllable-presentation interval and one syllable inter-presentation interval of a syllable-presentation cycle of such massed practice.

D33'. $\text{prtm} = \hat{t}\hat{T}[t = \text{nd}'\text{B}'\text{Cnv}'\text{B}'\breve{T} + \text{slpnin}'\text{B}'T$
$+ \text{slinpnin}'\text{B}'T . T \in \text{mspr}]$

D34. A *distributed practice* (dspr) is a rote practice in which the inter-cycle interval is greater than the sum of one syllable-presentation interval and one syllable inter-presentation interval.

D34'. $\text{dspr} = \hat{T}[T \in \text{rtpr} : (P, Q) : \text{R}'(P, T) + 1 = \text{R}'(Q, T) .$
$\supset . \text{bg}'\text{B}'Q - \text{nd}'\text{B}'\breve{P} > \text{slpnin}'P + \text{slinpnin}'P]$

D35. An *evenly distributed practice* (evdspr) is a distributed practice in which the inter-cycle intervals are equal throughout.

D35'. $\text{evdspr} = \hat{T}[T \in \text{dspr} : (\exists x)(P, Q) : \text{R}'(P, T) + 1 = \text{R}'(Q, T) .$
$\supset . \text{bg}'\text{B}'Q - \text{nd}'\text{B}'\breve{P} = x]$

D36. The *syllable reaction number* (slrnnm) of a syllable reaction of a subject during a syllable-presentation cycle (involving the subject in question) is the syllable presentation number of the syllable presentation which involves a syllable exposure congruent to the reaction in question.

D36'. $\text{slrnnm} = \hat{n}\hat{a}\hat{P}[(\exists b, c, d) : b \in \text{slex} . c \in \text{sb} .$
$a \text{ rn } c . a \text{ slcg } b . d \text{ cn}(b, c) . n = \text{slpnnm}'(d, P)]$

D37. An *anticipatory reaction* (anrn) during a syllable-presentation cycle is a syllable reaction beginning during a syllable presentation, or during the next succeeding syllable-inter-presentation interval, the number of which is less than the number of the syllable reaction in question.

D37'. $\text{anrn} = \hat{a}\hat{b}[(\exists P) : \text{slpnnm}'(b, P) < \text{slrnnm}'(a, P) .$

$\text{bg}'b \leq \text{bg}'\text{a} < \text{nd}'b + \text{slinpnin}'P]$

D38. A *correct reaction* (ctrn) during a syllable-presentation cycle is a syllable reaction beginning during a syllable presentation, or during the next succeeding syllable-inter-presentation interval, the number of which is one less than the number of the syllable reaction in question.

D38'. $\text{ctrn} = \hat{a}\hat{b}[(\exists P) : \text{slpnnm}'(b,P) + 1 = \text{slrnnm}'(a,P)\ .$

$\text{bg}'b \leq \text{bg}'a < \text{nd}'b + \text{slinpnin}'P]$

D39. A *perseverative reaction* (pern) during a syllable-presentation cycle is a syllable reaction beginning during a syllable presentation, or the next succeeding syllable inter-presentation interval, the number of which is greater than the number of the syllable reaction in question.

D39'. $\text{pern} = \hat{a}\hat{b}[(\exists P) : \text{slpnnm}'(b,P) > \text{slrnnm}'(a,P)\ .$

$\text{bg}'b \leq \text{bg}'a < \text{nd}'b + \text{slinpnin}'P]$

D40. A *reading reaction* (rdrn) during a syllable-presentation cycle is a syllable reaction beginning during a syllable presentation, or the next succeeding syllable inter-presentation interval, the number of which is the same as the number of the syllable reaction in question.

D40'. $\text{rdrn} = \hat{a}\hat{b}[(\exists P) : \text{slpnnm}'(b,P) = \text{slrnnm}'(a,P)\ .$

$\text{bg}'b \leq \text{bg}'a < \text{nd}'b + \text{slinpnin}'P]$

D41. The *degree of syllable reaction anticipation* (anrndg) during a syllable-presentation cycle is the difference obtained by subtracting the syllable-presentation number of the syllable presentation involved from the syllable-reaction number of the syllable reaction involved.

D41'. $\text{anrndg} = \hat{n}\hat{a}[(\exists P,b) : a \text{ anrn } b\ .\ n = \text{slrnnm}'(a,P)$

$- \text{slpnnm}'(b,P)]$

D42. The *degree of syllable reaction perseveration* (perndg) during a syllable-presentation cycle is the difference obtained by subtracting the syllable reaction number of the syllable reaction involved from the syllable presentation number of the syllable presentation involved.

D42'. $\text{perndg} = \hat{n}\hat{a}[(\exists P,b) : a \text{ pern } b\ .\ n = \text{slpnnm}'(b,P)$

$- \text{slrnnm}'(a,P)]$

D43. An *anticipatory error* (aner) is an anticipatory reaction of a degree greater than 1.

D43'. $\text{aner} = \hat{a}[a \in \text{D}'\text{anrn}\ .\ \text{anrndg}'a > 1]$

D44. A *perseverative error* (peer) is any perseverative reaction.

D44'. $\text{peer} = \text{D}'\text{pern}$

D45. A *false reaction* (fsrn) during a syllable-presentation cycle is a reaction which is not congruent with any syllable exposure involved in that syllable presentation cycle.

$$\text{D45}'.\quad \text{fsrn} = \hat{a}[(\exists P) : P \,\varepsilon\, \text{slpncy} \,.\, \text{bg}'\text{B}'P \leq \text{bg}'a \leq \text{nd}'\text{B}'\breve{P} + \text{slinpnin}'P \,.\, a \;\text{rn}(\text{slpnsb}'\text{B}'P) \,.\, \sim(\exists b) \,.\, a \;\text{slcg}\; b \,.\, \text{cn}'(b, \text{slpnsb}'\text{B}'P) \,\varepsilon\, \text{C}'P\,]$$

D46. A *presentation failure* (pnfl) in a syllable-presentation cycle is any syllable presentation (except the last) during which, and during the syllable-inter-presentation interval immediately following which, no correct reaction begins.

Presentation failures thus include both those syllable presentations during which false reactions occur and those during which there is failure to give any reaction whatever.

$$\text{D46}'.\quad \text{pnfl} = \hat{a}\hat{P}[P \,\varepsilon\, \text{slpncy} \,.\, a \,\varepsilon\, \vec{P}'\text{B}'\breve{P} \,.\, \sim(\exists b) \,.\, b \;\text{ctrn}\; a]$$

D47. The *critical syllable-learning cycle* (crsllncy_m) for any syllable presentation number n in a rote practice, with reference to the number m as a criterion of "complete" learning, is the first syllable-presentation cycle of that rote practice to be followed by m or more successive syllable-presentation cycles in each of which the nth syllable presentation is not a presentation failure.

$$\text{D47}'.\quad \text{crsllncy}_m = \hat{P}\hat{n}\hat{T}[(Q,a) : \text{R}'(P,T) + m \geq \text{R}'(Q,T) > \text{R}'(P,T) \,.\, n = \text{slpnnm}'(a,Q) \,.\, \supset \,.\, a \;\dot{-}\text{pnfl}\; Q :. \sim(\exists S) :. \text{R}'(S,T) < \text{R}'(P,T) : (Q,a) : \text{R}'(S,T) + m \geq \text{R}'(Q,T) > \text{R}'(S,T) \,.\, n = \text{slpnnm}'(a,Q) \,.\, \supset \,.\, a \;\dot{-}\text{pnfl}\; Q]$$

D48. The *first success* (ftsc) for any syllable-presentation number n in a rote practice is the ordinal number (R) of the first syllable-presentation cycle of that rote practice in which the nth syllable presentation is not a presentation failure.

$$\text{D48}'.\quad \text{ftsc} = \hat{k}\hat{n}\hat{T}[T \,\varepsilon\, \text{rtpr} : (\exists P,a) : P \,\varepsilon\, \text{C}'T \,.\, n = \text{slpnnm}'(a,P) \,.\, \sim(a \;\text{pnfl}\; P) \,.\, k = \text{R}'(P,T) \,.\, \sim(\exists Q,c) \,.\, \text{R}'(Q,T) < k \,.\, n = \text{slpnnm}'(c,Q) \,.\, \sim(c \;\text{pnfl}\; Q)]$$

D49. The *last failure* (ltfl_m) for any syllable-presentation number n in a rote practice, with reference to the number m as a criterion of "complete" learning is the ordinal number (R) of the critical syllable-presentation cycle for the syllable-presentation number n.

$$\text{D49}'.\quad \text{ltfl}_m = \hat{k}\hat{n}\hat{T}[(\exists P) : k = \text{R}(P,T) \,.\, P \;\text{crsllncy}_m\; (n,T)]$$

D50. A *reaction oscillation* (rnos) for any syllable-presentation number n in a rote practice is a sequence of successive syllable-presentation cycles the first part of which consists of one or more successive syllable-presentation cycles in which occur correct reactions whose reaction numbers are $n+1$, and the second part of which consists of one or more successive syllable-presentation cycles in which occur no correct reactions whose reaction numbers are $n+1$, the whole sequence not being part of a more inclusive sequence satisfying the above conditions.

D50′. rnos = $\hat{U}\hat{n}\hat{T}[T \in \mathrm{rtpr} : \dot{\exists}!U : U \dot{\cap} T^2 \subset U^2 :. (\exists P, Q)(S)$

$:. \mathrm{R}'(P,T) + 1 = \mathrm{R}'(Q,T) :. S \in \mathrm{C}'U : \supset : S\,T\,P \,.\, \sim(\exists a) \,.\, n$

$= \mathrm{slpnnm}'(a,S) \,.\, a \ \mathrm{pnfl}\ S \,.\, \mathrm{v} \,.\, Q\,T\,S \,.\, (\exists b) \,.\, n$

$= \mathrm{slpnnm}'(b,S) \,.\, b \ \mathrm{pnfl}\ S]$

D51. The *critical series-learning cycle* ($\mathrm{crsrlncy}_m$) of a rote practice, with reference to m as a criterion of "complete" learning, is the first syllable-presentation cycle of that rote practice to be followed by m or more successive syllable-presentation cycles in each of which no reaction failure occurs.

Practice varies somewhat among experimentalists as to the criterion (m) chosen to mark the completion of the learning of a rote series. A common practice is to give m the value of 2, i.e., the subject is required to continue the repetitions of the series as a whole until two successive repetitions have occurred in which no reaction failures take place.

D51′. $\mathrm{crsrlncy}_m = \hat{P}\hat{T}[(Q,a) : \mathrm{R}'(P,T) + m \geq \mathrm{R}'(Q,T)$

$> \mathrm{R}'(P,T) \,.\, a \in \mathrm{C}'Q \,.\, \supset \,.\, a \dot{-}\mathrm{pnfl}\ Q :. \sim(\exists S) :. \mathrm{R}'(S,T)$

$< \mathrm{R}'(P,T) : (Q,a) : \mathrm{R}'(S,T) + m \geq \mathrm{R}'(Q,T) > \mathrm{R}'(S,T) \,.$

$a \in \mathrm{C}'Q \,.\, \supset \,.\, a \dot{-}\mathrm{pnfl}\ Q]$

D52. A *completed rote practice* (cprtpr_m) or a *learned rote series*, with reference to the number m as a criterion of "complete" learning, is a rote practice in which the ordinal number (R) of the last syllable-presentation cycle is greater by m than the ordinal number of the critical series-learning cycle of that rote practice.

D52′. $\mathrm{cprtpr}_m = \hat{T}[T \in \mathrm{rtpr} \,.\, \mathrm{R}'(\mathrm{B}'\breve{T}, T)$

$= \mathrm{R}'(\mathrm{crsrlncy}_m{}'T, T) + m]$

D53. The *recall* (rcpb) or *recall probability* of the nth syllable presentation of the syllable-presentation cycles having the same value of N and the same ordinal position R in their respective rote practices, all such rote practices constituting an arbitrarily chosen set of rote practices, is the ratio of the number of such rote practices in which the nth syllable presentation of the Rth syllable-presentation cycle is not a presentation failure, to the total number of members of the set of rote practices in question.

D53′. rcpb = $\hat{x}\hat{n}\hat{m}\hat{k}\hat{\rho}\{\rho \subset \mathrm{rtpr} : (P,T) : T \in \rho \,.\, P \in \mathrm{C}'T \,.$

$\supset . \text{N}'P = m : (\exists i, j) : j = \text{Nc}'\rho : i = \text{Nc}'\hat{U}[U \varepsilon \rho : (S, a) : k = \text{R}'(S, U) . n = \text{slpnnm}'(a, S) . \supset . a \dot{-} \text{pnfl } S] : x = i \div j\}$

D54. The *syllable presentation of origin* (slpnor) of a syllable-presentation cycle is the syllable presentation of the syllable-presentation cycle in question, the beginning of which is concurrent with the beginning of the stimulus trace in question.

Throughout the mathematical portions of the present monograph the symbol s is used to designate the number of the syllable of origin of a stimulus trace.

D54'. slpnor = Cnv'tr

D55. A *stimulus trace segment* (${}_s\text{tr}_n$) of a stimulus trace of a syllable-presentation cycle is that portion of such stimulus trace originating in syllable s which is concurrent with the syllable-presentation interval, n, of such syllable-presentation cycle, where $n \geq s$.

D55'. ${}_s\text{tr}_n = \hat{Q}\hat{P}\{(\exists a, b) : s = \text{slpnnm}'(a, P) \leq n = \text{slpnnm}'(b, P) . Q = \hat{x}\hat{y}\hat{c}[x = \text{bg}'b \geq \text{bg}'c . y = \text{nd}'b + \text{slinpnin}'P \leq \text{nd}'c . c = \text{tr}'a] . \dot{\exists}!Q\}$

D56. The *trace-segment number* (trsgnm) of a stimulus trace of a syllable-presentation cycle is the number of the syllable-presentation interval of such syllable-presentation cycle with which the stimulus trace segment is concurrent.

D56'. $\text{trsgnm} = \hat{n}\hat{Q}[(\exists P, s) : Q = {}_s\text{tr}_n'P]$

D57. The *initial segment* (${}_s\text{tr}_s$) of a stimulus trace of a syllable-presentation cycle is that segment of such stimulus trace whose trace-segment number is the same as that of the syllable presentation of origin of the stimulus trace in question.

No symbolic logic definition is given here, because ${}_s\text{tr}_s$ is already defined in **D55'**.

D58. The *perseverative segments* (${}_s\text{tr}_{s+1\ldots N}$) of a stimulus trace originating in syllable presentation s of a syllable-presentation cycle are those segments of such stimulus trace whose trace-segment numbers extend from $s+1$ to N, inclusive.

D58'. ${}_s\text{tr}_{s+1\ldots N} = \hat{Q}\hat{P}[(\exists n) : s < n . Q = {}_s\text{tr}_n'P]$

D59. A *compound stimulus trace* ($\|{}_{0\ldots n}\text{tr}_n\|$) of a syllable-presentation cycle is that combination of stimulus-trace segments all of whose trace-segment numbers are the same.

D59'. $\|{}_{0\ldots n}\text{tr}_n\| = \hat{\lambda}\hat{P}\{\lambda = \hat{Q}[(\exists s) . Q = {}_s\text{tr}_n'P]\}$

D60. The *compound-trace number* (cptrnm) of a compound stimulus trace ($\|{}_{0\ldots n}\text{tr}_n\|$) is the trace-segment number common to the trace segments involved in the compound trace in question.

D60'. cptrnm = $\hat{n}\hat{\lambda}[(\exists P) : \lambda = \|_{0\ldots n}\mathrm{tr}_n\| \text{'} P]$

D61. The *reaction latency* (*T*) of a correct syllable reaction during a rote practice is the length of time elapsing between the beginning of a compound stimulus trace ($\|_{0\ldots n}\mathrm{tr}_n\|$) and the beginning of the syllable reaction in question.

In symbolic logic a non-italic T is used for this concept instead of the italic *T*, because the italic *T* is generally employed in the symbolic logic to represent a quite different concept, viz., an arbitrary relation between relations. This latter practice is in conformity with the symbolic usage of Woodger (*66*) and Whitehead and Russell (*65*).

D61'. $\mathrm{T} = \hat{x}\hat{a}[(\exists b) : b\ \mathrm{ctrn}\ a\ .\ x = \mathrm{bg}\text{'}\mathrm{tr}\text{'}a - \mathrm{bg}\text{'}b\ .$ $a \in \mathrm{slpn}]$

D62. The *cycle termination* (cytm) of a syllable-presentation cycle is that point of time which is later than the end of the last syllable presentation of the syllable-presentation cycle by an amount of time equal to the syllable-inter-presentation interval of the syllable-presentation cycle.

D62'. cytm = $\hat{t}\hat{P}[P \in \mathrm{slpncy}\ .\ t = \mathrm{nd}\text{'}\mathrm{B}\text{'}\breve{P} + \mathrm{slinpnin}\text{'}P]$

D62½. The *massed practice number* (msprnm) of a massed practice is the number of massed practices up to and including the massed practice in question which involve syllable-presentation cycles congruent with those of the latter.

D62½'. msprnm = $\hat{n}\hat{T}\{T \in \mathrm{mspr}\ .\ n = \mathrm{Nc}\text{'}\hat{U}[(\exists P, Q) :$ $U \in \mathrm{mspr}\ .\ P \in \mathrm{C}\text{'}T\ .\ Q \in \mathrm{C}\text{'}T\ .\ Q\ \mathrm{slpncycg}\ P\ .\ \mathrm{prtm}\text{'}U$ $\leq \mathrm{prtm}\text{'}T]\}$

D63. $\mathcal{E}_{n,r}$ in a syllable-presentation cycle is the total excitatory potential at any given instant for a reaction having reaction number r to occur during the syllable presentation in question or during the syllable-inter-presentation interval immediately following it, where $r = n+1$.

D63'. $\mathcal{E}_{n,r} = \hat{x}\hat{t}\hat{P}[x\ \mathrm{E}(t, n, P)\ .\ r = n+1]$

D64. The *increment of excitatory potential* (ΔE) is the difference between the amount of excitatory potential ($\mathcal{E}_{n,r}$) at the beginning of a syllable-presentation cycle and the amount at its cycle termination, provided that the syllable-presentation cycle in question occurs in the first massed practice of a rote learning.

D64.' $\Delta\mathrm{E} = (\iota x)[(\exists T, P, n) : \mathrm{msprnm}\text{'}T = 1\ .\ P \in \mathrm{C}\text{'}T\ .$ $n < \mathrm{N}\text{'}P\ .\ x = \mathcal{E}_{n,n+1}\text{'}(\mathrm{cytm}\text{'}P, P) - \mathcal{E}_{n,n+1}\text{'}(\mathrm{bg}\text{'}\mathrm{B}\text{'}P, P)]$

D65. ${}^{M,R,N}\mathrm{E}_{n,r}$ or, more briefly, $\mathrm{E}_{n,r}$, where $r = n+1$, is at any instant of time, the amount of excitatory potential due to R syllable-presentation cycles of the Mth massed practice of a rote learning containing N syllables, for reaction r to begin during the nth syllable-presentation interval of any such syllable-presentation cycle, provided that the instant of time in question falls on or after the termination of the Rth syllable-presentation cycle and before the termination of the $(R+1)$st syllable-presentation cycle, or, if there is no $(R+1)$st syllable-presentation cycle in the massed practice in question then before the termination of the massed practice. Quantitatively ${}^{M,R,N}\mathrm{E}_{n,r}$ is defined as equal to $R \times \Delta\mathrm{E}$.

In the symbolic logic the notation ${}^{i,j,\mathrm{N}'P}\mathrm{E}_{n,r}$ is used instead of ${}^{M,R,N}\mathrm{E}_{n,r}$.

D65.1' ${}^{i,j,\mathrm{N}'P}\mathrm{E}_{n,r} = \hat{x}\hat{t}\hat{P}\ [(\exists\, Q, T)\ .\ \mathbf{msprnm}'T = i\ .\ \mathrm{R}'(Q,T) = j\ .$ $Q\ \mathbf{slpncycg}\ P\ .\ \mathbf{cytm}'Q \leq t < \mathbf{cytm}'Q + (1+\mathrm{N}'P) \times \mathbf{slpnin}'Q\ .\ t < \mathbf{prtm}'T\ .\ x = \Delta\mathrm{E} \times \mathrm{R}'(Q,T) :\equiv: x \neq 0\ :.\ r = n+1]$

D65.2' $\mathrm{E}_{n,r} = \hat{x}\hat{t}\hat{P}[(\exists i, j) : x = {}^{i,j,\mathrm{N}'P}\mathrm{E}_{n,r}{}'(t,P)]$

D66. The *point of conditioning* (ptcd) of $\mathrm{E}_{n,r}$ is n.

D66'. $\mathbf{ptcd} = \hat{n}\hat{U}[(\exists r) : U = \mathrm{E}_{n,r}]$

D66½'. $\mathbf{invl} = \hat{u}\hat{t}\hat{P}\hat{U}[u \neq 0 \therefore \equiv \therefore \mathrm{C}'U \subset \mathbf{rtln}'P : u = \mathbf{prtm}'U - \mathbf{bg}'\mathrm{B}'\mathrm{B}'U\ .\ \mathbf{prtm}'U < t\ .\ \mathrm{v}\ .\ u = t - \mathbf{bg}'\mathrm{B}'\mathrm{B}'U\ .\ \mathbf{bg}'\mathrm{B}'\mathrm{B}'U < t \leq \mathbf{prtm}'U]$

D67. $\mathrm{E}_{n,r}(t)$ is the amount of excitatory potential due to a given massed practice at t units of time following the termination of that massed practice.

D67'. $\mathrm{E}_{n,r}(t) = \hat{x}\hat{P}\hat{T}[x \neq 0 :\equiv: t \geq 0\ .\ (\exists v)\ .\ v = t - \sum_{u\ \mathrm{invl}\ (t,P,T)} u\ .\ x = \mathrm{R}'(\mathrm{B}'\breve{T}, T) \times \Delta\mathrm{E} \times e^{-bv}\ .\ \mathrm{C}'T \subset \mathbf{rtln}'P \therefore b > 0]$

D68. The compound stimulus trace ($\|_{0\ldots n}\dot{\mathrm{tr}}_n\|$) associated with the excitatory potential ($\mathcal{E}_{n,r}$) of the Rth syllable-presentation cycle of a rote learning, is said to *evoke* (ev) the reaction r if that reaction begins during such compound stimulus trace of the Rth syllable-presentation cycle and if $r = n + 1$.

D68'. $\mathbf{ev} = \hat{\lambda}\hat{a}\,[(\exists b, P) : \mathbf{slpnnm}'(b,P) = \mathbf{cptrnm}'(\lambda, P)\ .\ a\ \mathbf{ctrn}\ b]$

D69. ${}_s\mathcal{I}_{n,r}$ in a syllable-presentation cycle is a component of the total inhibitory potential acting to diminish $\mathcal{E}_{n,n+1}$, the component in question being the one associated with the trace segment ${}_s\mathrm{tr}_n$ and arising as a consequence of the growth of $\mathcal{E}_{r-1,r}$.

D69'. ${}_s\mathcal{I}_{n,r} = \hat{x}\hat{t}\hat{P}[x\ \mathrm{I}\ (t, n, P, s, r)]$

D70. $\mathcal{I}_n$ in a syllable-presentation cycle is the total inhibitory potential acting at a given instant to reduce $\mathcal{E}_{n,n+1}$.

D70'. $\mathcal{I}_n = \hat{x}\hat{t}\hat{P}[x = \sum_{r=n+1}^{\mathrm{N}'P} \sum_{s=0}^{n} {}_s\mathcal{I}_{n,r}]$

D71. The *increment of inhibitory potential* ($\Delta\, {}^{N}_{s}\mathrm{I}_{n,r}$) is the difference between the amount of inhibitory potential at the beginning of a syllable-presentation cycle and the amount at its cycle termination, provided that the syllable-presentation cycle in question occurs in the first massed practice of a rote learning and provided that $s \leq n$ and $n \leq r-1$ and $r \leq N$.

D71'. $\Delta\, {}^{N'P}_{s}\mathrm{I}_{n,r} = (\iota x)[(\exists T, Q, n) : \mathrm{msprnm}'T = 1 \,.\, Q \in \mathrm{C}'T \,.\, s \leq n \leq r-1 \,.\, n \leq \mathrm{N}'Q \,.\, x = {}_{s}\mathcal{I}_{n,r}'(\mathrm{cytm}'Q, Q) - {}_{s}\mathcal{I}_{n,r}(\mathrm{bg}'\mathrm{B}'Q, Q) \,.\, Q \ \mathrm{slpncycg}\ P]$

D72. $\Delta\mathrm{K}$ is the amount of inhibitory potential associated with the initial segment (${}_{n}\mathrm{tr}_{n}$) of the stimulus trace (${}_{n}\mathrm{tr}_{n\ldots N}$), the inhibitory potential in question resulting from one syllable-presentation cycle involving syllable reaction r, where $r = n+1$, i.e., $\Delta\mathrm{K} = \Delta\, {}^{N}_{n}\mathrm{I}_{n,n+1}$.

D72'. $\Delta\mathrm{K} = (\iota x)[(\exists n, P) : x = \Delta\, {}^{N'P}_{n}\mathrm{I}_{n,n+1}]$

D73. F is the factor of reduction according to which the amount of inhibitory potential associated with the several component stimulus-trace segments, ${}_{n}\mathrm{tr}_{n}$, ${}_{n-1}\mathrm{tr}_{n}$, ${}_{n-2}\mathrm{tr}_{n}$, ... ${}_{n-n}\mathrm{tr}_{n}$ of a compound stimulus trace ($\|_{0\ldots n}\mathrm{tr}_{n}\|$) decreases from the basic value $\Delta\, {}^{N}_{n}\mathrm{I}_{n,n+1}$, i.e., $\mathrm{F} = \Delta\, {}^{N}_{s}\mathrm{I}_{n,n+1} \div \Delta\, {}^{N}_{s-1}\mathrm{I}_{n,n+1}$, where $s \leq n$.

D73'. $\mathrm{F} = (\iota x)\{(\exists s, n, P) : x = [\Delta\, {}^{N'P}_{s}\mathrm{I}_{n,n+1} \div \Delta\, {}^{N'P}_{s-1}\mathrm{I}_{n,n+1}] \,.\, x > 0 \,.\, s < n+1 \leq 100\}$

D74. $$\Delta\, {}^{N}_{s}\mathrm{I}_{n} = \begin{cases} \|\Delta\, {}^{N}_{s}\mathrm{I}_{n,n+1\ldots N}\| \\ \sum\limits_{r=n+1}^{N} \Delta\, {}^{N}_{s}\mathrm{I}_{n,r} \end{cases}$$

D74'. $$\Delta\, {}^{N'P}_{s}\mathrm{I}_{n} = \sum_{r=n+1}^{N'P} \Delta\, {}^{N'P}_{s}\mathrm{I}_{n,r}$$

D75. $$\Delta\, {}^{N}\mathrm{I}_{n} = \begin{cases} \|\Delta\, {}^{N}_{0\ldots n}\mathrm{I}_{n,n+1\ldots N}\| \\ \sum\limits_{s=0}^{n} \sum\limits_{r=n+1}^{N} \Delta\, {}^{N}_{s}\mathrm{I}_{n,r} \\ \sum\limits_{s=0}^{n} \Delta\, {}^{N}_{s}\mathrm{I}_{n} \end{cases}$$

D75'. $$\Delta\, {}^{N'P}\mathrm{I}_{n} = \sum_{s=0}^{n} \Delta\, {}^{N'P}_{s}\mathrm{I}_{n}$$

D76. $${}^{N}\mathrm{J}_{n} = \frac{\Delta\, {}^{N}\mathrm{I}_{n}}{\Delta\mathrm{K}}$$

D76'. $${}^{N'P}\mathrm{J}_{n} = \Delta\, {}^{N'P}\mathrm{I}_{n} \div \Delta\mathrm{K}$$

D77. ${}^{M,R,N}\mathrm{I}_n$, or, more briefly, I_n, is at any instant of time the amount of inhibitory potential due to R syllable-presentation cycles of the Nth massed practice in a rote learning of N syllables, which opposes the beginning of reaction r during the nth syllable-presentation interval of any such syllable-presentation cycle, where $r = n+1$; provided that the instant of time in question falls on or after the termination of the Rth syllable-presentation cycle and before the termination of the $(R+1)$st syllable-presentation cycle, or, if there is no $(R+1)$st syllable-presentation cycle, in the massed practice in question, then before the termination of the massed practice. Quantitatively ${}^{M,R,N}\mathrm{I}_n$ is defined as equal to $R \times \Delta\,{}^{N}\mathrm{I}_n$.

In the symbolic logic the notation ${}^{i,j,\mathrm{N}'P}\mathrm{I}_n$ is used instead of ${}^{M,R,N}\mathrm{I}_n$.

D77.1' ${}^{i,j,\mathrm{N}'P}\mathrm{I}_n = \hat{x}\hat{t}\hat{P}[(\exists Q,T)\ .\ i = \mathrm{msprnm}'T\ .\ Q \in \mathrm{C}'T\ .\ j = \mathrm{R}'(Q,T)\ .\ Q\ \mathrm{slpncycg}\ P\ .\ \mathrm{cytm}'Q \leq t < \mathrm{cytm}'Q + (1+\mathrm{N}'P) \times \mathrm{slpnin}'Q\ .\ t < \mathrm{prtm}'T\ .\ x = \Delta^{\mathrm{N}'P}\mathrm{I}_n \times \mathrm{R}'(Q,T) : \equiv : x \neq 0]$

D77.2'. $\mathrm{I}_n = \hat{x}\hat{t}\hat{P}[(\exists i,j) : x = {}^{i,j,\mathrm{N}'P}\mathrm{I}_n]$

D78. $\mathrm{I}_n(t)$ is the amount of inhibitory potential due to a given massed practice at t units of time following the termination of that massed practice.

D78'. $\mathrm{I}_n(t) = \hat{x}\hat{P}\hat{T}[x \neq 0 :\equiv: t \geq 0\ .\ (\exists v)\ .\ v = t - \sum_{u\ \mathrm{invl}\ (t,P,T)} u\ .\ x = \mathrm{R}'(\mathrm{B}'\breve{T},T) \times \Delta^{\mathrm{N}'P}\mathrm{I}_n \times e^{-dv}\ .\ \mathrm{C}'T \subset \mathrm{rtln}'P\ .\!: \mathrm{d} \times \Delta^{\mathrm{N}'P}\mathrm{I}_n > \mathrm{b} \times \Delta\mathrm{E}]$

D79. The *effective excitatory potential* ($\bar{\mathrm{E}}_{n,r}$) or, more briefly, $\bar{\mathrm{E}}$ of a given rote series with a given subject, is the difference between $\mathrm{E}_{n,r}$ and I_n, i.e., $\bar{\mathrm{E}}_{n,r} = \mathcal{E}_{n,r} - \mathcal{I}_n$.

D79.1' $\bar{\mathrm{E}}_{n,r} = \hat{x}\hat{t}\hat{P}[\mathcal{E}_{n,r}{}'(t,P) - \mathcal{I}_n{}'(t,P)]$

D79.2' $\bar{\mathrm{E}} = \hat{x}\hat{t}\hat{n}\hat{P}[x = \bar{\mathrm{E}}_{n,n+1}{}'(t,P)]$

D80. *Reminiscence* (rm) is a net increase in excitatory potential $\bar{\mathrm{E}}_{n,r}$ during a period of time in which no syllable presentations occur to the subject in question.

D80'. $\mathrm{rm} = \hat{x}\hat{t}_1\hat{t}_2\hat{n}\hat{P}[t_1 < t_2\ .\ x = \bar{\mathrm{E}}_{n,n+1}{}'(t_2,P) - \bar{\mathrm{E}}'(t,P) > 0 : \sim(\exists a,b) : a\ \mathrm{slpnsb}'\mathrm{B}'P\ .\ a\ \mathrm{slpnsb}\ b\ .\ t_1 < \mathrm{nd}'b\ .\ \mathrm{v}\ .\ \mathrm{bg}'b < t_2]$

D81. The *mean reaction threshold* (mernth) of a class of syllable presentations is the mean of such reaction thresholds.

In the mathematical portions of the present monograph, the symbol ℓ is used in place of the logical symbol rnth to indicate the value of the reaction threshold which, according to Postulate 15, varies from syllable presentation to syllable presentation according to the normal probability integral.

D81'. $\text{mernth} = \hat{x}\hat{\alpha}[x = (\sum_{a\varepsilon\alpha} \text{rnth}'a) \div \text{Nc}'\alpha]$

D82. L is the limit approached by the mean reaction thresholds of successively larger and larger classes of syllable presentations, each such class including all syllable presentations beginning over some stretch of time.

The existence of such a limit is assumed by Postulate 16.

D82'. $\text{L} = (\imath x)\{(y)(\exists j)(\alpha,z,w,v) : \text{Nc}'\alpha > j \,.\, \alpha$

$\hat{a}[a \; \varepsilon \; \text{slpn} \,.\, z < \text{bg}'a < w] \,.\, v = \text{mernth}'\alpha \,.\, y > 0 \,.$

$\supset \,.\, v-y < x < v+y\}$

D83. The *standard deviation of reaction threshold* (sdrnth) of a class of syllable presentations is the standard deviation of their reaction thresholds based on the mean reaction threshold of the class.

D83'. $\text{sdrnth} = \hat{x}\hat{\alpha}\{x^2 = [\sum_{a\varepsilon\alpha} (\text{rnth}'a - \text{mernth}'\alpha)^2]$

$\div \text{Nc}'\alpha \,.\, x > 0\}$

D84. σ is the *limit approached by the standard deviation of reaction thresholds* of successively larger classes of syllable presentations, each such class including all syllable presentations beginning over some stretch of time.

The existence of such a limit is assumed by Postulate 17.

D84'. $\sigma = (\imath x)\{(y)(\exists j)(\alpha,z,w,v) : \text{Nc}'\alpha > j \,.\, \alpha$

$= \hat{a}[a \; \varepsilon \; \text{slpn} \,.\, z < \text{bg}'a < w] \,.\, v = \text{sdrnth}'\alpha \,.\, y > 0 \,.$

$\supset \,.\, v - y < x < v + y\}$

D85. The *frequency* (fq) of a class of syllable presentations with respect to v and u is the number of members of that class which have reaction thresholds differing from v-L by less than u.

D85'. $\text{fq} = \hat{h}\hat{v}\hat{u}\hat{\alpha}\,\{(\exists x,y) : \alpha = \hat{a}[a \; \varepsilon \; \text{slpn} \,.\, x < \text{bg}'a$

$< w] \,.\, h = \text{Nc}'\hat{b}\,[b \; \varepsilon \; \alpha \,.\, v\text{-L-}u < \text{rnth}'b < v\text{-L+}u] \,.$

$u > 0\}$

D86. The *functional combination* (Σ) of excitatory potentials (E) and of inhibitory potentials (I), is that method of combination whereby they summate in determining, in conjunction with other conditions, whether under these conditions the syllable reaction (r) involved shall occur, and in determining, in case it does occur, the latency (T) of such reaction.

Postulates

POSTULATE 1

A.[2] *During all syllable-presentation cycles of the learning of any rote series, stimulus traces* $(\| {}_{s}\mathrm{tr}_{s \ldots N} \|)$ U6 *extend from the beginning of every syllable presentation (s) through the remainder of the syllable-presentation cycle but no further.*

A'. $P \in \mathrm{slpncy} \,.\, a \in C'P \,.\, \supset \,.\, E!\mathrm{tr}'a \,.\, \mathrm{bg}'\mathrm{tr}'a = \mathrm{bg}'a \,.\, \mathrm{nd}'\mathrm{tr}'a = \mathrm{cytm}'P$

B.[3] Let us suppose that we have the rote series:

Stimulus syllable:	KEM,	FAP,	ZIT,	YOD,	NAR,	JUB,	etc.
Number of syllable:	0	1	2	3	4	5	...

Each syllable (e.g., *ZIT*) as a bit of print is first of all primarily a visual stimulus object. Secondly, to persons who have acquired reading habits, this syllable as a visual stimulus gives rise to a vocal reaction. This vocal reaction, while totally different from the visual stimulation, bears within a given culture a stable relationship to it; the two are said to be congruent (U5). Accordingly we say that the syllable reaction of saying "ZIT" corresponds to (is congruent with) the printed stimulus consisting of the juxtaposed letters Z, I, and T. In order to avoid ambiguity in referring to these two distinct but intimately related variables, the printed syllable as a visual stimulus presented by the memory drum will be italicized, thus: *ZIT*; whereas the corresponding syllable reaction will be represented by the syllable placed within quotation marks, thus: "ZIT". The syllable-of-origin of stimulus traces will be indicated by the letters of the syllable without either underscoring or quotation marks, e.g., ZIT.

The concrete meaning of this postulate is shown diagrammatically in Figure 1. Observe, first, that during the

2. Each postulate is written out informally in ordinary word language under A, and in a formal and precise manner in the notation of symbolic logic under A'. The statements in symbolic logic have been included with the object of demonstrating its use in scientific theory. For the reader unacquainted with symbolic logic, the A-sections will be quite adequate for an understanding of the development of the system as a whole, and the A'-sections should be completely disregarded.

3. In the B-portion of this and several of the following postulates, evidence of various kinds is presented relative to the independent status of the postulates in question, chiefly with the object of orienting the reader. This should not be mistaken as an attempt to deduce the postulate; a theorist cannot be asked to deduce his postulates.

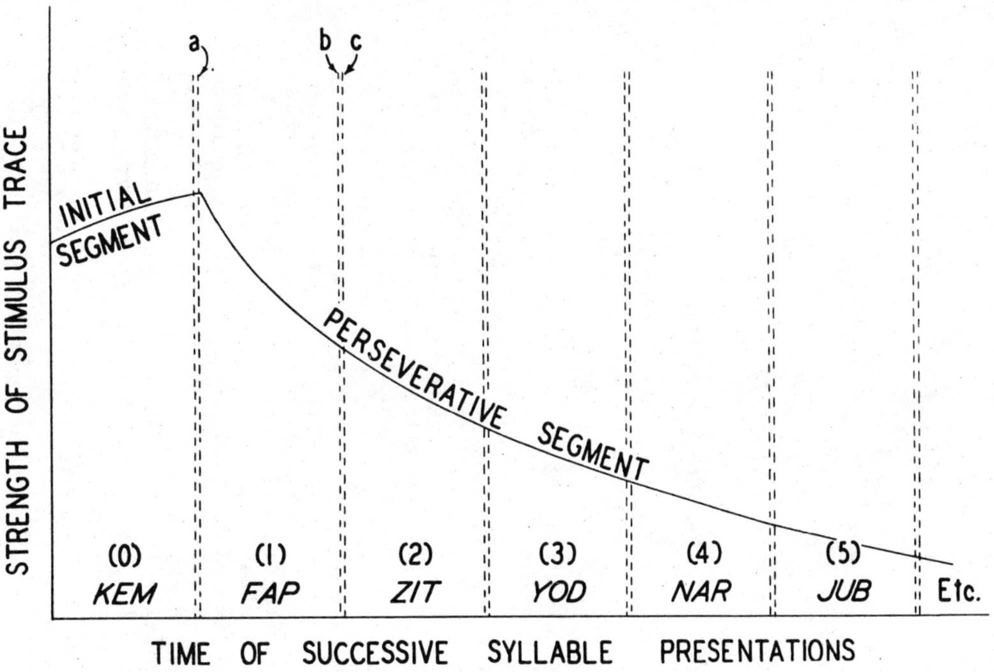

Figure 1. Diagrammatic representation of a single stimulus trace (${}_{0}tr_{0...5...}$) the syllable of origin being *KEM*. Exactly identical stimulus traces are assumed to originate at all the other syllables of the rote series but, in the interest of simplicity of exposition, they have not been represented here.

From *a* to *b* is one *syllable exposure*, that of *FAP* (see U1) or, if a subject is present, a *syllable presentation* (D17); from *b* to *c* is one *syllable-inter-presentation interval* (D23); from *a* to *c* is one *syllable-presentation interval* (D24). In subsequent diagrams of this type, the syllable-inter-presentation interval, because of its practical unimportance, will not be explicitly represented.

presentation of the stimulus syllable *KEM*, the *initial* segment of the trace increases in strength but with a negative acceleration. Secondly, after the termination of the presentation of the stimulus syllable, the stimulus trace persists but with a progressively weakening intensity, the rate of diminution being negatively accelerated. Postulate 1 assumes that stimulus traces exactly similar to that shown in Figure 1 for *KEM* are initiated by *FAP*, *ZIT*, *YOD*, and all the remaining stimulus syllables of the rote series. (See Figure 2.)

The notion of the stimulus trace seems to have been originated by Pavlov as early as 1906 (*48*, 93). In an attempt to explain the phenomena of what he called *trace conditioned reflexes*, Pavlov remarks:

> "Any convenient external stimulus is applied to the animal and continued for 1/2 to 1 minute. After another definite interval of 1-3 minutes, food or a rejectable substance is introduced into the mouth. It is found that after several repetitions of this routine the stimulus will not itself evoke any reaction; neither will its disappearance; but the appropriate reaction will occur after a definite interval, the after-effect of the excitation caused by the stimulus being the operative factor . . . . every stimulus must leave a trace on the nervous system for a greater or less time - a fact which has long been recognized in physiology under the name of after effect." (*47*, 39-40.)

As was his custom, Pavlov left the concept of the stimulus trace in a non-quantitative state, which places great limitations in the way of the determination of its scientific implications. Recently others (*31*, 16) have sharpened the concept by formulating it as a negatively accelerated falling gradient, much as it appears in Figure 1.[4]

The concept of the stimulus trace as formulated by Pavlov comprised only the "perseverative segment" (Figure 1). The present formulation explicitly includes, in addition, the "initial segment" (Figure 1) and also postulates the general shape of the gradients of both segments. It is here supposed that the initial segment of the stimulus trace would be a rising gradient because of the presumptive summation of the traces left from the early instants of stimulation with the actual results of stimulation at each later instant. Presumably this gradient would approximate the positive growth variety represented by the equation, $X = A - Ae^{-ft}$. It will be noted from the form of this equation that the initial segment begins at zero rather than with a comparatively large value, as suggested in Figure 1. Analogously, the perseverative segment would approximate a negative growth gradient of the form, $X = Ae^{-ft}$.

The ground for supposing that the stimulus trace, following the termination of the stimulus, will diminish according to a negative growth function is mainly analogical. A large

4. This quantitative refinement of the concept is not a part of the present system, but it is believed that it will be required in subsequent developments of the theory of rote learning. A very similar hypothesis was put forward by Köhler in 1923. (See Köhler, W., Zur Theorie des Sukzessivvergleichs und der Zeitfehler, *Psychol. Forsch.*, 1923, *4*, 115-175.

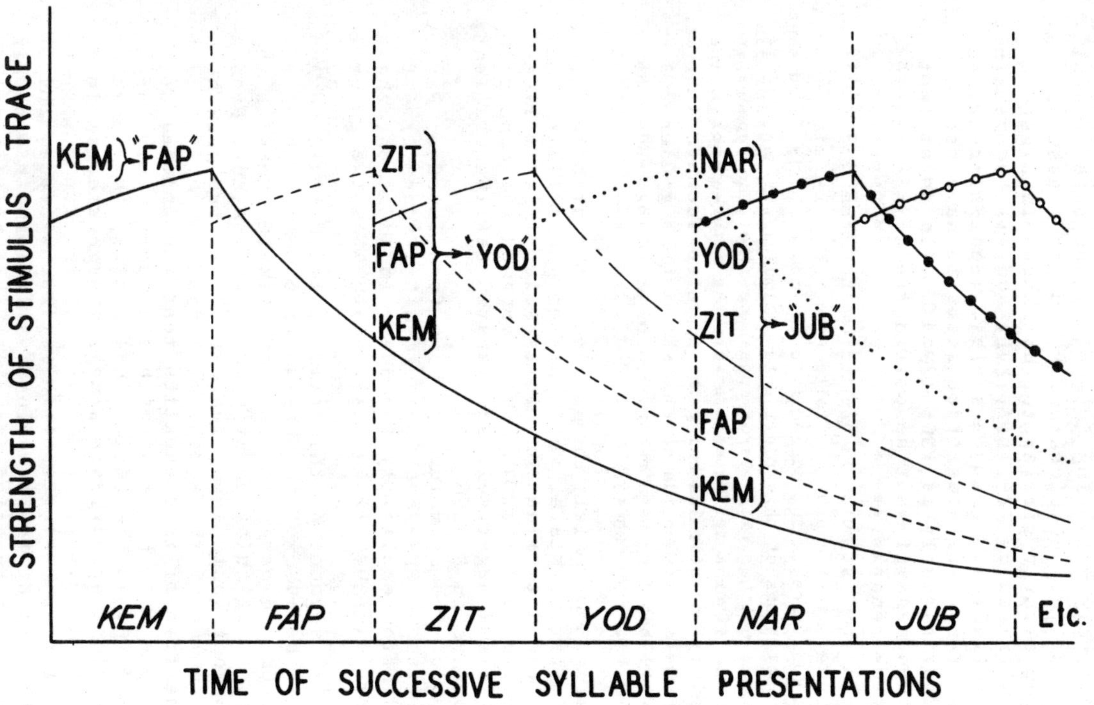

Figure 2. Diagrammatic representation of the overlapping stimulus traces of the first few syllables of a rote series in process of being learned. The braces indicate the phases of the several stimulus traces making up the compound stimulus trace to which three typical syllable reactions are finally conditioned (associated). The detailed indication of the compound stimulus traces to which the remaining syllable reactions are conditioned are omitted from the diagram in the interest of simplicity of exposition.

number of processes in which the amount of loss appears to be a constant proportion of the portion yet remaining have followed rather closely a course represented by the above equation. No hypothesis is offered at this time as to the physical nature of the trace itself, though the present type of analysis ought to suggest promising modes of experimental attack on the latter problem.

The hypothesis that the stimulus trace increases in intensity during the action of the stimulus offers a plausible explanation of the well-known fact that weak excitatory potentials yield longer reaction times than strong. It is highly probable that excitatory potential is a joint function of (1) the degree of conditioning, and (2) the intensity of the evoking stimulus trace. On this assumption, if the original conditioning has been weak it may not suffice, in combination with a weak stimulus trace, to produce an excitatory potential great enough to exceed the reaction threshold and, consequently, no reaction will occur. If, however, the gradient of the stimulus trace rises during the incidence of the stimulus, it follows that if reaction does not occur at the beginning of the stimulus the excitatory potential will rise continuously and thus may exceed the reaction threshold before the termination of the stimulus at which time reaction will occur. Thus there would come about an association between weak conditioning and long reaction time, which is the substance of Postulate 18.

POSTULATE 1, COROLLARY 1

A. *In the learning of any rote series, concurrently with each syllable-presentation interval (n) of each syllable-presentation cycle, there is a compound stimulus trace* ($\| {}_{0\ldots n}tr_n \|$) *which consists of all those stimulus-trace segments, each of which is the nth stimulus-trace segment of a stimulus trace originating in a syllable presentation whose syllable presentation number lies between zero and n inclusive.*

Proof:

This proposition follows directly from Postulate 1 and D59.

B. The meaning of this corollary is shown clearly in Figure 2, where the constituent traces of all the compound stimulus traces are shown graphically. For example, the trace segments making up the compound stimulus trace at ZIT are seen to be those of ZIT itself, together with those of all the syllables preceding ZIT, i.e., FAP and KEM.

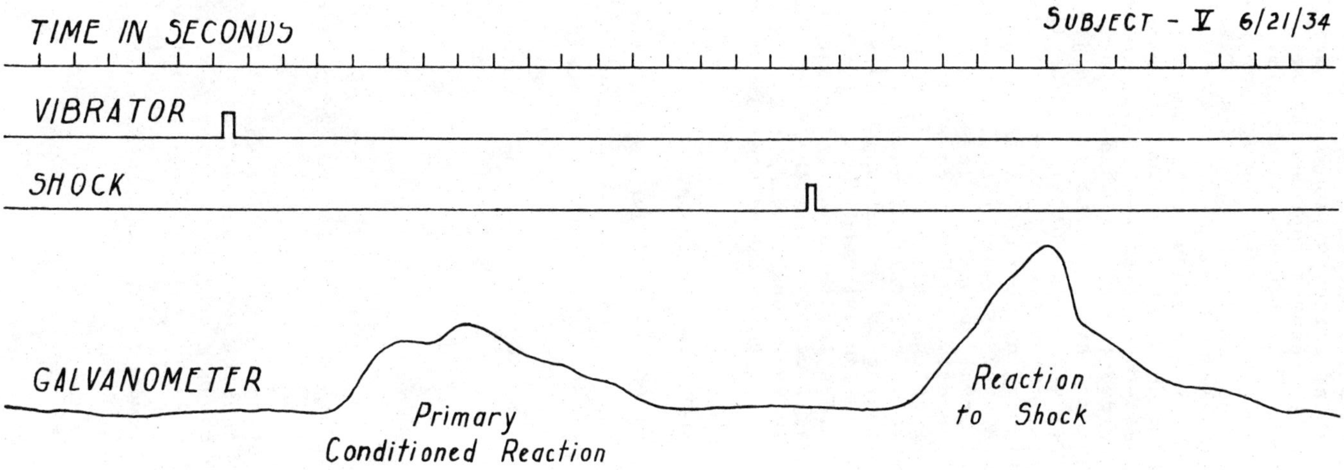

Figure 3. Diagrammatic representation of the setting up (reinforcement) of a simple trace conditioned reaction. Read the record from left to right. Note that the conditioned stimulus (vibrator) in the trace conditioned reinforcement terminates long before the onset of the reinforcing stimulus (shock). The intervening interval is empty so far as these two stimuli are concerned, and is held by Pavlov to be charged with inhibitory potentiality. It will be observed that the period of delay of the primary conditioned reaction here shown is but little more than the normal latency of the galvanic skin reaction conditioned.

POSTULATE 2

A. *In the learning of any rote series, at the termination of each syllable-presentation cycle there is an addition of the amount,* $\Delta E_{n,r}$, *to the excitatory potential,* $\mathcal{E}_{n,r}$, *existent at that instant, where* $n = r-1$ *and* $1 \leq r \leq N$.

A'. $P \; \varepsilon \; \mathrm{slpncy} \; . \; n < \mathrm{N}'P \; . \supset . \; \mathcal{E}_{n,n+1}'(\mathrm{cytm}'P, P)$

$> \mathcal{E}_{n,n+1}'(\mathrm{bg}'\mathrm{B}'P, P)$

B. Thus in the rote series here used for illustrative purposes, each repetition of the series as a whole produces a finite increment to the potentiality for each syllable to be spoken during the presentation of the syllable preceding it; i.e., for "FAP" to be spoken during the presentation of *KEM*, for "ZIT" to be spoken during the presentation of *FAP*, for "YOD" to be spoken during the presentation of *ZIT*, and so on. Because no theorems in the present monograph require a postulate stating the stimulus which initiates or evokes the several reactions in a completely learned rote series, no such postulate is given. There is, however, the implicit assumption that that which evokes the reaction (r) in a correctly learned rote series is the compound stimulus trace (${}_{0\ldots r-1}\mathrm{tr}_{r-1}$), i.e., syllable reaction r is "conditioned" (D66) to compound stimulus trace, ${}_{0\ldots r-1}\mathrm{tr}_{r-1}$, and the compound stimulus trace, ${}_{0\ldots r-1}\mathrm{tr}_{r-1}$, is said to "evoke" (D68) reaction r. Thus, reaction "YOD" (r=4) will be conditioned to the segments of the traces originating in syllables KEM, FAP, and ZIT, which coincide with the presentation of *ZIT*; i.e., to the segments of traces of syllables 0, 1, and 2 at point 2 since $r-1 = 2$. In future elaborations and extensions of the present system which presumably will appear, such a postulate will need to be made explicitly. As a partial anticipation of this development the following explanation is given.

Postulate 2 represents in general a version of Lepley's hypothesis (*36*). This hypothesis states, in substance, that rote learning consists of the concurrent formation of a large number of trace conditioned reactions. It thus comes about that not only is each reaction conditioned to a multiplicity of traces, but each of the traces so conditioned originates at a different place in the series. No reports have been found of such compound trace conditioning in the ordinary conditioned reaction investigations, though it would seem that such an experiment would be entirely feasible. In order to refresh the reader's memory, a record of the reinforcement of a typical trace conditioned reaction of the ordinary variety is reproduced as Figure 3. The reaction in this case (the galvanic skin reaction) is conditioned to the trace originated by a brief vibratory stimulation on the skin (*2*,49). At the stage of training represented by this record, the period of delay attained by the conditioned reaction may be seen to be less than four seconds, a portion of which must be attributed to the natural latency of this rather sluggish reaction. Experiments have shown abundantly (*47*; *55*) the great difficulty of attaining delays of any considerable length in trace conditioned reactions.

The braces in the diagram of Figure 2 indicate the phases of the particular stimulus traces making up the compound traces to which the representative syllable reactions "FAP", "YOD", and "JUB" are conditioned. Note that the number of stimulus traces found in the compound trace to which each succeeding syllable reaction is conditioned increases by one. Thus, "FAP" is conditioned to but a single stimulus trace (that originating in *KEM*), "ZIT" is conditioned to a compound of two stimulus traces (those originating in *KEM* and *FAP*), "YOD" is conditioned to a compound comprised of three stimulus traces (those originating in *KEM*, *FAP*, and *ZIT*), "JUB" is conditioned to a compound composed of five stimulus traces (those originating in *KEM*, *FAP*, *ZIT*, *YOD*, and *NAR*), and so on. It may also be observed that the strengths of the perseverative segments of traces in any given conditioned compound trace make up a quantitative hierarchy in which the more remote the syllables of origin, the weaker the trace. Moreover, the differences between the strengths of successive perseverative segments of traces within a given hierarchy, when taken in a descending order, are progressively smaller.

It is to be observed that each stimulus trace may have conditioned to it a number of different syllable reactions, each reaction conditioned to a distinct phase or portion of the trace. For example, the initial segment of the stimulus trace originating in *KEM* is conditioned to the syllable reaction "FAP", the first section of the perseverative segment is conditioned to "ZIT", the second section of the perseverative segment is conditioned to "YOD", and so on with "NAR", "JUB", and all the remaining syllable reactions of the rote series.

The marked difficulty of setting up long delays in trace conditioned reactions has been mentioned. It may now be noted that the conditions of the typical rote-learning experiment (instructions to the subject, etc.) bring it about that no period of delay whatever is required so far as the shortest (and presumably dominant) trace in the compound is concerned. The subject is told to make each syllable reaction as soon as possible after the presentation of the stimulus syllable immediately preceding it. Thus the instructions are that "JUB" (Figure 3) is to be spoken by the subject at the beginning of the initial segment of the stimulus trace of *NAR* with no delay except that of ordinary latency. This itself is possible, presumably, because "JUB" is *originally* conditioned to the trace of *NAR* at that phase which is concurrent with stimulus syllable *JUB*. Thus when reaction "JUB" occurs during the presentation of *NAR*, it is a trace conditioned reaction, but with no delay so far as the stimulus trace *NAR* is concerned. This means, of course, that the only periods of delay which need to be developed are those associated with the remaining (and weaker) traces constituting the conditioned compound. The fact that no delay needs to be set up for what is probably the dominant component of the compound trace conditioned, probably accounts for the relative ease of developing the necessary delays in rote series, as compared with simple trace conditioned reactions.

POSTULATE 3

A. *In the first massed practice of the learning of any rote series, the increments of excitatory potential to correct reactions are alike in amount for every reaction at every repetition of series of every length, i.e.,*

$$\Delta^{R,N}E_{0,1} = \Delta^{R,N}E_{1,2} = \ldots\ldots = \Delta^{R,N}E_{N-1,N}$$

and

$$\Delta^{R,6}E_{r-1,r} = \Delta^{R,7}E_{r-1,r} = \Delta^{R,8}E_{r-1,r} = \ldots$$

and

$$\Delta^{1,N}E_{r-1,r} = \Delta^{2,N}E_{r-1,r} = \Delta^{3,N}E_{r-1,r} = \ldots$$

A'. $\exists ! \Delta E$

B. This postulate means (1) that from any given amount of practice each syllable reaction, such as "FAP" and "ZIT" of Figure 2, receives the same amount of absolute excitatory potentiality as the other syllable reactions, such as "YOD", "NAR", "JUB", etc.; (2) that a given syllable reaction at one stage of practice receives the same amount of excitatory potentiality (ΔE) from a single repetition as at any other stage of practice; and (3) that syllable reactions in long series receive the same amount of excitatory potential as in short series.

While the above propositions are almost certainly not strictly accurate, there is some reason to believe that they are fair first approximations. Hull has shown that (*29*) after about twenty hours the capacity to recall the several syllables of a series approaches fairly close to equality, which fact argues rather strongly for the first proposition. These results are supported by an experiment by Hovland (*25*) which shows that this change in recall is a direct function of the time interval between learning and recall. His results confirm Hull's at a 24-hour interval, confirm Ward's of 20 minutes, and those of Robinson and Brown immediately following learning.

It is possible, however, that where there is a very large amount of inhibition (interference) at a point of conditioning, such as the present system presumes occurs near the middles of series, the amount of absolute excitatory potential which is assumed to be acquired as the result of a single repetition may be somewhat less than would be the case if only a little, or no inhibition whatever, were present. On this assumption it would be plausible to expect that as more and more inhibition accumulates in the course of learning, e.g., when learning occurs by massed repetitions, the increment of excitatory potential ($\Delta E_{r-1,N}$) at each successive repetition might suffer progressive diminution. This whole question calls urgently for direct experimental investigation.[5]

A somewhat different question is involved in the assumption that $\Delta E_{r-1,r}$ is constant throughout the repetitions required to produce a perfect recall of the series as a whole. Analogy with most empirical curves of learning would suggest that $\Delta E_{r-1,N}$, at least after the early stages of practice, should progressively diminish to zero as learning progresses. Moreover, it is difficult to believe that some diminution in $\Delta E_{r-1,r}$ will not occur as the physiological limit of the capacity of the organism to acquire the particular excitatory potential is approached.

In opposition to the above considerations, and in support of the assumptions of the present form of the postulate, it may be said:

1. The curve of learning in terms of excitatory potential need not necessarily take the same form as empirical learning curves; the learning curve in terms of $E_{r-1,r}$ is a distinctly different concept.

2. It is conceivable that the flattening out of empirical learning curves which usually occurs as the function in question approaches the physiological limit may not begin, in the present instance, until this limit is approached rather closely. It is possible that this negative acceleration may not begin appreciably in ordinary rote series until after the stage of perfect recall has been reached.

Actually, however, it makes no difference for the theorems involved in the present monograph, whether the curve of acquisition of $\mathcal{E}_{r-1,r}$ is assumed to be a straight line (as here) or whether it takes the more conventional form of a negatively accelerated positive growth curve of the type ($X = A - Ae^{-fR}$). The theorems included in the present monograph follow on either assumption. A complete set of duplicate mathematical proofs for critical theorems of the system, on the assumption that $\Delta E_{r-1,r}$ grows progressively less as learning advances, was worked out by Dr. Perkins but the proofs are not reproduced because of the expense of publication. Sooner or later, presumably, one or more phenomena will be encountered which can be deduced on the assumption of one of the above possibilities but not the other. When that stage of development is reached it will be a simple matter to discard the postulational alternative which has thus shown itself inadequate.

5. A postulate incorporating this conjectured dependence of $\Delta E_{r-1,r}$ upon the amount of inhibitory potential present at the point of final conditioning was tried late in the formulation of the present system in order to bring the theory into line with the known facts of Jost's law. Computations according to this postulate based on a set of typical concrete values yielded Jost's law in some detail. We were forced to abandon the postulate, however, because when used for the deduction of general theorems it developed such unwieldy equations that the differentiation of one of them would have covered forty or fifty pages.

POSTULATE 4

A. *In the first massed practice of the learning of any rote series, at the instant of the addition of* $\Delta E_{n,r}$ *to* $\mathcal{E}_{n,r}$ *(where* $n + 1 = r$*) there are also added finite inhibitory potentials, one of which is active concurrently with each of the stimulus-trace segments preceding* r*, of each of the stimulus traces originating at syllable presentations antecedent to* r*, i.e.,*

$$\text{if} \quad \Delta E_{n,r}, \quad \text{then} \quad \| \Delta\, {}_{0 \ldots n}I_{s \ldots n,r} \| > 0$$

for all combinations of integral values of s *and* n*, such that* $s \leq n \leq r - 1 \leq N - 1$.

A'. $s \leq n \leq r \leq N'P \,.\supset .\, \exists ! \Delta\, {}^{N'P}_{s}I_{n,r} \,.\, 0 < \Delta\, {}^{N'P}_{s}I_{n,r}$

B. Consider the conditioning of the syllable reaction "YOD" at final conditioning point *ZIT*. The various stimulus traces involved in this conditioning were shown by Figure 2.

Such empirical justification as exists for Postulate 4 lies mainly in the experimental work on conditioned reactions. Pavlov (*47*, 88) has reported from experiments with dogs "inhibition of delay" during the period of delay in delayed conditioned reactions. He has also reported what might be called "inhibition of reinforcement" (*47*, 248) where little or no delay occurred between the conditioned stimulus (e.g. the vibrator, Figure 3) and either the reinforcement (e.g. the shock, Figure 3) or the occurrence of the conditioned reaction. Rodnick (*56*), working with human subjects, has reported inhibitions of delay in delayed conditioned reactions, though there is some uncertainty about the interpretation of his results.

While the present systematic approach assumes the inhibitory potentials as stated in this postulate, it is recognized that sooner or later these postulates themselves may be deducible from other still more basic postulates. As an informal anticipation of this latter inquiry, we may ask why conditioned reactions should display inhibition at the point of conditioning; or, indeed, anywhere else. In considering this matter it may be well to recall that the type of inhibition here under consideration manifests itself as at least a temporary neutralization of excitatory potentials which otherwise would exhibit themselves in action at the point in question; no direct manifestation of inhibitory potential is ever observed. Pavlov interpreted these phenomena in terms of a neurological theory of inhibition which, however, has not been very generally accepted and is not accepted by the present writers.

A somewhat more plausible hypothesis has been advanced independently by Hovland (*17*) and by Eleanor J. Gibson (*13*). This is to the effect that if a stimulus has excitatory potentialities to two or more incompatible reactions, the resulting clash, struggle, or competition between these reaction potentialities tends to neutralize each, thus materially reducing the excitatory potential available for the evocation of even the dominant tendency. For example, suppose that in a stimulus-reaction situation the stimulus complex *A* tends to evoke reaction *X* to the extent of 10 units, and at the same time it tends to evoke the reaction *Y* to the extent of 8 units. On the above assumption, the excitatory potential available for the evocation of the dominant reaction *X* would be 10-8, or only 2 units in amount. If, now, 3 units of excitatory potential are required to pass the reaction threshold, this conflict of excitatory potentials would result in an effectual inhibition of both reaction potentialities. No attempt has yet been made to work out in detail the dynamics of this hypothesis. It would seem probable, however, that, other things equal, the greater the number of competing excitatory potentials, the greater the amount of inhibition that would result.

POSTULATE 5

A. *In the first massed practice of the learning of any rote series, the increments of inhibitory potential of the type* ${}_{r-1}I_{r-1,r}$, *where both the syllable-of-origin of the trace and the number of its segment are one less than the number of the syllable reaction involved, are alike in amount for every reaction* (*r*) *at every repetition* (*R*) *in series of every length* (*N*) *i.e.,*

$$\Delta K = \begin{cases} \Delta\,{}^{R,N}_{0}I_{0,1} = \Delta\,{}^{R,N}_{1}I_{1,2} = \ldots\ldots = \Delta\,{}^{R,N}_{N-1}I_{N-1,N} \\ \Delta\,{}^{R,6}_{r-1}I_{r-1,r} = \Delta\,{}^{R,7}_{r-1}I_{r-1,r} = \Delta\,{}^{R,8}_{r-1}I_{r-1,r} = \ldots\ldots \\ \Delta\,{}^{1,N}_{r-1}I_{r-1,r} = \Delta\,{}^{2,N}_{r-1}I_{r-1,r} = \Delta\,{}^{3,N}_{r-1}I_{r-1,r} = \ldots\ldots \end{cases}$$

A'. $\exists!\,\Delta K$

B. This postulate states in effect that all ΔK's are alike in magnitude. A comparison of Postulate 5 with Postulate 3 shows a close parallelism.

POSTULATE 6

A. *In the first massed practice of the learning of any rote series, the series of increments of inhibitory potential added concurrently with the addition of* $\Delta E_{r-1,r}$ *and active concurrently with the series of syllable trace segments making up the compound trace* ($\|\,{}_{0\ldots r-1}tr_{r-1}\,\|$) *is as follows: when the difference between the number of the syllable-of-origin of a trace* (*s*) *and the number of the*

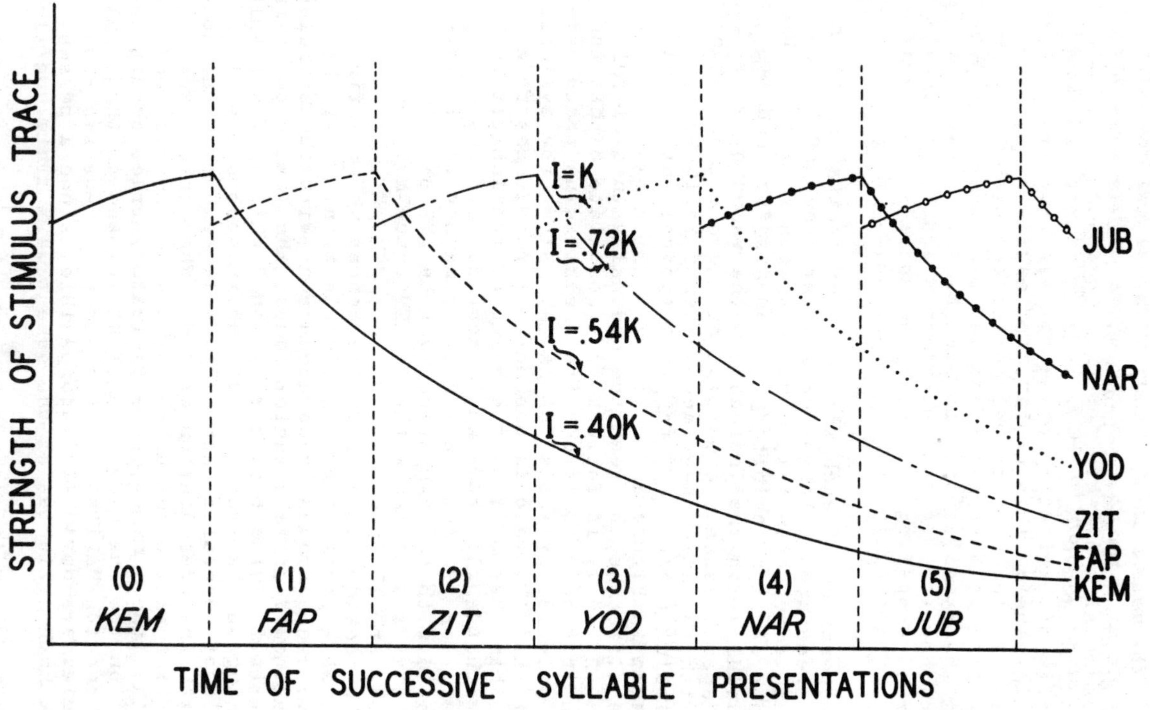

Figure 4. Diagram showing the amounts of inhibitory potential generated in connection with the learning of a syllable reaction "NAR" at its point of conditioning (D66). In this case, F is taken as 1.37. In this and subsequent diagrams of a similar nature, in order not to clutter the available space unduly, the subscripts are omitted from the I's, and the Δ's are omitted from both the I's and the K's.

syllable reaction (r) is 1, *the inhibitory potential associated with the trace segment has a basic value of* ΔK, *and for each successive additional unit of remoteness in the origin (s) of the stimulus trace from r, the inhibitory potential associated with that trace segment decreases by a constant factor of proportionality* (F), *where* $F > 1$, *i.e.*,

$$\Delta\,{}_{r-1}^{N}I_{r-1,r} = \Delta K, \quad \Delta\,{}_{r-2}^{N}I_{r-1,r} = \frac{\Delta K}{F}, \quad \Delta\,{}_{r-3}^{N}I_{r-1,r} = \frac{\Delta K}{F^2},$$

$$\ldots\ldots \quad \Delta\,{}_{r-r}^{N}I_{r-1,r} = \frac{\Delta K}{F^{r-1}}$$

A'. E!F

B. The generation of each increment of excitatory potential ($\Delta E_{r-1,r}$) generates a larger or smaller group of increments of inhibitory potential, dependent upon the value of r; the number of such inhibitory increments for each $\Delta E_{r-1,r}$ increases rapidly with the increase in the value of N. Postulate 6 specifies only a portion of these values - those associated with the stimulus-trace segments included in the compound stimulus trace, ${}_{0\ldots r-1}tr_{r-1}$. A typical set of such values is shown in Figure 4.

Since the difference between the ordinal values of *YOD* and *NAR* is 4 - 3, or 1, it follows from Postulate 6 that the amount of inhibition associated with the stimulus trace originating in stimulus syllable *YOD* is ΔK. Since the syllable of origin of the ZIT trace is one unit of remoteness from *NAR* greater than that of YOD, the value of the inhibition associated with the ZIT trace must, according to Postulate 6, be ΔK ÷ 1.37, or .72 ΔK. In a similar manner the inhibition associated with the FAP trace must be .72 ΔK ÷ 1.37, or .54 ΔK, and that associated with the KEM trace must be .54 ΔK ÷ 1.37, or .40 ΔK.

We may now consider in a tentative manner why the final conditioning of a particular syllable reaction, e.g., "NAR" (Figure 4) to the compound trace concurrent with the stimulus syllable *YOD* should produce inhibition at the latter point in the series. It is evident that reaction "NAR", as the result of conditioning to the trace of *YOD* and in accordance with the principle of generalization (*47*), will tend to occur concurrently with the presentation of *YOD*. When this takes place a conflict will occur between the simultaneous tendency to speak "NAR" (the permissible anticipatory trace conditioned reaction) on the one hand, and "YOD" (evoked as the result of ordinary reading habits), on the other. These two reaction tendencies are obviously incompatible, since a person cannot speak both syllables at the same time.[6] This conflict

6. It seems probable that the mechanism here sketched may account for the "inhibition of reinforcement" in ordinary conditioning situations. An interesting series of corollaries dependent on this hypothesis offers a promising field for experimental investigation.

between incompatible excitatory potentialities should tend to a mutual neutralization of excitatory potentials and thus produce what we have called inhibition.

We have still to consider the assignment of different amounts of this hypothetical conflict-generated inhibition to the several traces in the compound trace concurrent with stimulus syllable *YOD*. It is plausible to suppose (*19*;*47*) that in any compound conditioned stimulus, the weaker stimuli will be more feebly conditioned to the reaction than will the stronger stimuli. It follows that syllable reaction "NAR" will be conditioned less strongly to the trace of *ZIT* than to that of *YOD*, less strongly to the trace of *FAP* than to that of *ZIT*, and less strongly to the trace of *KEM* than to that of *FAP*. Accordingly the bringing forward of "NAR" into conflict with the reading reaction "YOD" occurring during the presentation of stimulus syllable *YOD* is most of all dependent on the stimulus trace originating in *YOD*, somewhat less on the trace of *ZIT*, still less on the trace of *FAP*, and least of all on the trace of *KEM*. Assuming that the inhibition attributable to each trace is proportional to the conflict of reaction dependent upon each, it follows that the trace of *YOD* is responsible for more inhibition than any of the others, that the trace of *ZIT* is responsible for somewhat less, that the trace of *FAP* is responsible for less still, and that the trace of *KEM* is responsible for the least of all, exactly as shown in Figure 4. It seems probable that in a future development something like this will be incorporated into the explicit theory of rote learning. This, of course, will amount to a deduction of some of the present postulates from still lower-level principles. At present, however, these are merely suggestions thrown out incidentally, looking to future developments, and are not a part of the present systematic structure.

POSTULATE 7

A. *In the learning of any rote series, the series of increments of inhibitory potential,* $\| \Delta\, {}_{s}I_{s\ldots n,r} \|$, *added at the instant of the addition of the increment of excitatory potential,* $\Delta E_{r-1,r}$, *and active concurrently with the several segments,* ${}_{s}tr_{s\ldots r-1}$, *antecedent to* r *of any stimulus trace,* ${}_{s}tr_{s\ldots N}$, *decrease in amount at a uniform rate from the amount occurring at segment* $r-1$ *to a zero value at a position just preceding the syllable-of-origin of the trace in question, i.e.,*

$$\Delta\, {}^{R,N}_{s}I_{s-1,r} = \frac{(s-1)-(s-1)}{(r-1)-(s-1)} \times \frac{\Delta K}{F^{x}} = 0 \times \frac{\Delta K}{F^{x}} = 0$$

$$\Delta\, {}^{R,N}_{s}I_{s,r} = \frac{s-(s-1)}{(r-1)-(s-1)} \times \frac{\Delta K}{F^{x}} = \frac{1}{(r-1)-(s-1)} \times \frac{\Delta K}{F^{x}}$$

$$\Delta^{R,N}{}_{s}I_{s+1,r} = \frac{(s+1)-(s-1)}{(r-1)-(s-1)} \times \frac{\Delta K}{F^x} = \frac{2}{(r-1)-(s-1)} \times \frac{\Delta K}{F^x}$$

. . . .

$$\Delta^{R,N}{}_{s}I_{r-1,r} = \frac{(r-1)-(s-1)}{(r-1)-(s-1)} \times \frac{\Delta K}{F^x} = 1 \times \frac{\Delta K}{F^x} = \frac{\Delta K}{F^x}$$

where $s \leq n < r \leq N$ *and* $x = r - s - 1$.

A'. $s \leq n < r < 100 \,.\supset.\, \Delta^{N'P}{}_{s}I_{n,r} = (n-s+1) \times \Delta^{N'P}{}_{s}I_{r-1,r} \div (r-s)$

B. The meaning of this postulate is illustrated by Figure 5, which is merely an extension of Figure 4. Take, for example, stimulus trace *FAP* (broken line). We have seen (Figure 4) that this trace bears an inhibitory potential of .54 ΔK at the point of conditioning. Postulate 6 states that all points on this trace preceding *YOD* will also show inhibition, the amounts decreasing uniformly to zero at the stimulus syllable preceding the syllable of origin of the trace. Since *FAP* is the syllable of origin of this trace, this progression must reach a value of zero at syllable *KEM*. Since the difference between the ordinal numbers of *YOD* and *KEM* is 3 - 0, or 3, the decrement at each step forward from *YOD* must be .54 ΔK ÷ 3, or .18 ΔK. Accordingly the inhibition arising from this conditioning, and attributable to the trace originating in *FAP*, has a value of .54 ΔK - .18 ΔK, or .36 ΔK at *ZIT*, and one of .36 ΔK - .18 ΔK, or .18 ΔK at *FAP*. The inhibitions associated with the traces originating in syllables *KEM* and *ZIT* have been computed in an exactly analogous manner and are also shown in Figure 5. A complete set of inhibitory potentials generated in association with the generation of the learning of syllable reaction "JUB" is shown in Figure 12. There it will be noted that the inhibitory potential has increased by 5.

If we care to go behind Postulate 7 as we did in the case of Postulate 6, we find a certain amount of *a priori* justification for the assumption made. The reasoning in this case is much like that given in connection with Postulate 6. We have seen (Figure 4) that reaction "NAR", originally conditioned to the trace originating in stimulus syllable *KEM*, through the principle of generalization, tends to be evoked not only concurrently with the presentation of stimulus syllable *YOD*, but at all of the other preceding syllables, i.e., *ZIT*, *FAP*, and *KEM*. This tendency for "NAR" to appear at these various points in advance of its proper position will, of course, produce interference with the excitatory tendencies at these points and, therefore, inhibition. But here we must note an important characteristic of generalization. This is that, other things equal, the generalized reaction decreases progressively in strength as the trace evoking it differs from that conditioned. It follows that, other things equal, reaction "NAR" will be evoked less

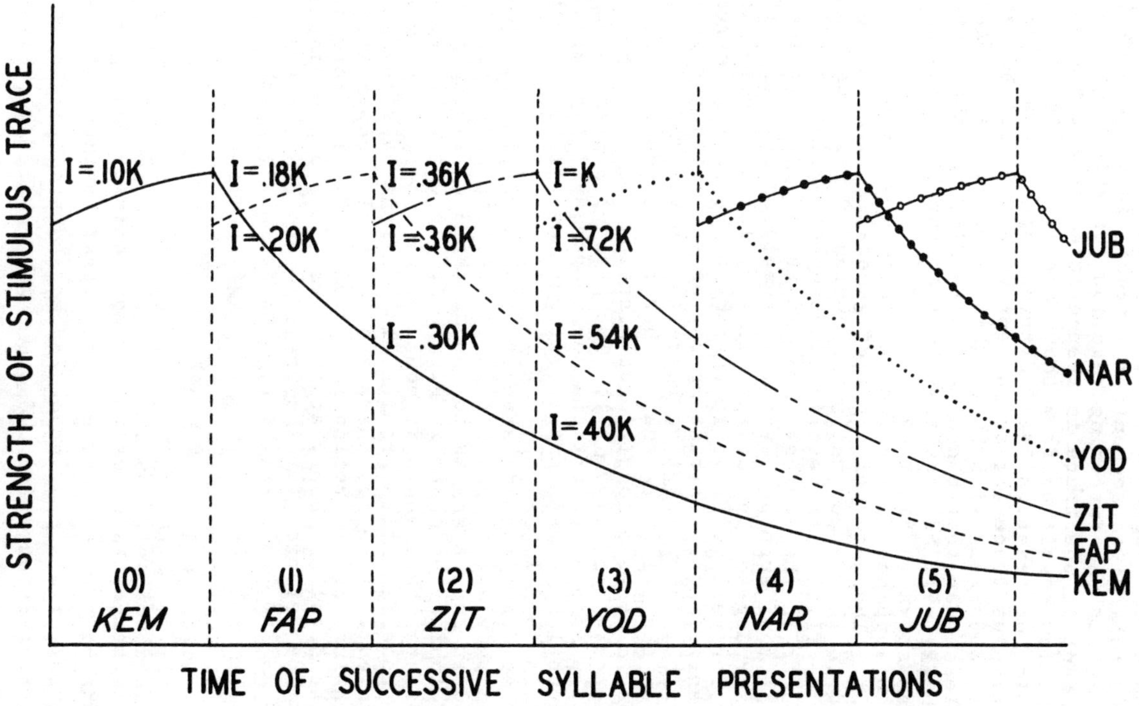

Figure 5. Diagram showing the complete set of inhibitory potentials generated in connection with the learning of the syllable reaction "NAR". As in Figure 4, the value of F is taken at 1.37.

and less strongly by segments of the *KEM* trace the more remote the evoking segment from the segment originally conditioned to the reaction. But, assuming that the degree of conflict with the other reaction tendencies active at that point, and so the degree of the resulting inhibition to be attributed to a given segment of a trace, is proportional to the strength of the intruding reaction for which the particular segment of the trace is responsible, it follows that the inhibition resulting from the generalized effects of the conditioning of reaction "NAR" attributable to the segment of the stimulus trace *KEM* concurrent with *ZIT* will be less than that concurrent with *YOD*, that concurrent with *FAP* will be less than that concurrent with *ZIT*, and that concurrent with *KEM* will be less than that concurrent with *FAP*, exactly as shown in Figure 5.

Certain supplementary observations need to be made at this point. The same principle of generalization should also lead to the *perseverative* intrusion of reaction "NAR" so as to conflict with reaction tendencies active at syllables *later* in the series. As a matter of fact, Patten has shown (*46*) that such perseverative reactions do occur, though to a lesser degree than do anticipatory intrusions. This tendency for perseverative intrusions to occur less strongly than do anticipatory intrusions may be accounted for in part on the principle of relative refractory phase (*57*; *6*), and in part on the principle that, other things equal, *a weak trace should evoke a conditioned reaction less strongly than would a strong trace* (*47* ; *1*). The present system takes no notice of possible inhibitions due to perseverative errors though doubtless subsequent investigations will need to work this out in detail.

It is evident that the strength-of-trace principle just mentioned should also operate in the case of antedating reactions and in such a way as to oppose the diminishing gradients of inhibition shown in Figure 5. In order to retain the gradients there shown it would be necessary to assume that the generalization gradient is dominant over the intensity gradient. Even so, there should result complications not taken into consideration in Postulate 7. The working out of the detailed theory at the level here under discussion must be deferred for the present, and Postulate 7 accepted as the best first approximation now available.

POSTULATE 8

A. *In the learning of any rote series by massed practice, all increments of excitatory potential* ($\Delta^{R,M}E_{r-1,r}$) *due to massed practice and all concurrent residues of excitatory potentials from preceding massed practices of the rote learning in question* [$E_{r-1,r}(t_1)$, $E_{r-1,r}(t_2)$ *etc.*] *summate functionally* (D86) *by simple addition.*

Postulate 8 may be restated in somewhat more mathematical terms as follows:

If $E_{r-1,r}(t_1) \dotplus E_{r-1,r}(t_2) \dotplus \ldots$ means the total *functional* (D86) excitatory potential relative to reaction r in the massed practices preceding a given instant of a given rote learning, then $\mathcal{E}_{r-1,r} = E_{r-1,r}(t_1) \dotplus E_{r-1,r}(t_2) \dotplus \ldots = E_{r-1,r}(t_1) + E_{r-1,r}(t_2) + \ldots$ Or, more generally, if the symbol, $\dot{\Sigma}$, designates the functional sum,

$$\mathcal{E}_{r-1,r} = \dot{\sum_{i=1}^{*}} E_{r-1,r}(t_i) = \sum_{i=1}^{*} E_{r-1,r}(t_i)$$

Likewise for the R completed repetitions of the massed practice in which the instant in question falls,

$$\dot{\sum_{i=1}^{R}} \Delta^{i,N} E_{r-1,r} = \sum_{i=1}^{R} \Delta^{i,N} E_{r-1,r} \, .$$

Suppose, for example, there is massed practice of the same rote learning (D29) on three successive days. Let t designate the time (in days) elapsed up to the time of the massed practice on the third day. Then $t_1 = 2$, $t_2 = 1$, $t_3 = 0$, and the total *functional* excitatory potential, for a reaction having reaction number r, at the termination of the massed practice on the third day is $\mathcal{E}_{r-1,r} = E_{r-1,r}(2) + E_{r-1,r}(1) + E_{r-1,r}(0)$. Furthermore, if there are R repetitions on the third day,

$$\mathcal{E}_{r-1,r} = E_{r-1,r}(2) + E_{r-1,r}(1) + \sum_{i=1}^{R} \Delta^{i,N} E_{r-1,r} \, .$$

A'. $n < N{}^{\backprime}P \,.\supset.\, \mathcal{E}_{n,n+1}{}^{\backprime}(t,P) = E_{n,n+1}{}^{\backprime}(t,P) + \sum_{\mathrm{prtm}{}^{\backprime}T \le t} E_{n,n+1}(t - \mathrm{prtm}{}^{\backprime}T){}^{\backprime}(P,T)$

B. This postulate states, in effect, that the *functional* sum of the excitatory potentials to the evocation of a reaction (r) at point $r-1$ at a given instant is the same as the sum produced by simple addition of those excitatory potentials. This means that the excitatory potentials from different sources combine jointly to mediate reaction (see D86) by simple addition rather than in innumerable other manners which would be conceivable. For example, it might easily be imagined that excitatory potentials might summate functionally in such a way as to be less than the sum obtained by simple addition. Indeed, this latter mode of functional combination would probably be the first choice of most students of behavior, and this may turn out to be the fact. There seems to be no direct test of these possibilities; presumably the only means of verification is the experimental testing of theorems deduced from the various conceivable possibilities taken in conjunction with the other postulates of a system. One reason for the choice of the present one of numerous obvious possibilities is its relative simplicity and consequent ease of mathematical manipulation.

POSTULATE 8, COROLLARY 1

A. *In any massed practice ($\dot{M}$) of the learning of any rote series,*

$$^{R,\dot{M}}E_{r-1,r} = R\Delta {}^{\dot{M}}E_{r-1,r} \, .$$

Proof:

This follows directly from Postulates 3 and 8, and Definitions D64 and D65.

B. Since in the present treatment $\Delta E_{r-1,r}$ is taken as the basic unit of measurement, $^{R,\dot{M}}E_{r-1,r} = R \times 1$, or R. It thus comes about that the value of $E_{r-1,r}$ is always the same as R. For example, if the first massed practice of a rote series has just completed 6 repetitions, the excitatory potential to the evocation of every syllable reaction in the series must be 6.

POSTULATE 9

A. *During any massed practice of any rote learning, all increments of inhibitory potential ($\Delta {}^{\dot{M}}_{s}I_{n,r}$) of the massed practice in question, and all concurrent residues of inhibitory potentials from preceding massed practices of the same rote learning [${}^{\dot{M}}_{s}I_{n,r}(t_i)$] summate functionally by simple addition.*

Postulate 9 may be restated in more mathematical terms in a manner analogous to the restatement of Postulate 8, as follows:

$$\mathcal{I}_n = \dot{\sum_{i=0}^{m}} {}^{\dot{M}}_{s}I_{n,r}(t_i) = \sum_{i=0}^{m} {}^{\dot{M}}_{s}I_{n,r}(t_i) \, .$$

Also

$$\dot{\sum_{i=1}^{N}} \Delta {}^{i}_{s}I_{n,r} = \sum_{i=1}^{N} \Delta {}^{i}_{s}I_{n,r}$$

and

$$\dot{\sum_{i=1}^{N-n}} \Delta {}^{R,\dot{M}}_{s}I_{n,n+i} = \sum_{i=1}^{N-n} \Delta {}^{R,\dot{M}}_{s}I_{n,n+i}$$

and

$$\dot{\sum_{i=0}^{n}} \Delta {}^{\dot{M}}_{i}I_{n,r} = \sum_{i=0}^{n} \Delta {}^{\dot{M}}_{i}I_{n,r} \, .$$

A'. $n < N'P \, . \supset . \, \mathcal{I}_n'(t,P) = I_n'(t,P) + \sum_{\text{prtm}'T \le t} I_n(t - \text{prtm}'T)'(P,T)$

B. The *functional* summation (D86) of inhibitory potentials may be illustrated by reference to Figure 5. According to this postulate the functional aggregate of inhibition resulting from the conditioning of reaction "NAR" is .10 ΔK at *KEM*, .18 ΔK + .20 ΔK, giving a total of .38 ΔK, at *FAP*, .36 ΔK + .36 ΔK + .30 ΔK, giving a total of 1.02 ΔK, at *ZIT*, and ΔK + .72 ΔK + .54 ΔK + .40 ΔK, giving a total of 2.66 ΔK at *YOD*.

It is to be observed that the above example illustrates the functional summation of only one class of inhibitory potentials; there are several others, as indicated in the restatement of Postulate 9.

POSTULATE 9, COROLLARY 1

A. *In any massed practice of any rote learning,*

$$^{R,N}K_n = R\Delta\,^{N}K_n$$

Proof:

1. By Postulate 5, the ΔK's for all repetitions are equal.
2. By (1) and Postulate 9, the corollary follows.

B. Consider, for example, the inhibition (Figure 4) associated with the initial segment of the stimulus trace *YOD* ($n = 3$) resulting from the conditioning of the syllable reaction "NAR" ($r = 4$). Each repetition adds an increment of inhibition (ΔK) to the accumulation of this inhibition. If $R = 6$, i.e., if six repetitions have occurred, it follows that,

$$^{6,N}K_3 = 6\,\Delta\,^{N}K_3$$

POSTULATE 9, COROLLARY 2

A. *In any massed practice of any rote learning,*

$$^{R,N}I_n = R\,\Delta\,^{N}I_n$$

Proof:

1. By Postulates 6 and 7, at any given repetition the various increments of inhibitory potential at any given syllable presentation are constant functions of ΔK.
2. By Postulate 5, the ΔK's for all repetitions at all positions are equal.
3. From (1), (2), and Postulate 9, it follows that the functional (D86) sum of the various increments of inhibition active at any given syllable presentation must be the same at each repetition.
4. From (3), D75, D77, and Postulate 9, the corollary follows.

B. It is evident from the foregoing and Figures 4 and 5 that, according to the present set of postulates, at each repetition in rote learning there is set up at any given point of conditioning (n or $r - 1$) a whole series of inhibitory potentials. Since these are all various functions of ΔK, and since (Postulate 5) ΔK is the same at all repetitions, it follows that the amount of aggregate inhibitory potential at a given point of conditioning (ΔI_n) resulting from the repetition of a rote series as a whole must be the same for each repetition. Suppose that in a given series (ΔI_n) - 4.65 ΔK, and ΔK = .09 ΔE. Substituting, we have,

$$\Delta I_n = 4.65 \times .09\ \Delta E = .42\ \Delta E$$

Substituting this in the equation of Corollary 1, we have

$$^{R,N}I_n = R.42\ \Delta E$$

If, now, we assume that 6 repetitions have occurred, this equation becomes,

$$^{6,N}I_n = 6 \times .42\ \Delta E = 2.52\ \Delta E$$

POSTULATE 10

A. *In the learning of any rote series, the total excitatory potential functioning at a given instant and the total inhibitory potential functioning at the same instant, summate by simple subtraction, i.e.,*

$$\bar{E}_{n,r} = \mathcal{E}_{n,r} - \mathcal{I}_n$$

where $r = n + 1$.

A'. $\bar{E}_{n,n+1}{}'(t,P) = \mathcal{E}_{n,n+1}{}'(t,P) - \mathcal{I}_n{}'(t,P)$

In the symbolic logic this postulate is a consequence of D79.1'.

B. In order to illustrate this negative summation it must be explained that, because inhibition never shows itself except as it neutralizes an excitatory potential, ΔK is measured in terms of its ratio to ΔE. For example, in an empirical determination of the value of ΔK (Problem I) it was found that ΔK would neutralize .09 of ΔE, i.e., ΔK = .09 ΔE. Now, suppose that the series under consideration has had ten repetitions and that at the point of conditioning of a given syllable reaction, such as "NAR", ΔI = 5.58 ΔK. Substituting in this equation the value of K above, we have,

$$\Delta I = 5.58 \times .09\ \Delta E = .502\ \Delta E\,.$$

Taking the amount of excitatory potential from one repetition as the unit, it follows from this that ten repetitions would yield 10 units of excitatory potential and (by Corollary 1) .502 × 10 or 5.02 units of inhibitory potential. Postulate 10 states that the amount of effective excitatory potential at the conditioning point of syllable "NAR" would then be,

$$\bar{E}_{r-1,r} = 10\ \Delta E - 5.02\ \Delta E = 4.98\ \Delta E\ .$$

POSTULATE 11

A. *In any rote learning,*

$$\Delta E_{r-1,r} > \Delta I_{max} > 0$$

where ΔI_{max} *is the maximum* ΔI *associated with any compound trace* $({}_{0\ldots r-1}tr_{r-1})$ *of a syllable-presentation cycle.*

A'. $n < N'P\ .\supset.\ 0 < \Delta^{N'P}I_n < \Delta E$

B. This postulate may be illustrated concretely as follows. In a certain theoretical set-up (p. 109) consisting of 15 syllables, exclusive of the cue syllable, 8 repetitions would be required for complete learning and the inhibition would be maximal at reaction syllable No. 9, where $\Delta I_{max} = 7.69\ \Delta K$. Taking $\Delta K = .09\ \Delta E$ and substituting in the above equation, we have,

$$\Delta I_{max} = 7.69 \times .09\ \Delta E = .69\ \Delta E\ .$$

But, since ΔE, by definition, is equal to one, we have the concrete inequality,

$$1 > .69 > 0$$

i.e.,

$$\Delta E > \Delta I_{max} > 0$$

POSTULATE 11, COROLLARY 1

A. *In the process of rote learning by any massed practice the amount of excitatory potential* ($E_{r-1,r}$) *at the completion of learning by massed practice is greater than the amount of inhibition at its maximum point* (I_{max}) *in the series, i.e.,*

$$a > c_{max} > 0$$

where a *is the amount of excitatory potential appearing anywhere in the series, and* c *is the amount of inhibitory potential.*

Proof:

1. By Postulate 11, $\Delta^{N}E > \Delta^{N}I_{max} > 0$.

2. Multiplying through (1) by R, we have,

$$R\Delta{}^{N}E > R\Delta{}^{N}I_{max} > 0.$$

3. But by Postulate 8, Corollary 1,

$$R\Delta{}^{N}E = {}^{R,N}E.$$

4. Also by Postulate 9, Corollary 1,

$$R\Delta{}^{N}I_{max} = {}^{R,N}I_{max}.$$

5. Substituting (3) and (4) in (2), we have,

$${}^{R,N}E > {}^{R,N}I_{max} > 0,$$

i.e.,

$$a > c_{max} > 0.$$

B. This corollary may be illustrated with the same concrete data as that employed to illustrate the main postulate. In that example, 8 repetitions were required to learn the series as a whole. It follows from the definition of the unit of excitatory potential that in this case $E = 8$. Also by Postulate 9, Corollary 1, $I_{max} = .69 \times 8 = 6.14$. Accordingly we may write the inequality,

$$8 > 6.14 > 0$$

i.e.,

$${}^{R,N}E > {}^{R,N}I_{max} > 0.$$

POSTULATE 11, COROLLARY 2

A. *The effective excitatory potential* ($\bar{E}_{r-1,r}$) *at the termination of learning by massed practice at any given syllable-presentation interval is always positive.*

Proof:

1. By Postulate 10,

$$\bar{E}_{n,r} = \mathcal{E}_{n,r} - \mathcal{I}_n,$$

where

$$r = n + 1.$$

2. But by D65 and D77, during the first massed practice,

$$\mathcal{E}_{n,r} = E_{n,r},$$

and

$$\mathcal{I}_n = I_n.$$

3. Substituting (2) in (1), we have,

$$\bar{E}_{n,r} = E_{n,r} - I_n.$$

4. From Corollary 1 of Postulate 11,

$$E_{n,r} - I_{max} > 0.$$

5. Combining (3) and (4) the corollary follows.

B. This corollary may be illustrated by means of the same example used above. We have seen that in that case $E_{n,r} = 8$ and $I_{max} = 6.14$. From this we may write the inequality,

$$\bar{E}_{n,r} = 8 - 6.14 = 1.86 > 0,$$

i.e.,

$$\bar{E}_{n,r} = E_{n,r} - I_{max} > 0.$$

POSTULATE 12

A. *In the learning of any rote series, any excitatory potential ($E_{r-1,r}$) resulting from any given massed practice decreases with the passage of time following the termination of such massed practice, the rate of decrease at any time being proportional to the amount of the residue of such excitatory potential existent at that time, i.e.,*

$$E_{r-1,r}(t) = E_{r-1,r}\, e^{-bt}$$

where b is a constant > 0 and t is the time elapsed since the termination of the massed practice in question with the exception of that part of such time which may be occupied by congruent massed practice.

A′. $x = E_{n,n+1}(t)`(P,T) . \equiv :: x \neq 0 : \equiv : t \geq 0 . (\exists v) . v = t - \sum_{u\ \mathrm{invl}(t,P,T)} u .$
$x = R`(B`\breve{T}, T) \times \Delta E \times e^{-bv} . C`T \subset \mathrm{rtln}`P :. b > 0$

In the symbolic logic this postulate is a consequence of D67′.

B. This equation attempts to state a first approximation to a quantitative law of forgetting. The equation means that with the passage of time E decreases progressively according to a fraction of which the numerator is the amount of E at the termination of massed practice, and the denominator is e, i.e., 2.7183, raised to the bt power. Thus, suppose that E at the termination of learning is 8, and that b has a constant value of .1. What will be the value of E after the lapse of three units of time, i.e., $t = 3$? Substituting in the equation, we have,

$$E = \frac{8}{2.7183^{.1\times 3}} = \frac{8}{2.7183^{.3}}$$

The value of $2.7183^{.3}$ may be found by the aid of logarithms. The logarithm of 2.7183 is .43425. Multiplying this by .3, we have .13027. Looking up the value of .13027 in a table of logarithms we find it corresponds to a value of 1.35,

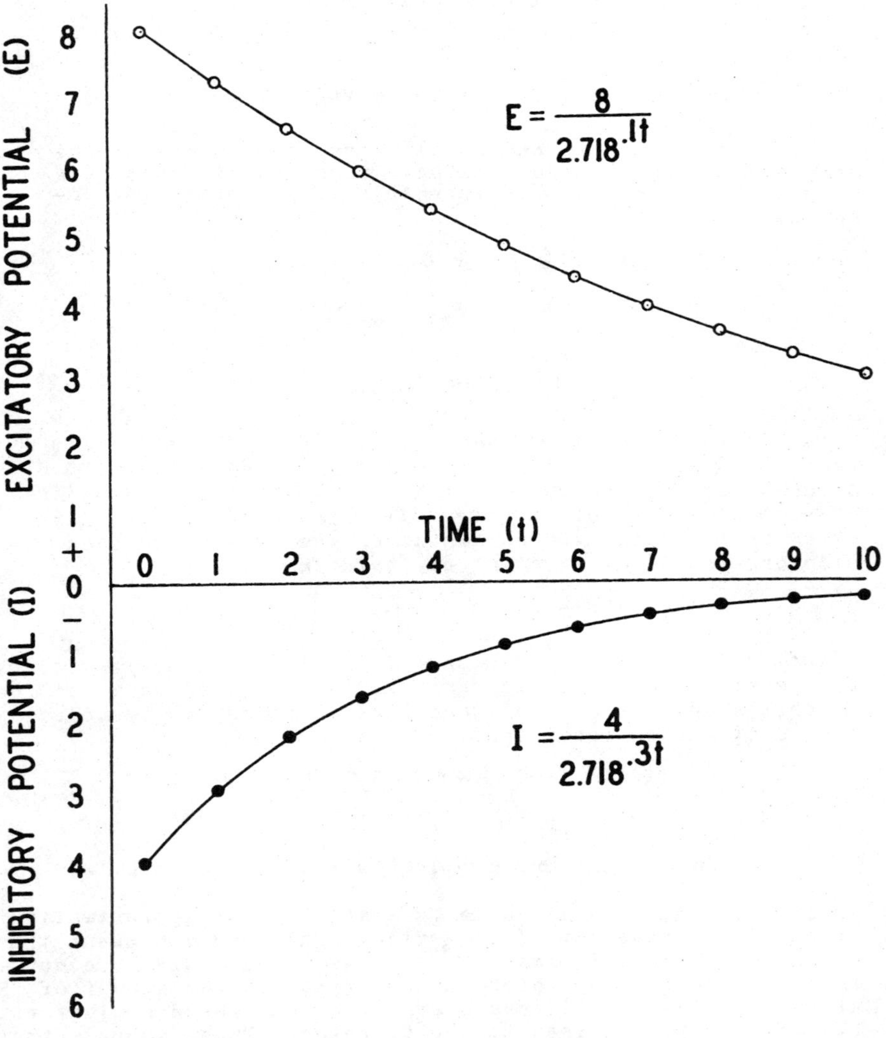

Figure 6. The upper graph of this figure shows the progressive diminution in the strength of an excitatory potential with the passage of time since the termination of active learning according to the equation

$$E = \frac{8}{2.718^{.1t}}$$

The lower graph of the figure shows the progressive diminution in the strength of an inhibitory potential with the passage of time since the termination of the conditions producing the inhibition, according to the equation

$$E = \frac{8}{2.718^{.3t}}$$

Note that at the outset the amount of I lost per unit time is greater than the amount of diminution in E.

which is the .3 power of 2.7183. Substituting this in the formula, we have,

$$E = \frac{8}{1.35} = 5.93$$

i.e., during the passage of 3 units of time, E has fallen from 8 to 5.93.

A series of values for E has been computed on the above assumptions where t has been varied systematically from 0 to 10. The results from this series of computations are shown graphically in Figure 6. An examination of this graph will show that it falls progressively more slowly with the passage of time. This has long been known (*8;49;11*) to be a characteristic of forgetting of rote material (Figure 7), though this is not true of the retention of simple conditioned reactions (*68;16*).

POSTULATE 13

A. *In the learning of any rote series, any inhibitory potential* (I_n) *resulting from any given massed practice decreases with the passage of time* (t) *following the termination of such massed practice, the rate of decrease at any time being proportional to the amount of inhibitory potential existent at that time, i.e.,*

$$I_n (t) = I_n e^{-dt}$$

where d *is a constant* > 0, t *is the time elapsed since the termination of the massed practice in question with the exception of that part of such time which may be occupied by congruent massed practice, and*

$$d \Delta {}^{N}I_n = b \Delta {}^{N}E_{n,n+1}.$$

A'. $x = I_n (t)`(P, T) . \equiv :: x \neq 0 : \equiv : t \geq 0 . (\exists v) . v \cdot = t - \sum_{u \text{ invl } (t,P,T)} u . x = R`(B`\breve{T}, T) \times \Delta^{N`P} I_n \times e^{-dv} . C`T \subset \text{rtln}`P :. d \times \Delta^{N`P} I_n > b \times \Delta E$

In the symbolic logic this postulate is a consequence of D78'.

B. This equation means (in exact analogy to Postulate 12) that with the passage of time following the termination of a given massed practice, the inhibition (I_n) decreases progressively according to the value of a fraction of which the numerator is the amount of I_n at the termination of the massed practice in question (where $t = 0$) and the denominator is e, i.e., 2.7183, raised to the dt power. Thus, suppose that I_n is 4, that d is .3, and that t is 3. Substituting these values in the above formula, as in the example worked out for Postulate 12, we have,

$$I_n (3) = \frac{4}{2.7183^{.3 \times 3}}$$

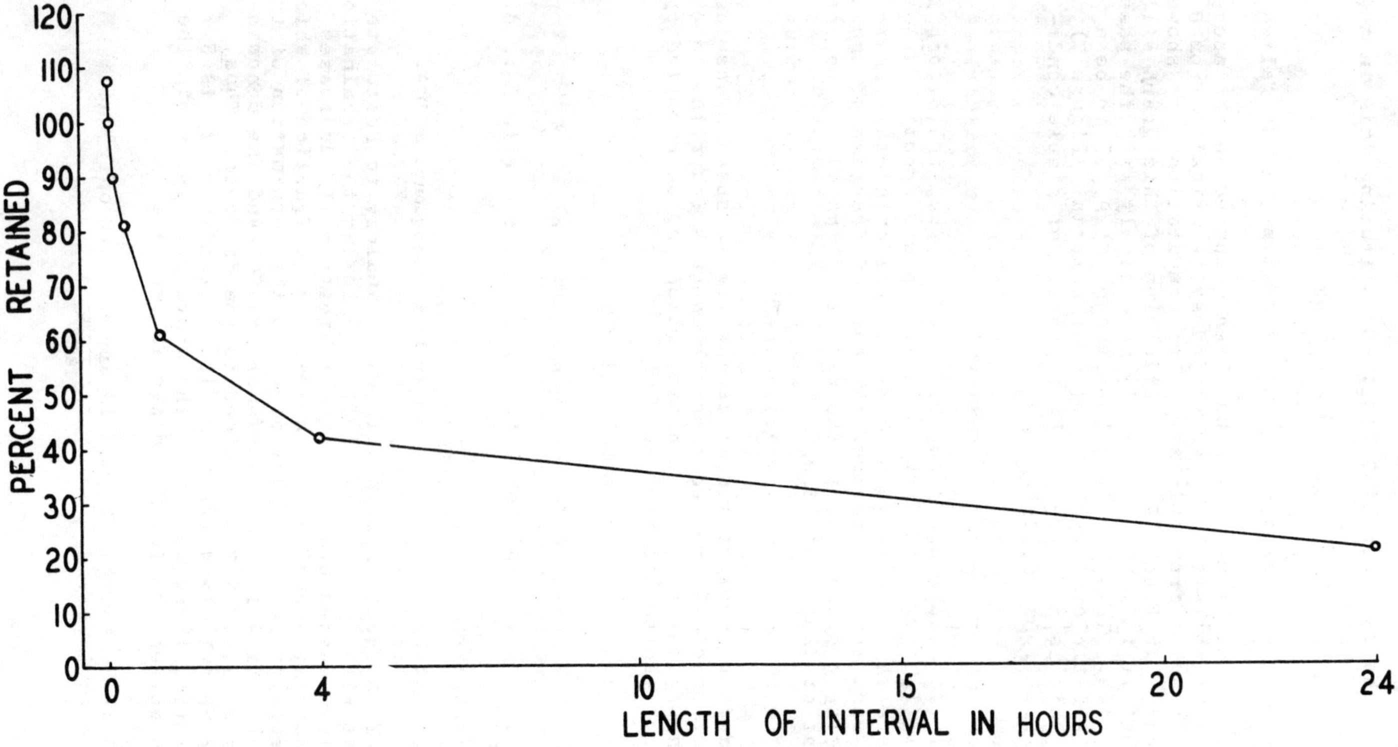

Figure 7. Curve of forgetting of material learned by rote. This is a composite of the results of Luh (38) and of Ward (64).

$$= \frac{4}{2.7183^{.9}}$$

$$= \frac{4}{2.46}$$

$$= 1.62$$

i.e., during the passage of three units of time, I_n should fall from its original value of 4 to 1.62.

A series of I-values has been computed on the above assumption where t has been varied systematically from 0 to 10. The results from this series of computations are shown graphically in Figure 6. A simple inspection of this graph will show that it decreases with the passage of time, but more rapidly at first than later. It will be noted that these inhibitory values have been plotted against a negative extension of the scale of excitatory values. The general shape of this curve agrees very well with its characteristics as determined experimentally by Ellson (Figure 8).

Finally, we may notice the portion of Postulate 13 which states that

$$d\,\Delta\, {}^{N}\mathrm{I}_n > b\,\Delta\, {}^{N}\mathrm{E}_{n,n+1}$$

This statement of inequality is merely a mathematical expression of Pavlov's empirical generalization that inhibitory effects are more rapidly dissipated with the passage of time than are those of an excitatory nature (*47*, 99). In the examples chosen above we have $d = .3$, $\mathrm{I} = 4$, $b = .1$, and $\mathrm{E} = 8$. Substituting in the inequality we have,

$$.3 \times 4 > .1 \times 8$$

i.e.,

$$1.2 > .8$$

The effect of this inequality may be seen in the more rapid diminution at the beginning of the I-curve than that of the E-curve, as shown in Figure 6.

POSTULATE 13, COROLLARY 1

A. *At the termination of any syllable-presentation cycle of the first massed practice of any rote practice,*

$${}^{R,N}\overline{\mathrm{E}}_{n,n+1} = {}^{R,N}\mathrm{E}_{n,n+1} - {}^{R,N}\mathrm{I}_n$$

Proof:

1. By Postulate 10,

$${}^{R,N}\overline{\mathrm{E}}_{n,n+1} = \mathcal{E}_{n,n+1} - \mathcal{I}_n$$

2. By Postulates 12 and 13 and D65, if only one massed practice is involved,

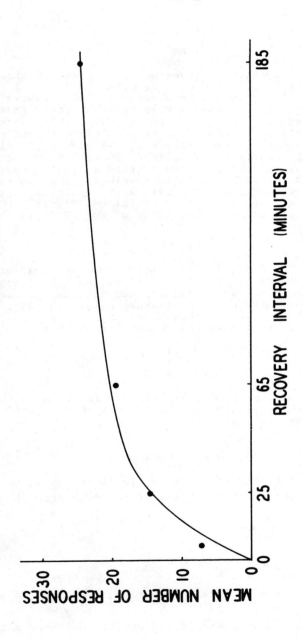

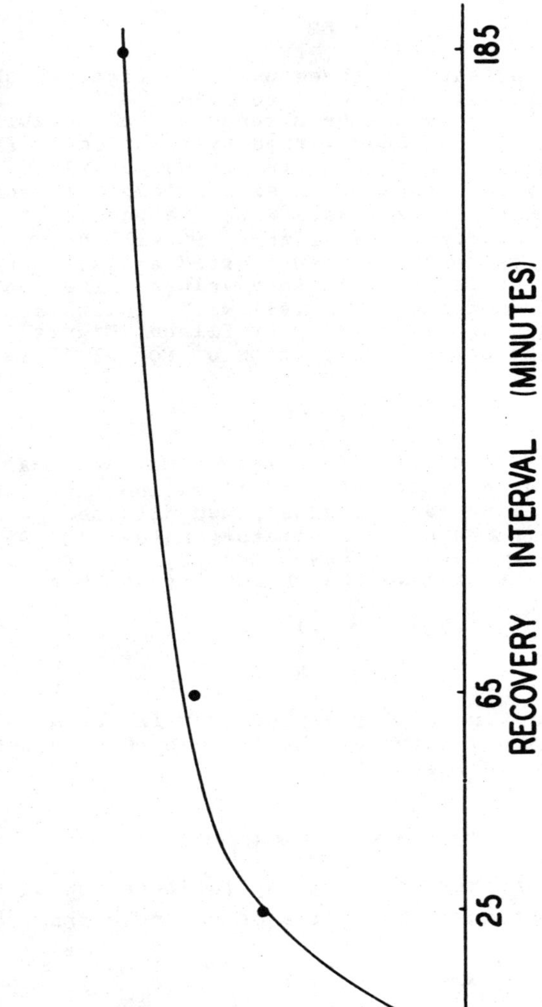

Figure 8. Graph showing the spontaneous recovery of an excitatory potential from the inhibition resulting from experimental extinction. The subjects were albino rats and the reaction was the pressure of a metal bar to get a pellet of food. The mean number of reactions required to produce experimental extinction preceding the extinctions here shown was 49.7 repetitions. (After Ellson, 10)

$$\mathcal{E}_{n,n-1} = {}^{R,N}E_{n,n+1}\, e^{-bt}$$

and

$$\mathcal{I}_n = {}^{R,N}I_n\, e^{-dt}$$

3. But, by Postulates 12 and 13, at any time preceding the termination of the first massed practice, $t = 0$.

4. From (2) and (3) it follows that during the first massed practice,

$${}^{R,N}E_{n,n+1}\, e^{-bt} = {}^{R,N}E_{n,n+1}$$

and

$${}^{R,N}I_n\, e^{-dt} = {}^{R,N}I_n$$

5. Substituting (4) in (2) we have,

$$\mathcal{E}_{n,n+1} = {}^{R,N}E_{n,n+1}$$

and

$$\mathcal{I}_n = {}^{R,N}I_n$$

6. Substituting (5) in (1) we have,

$${}^{R,N}\bar{E}_{n,n+1} = {}^{R,N}E_{n,n+1} - {}^{R,N}I_n$$

from which the corollary follows.

POSTULATE 13, COROLLARY 2

A. *At the termination of any syllable-presentation cycle of the first massed practice of any rote practice,*

$${}^{R,N}\bar{E}_{n,n+1} = R\,\Delta\, {}^{N}\bar{E}_{n,n+1}$$

where

$$\Delta\, {}^{N}\bar{E}_{n,n+1} = \Delta\, {}^{N}E_{n,n+1} - \Delta\, {}^{N}I_n$$

Proof:

1. By Postulate 8, Corollary 1,

$${}^{R,N}E_{n,n+1} = R\,\Delta\, {}^{N}E_{n,n+1}$$

2. By Postulate 9, Corollary 1,

$${}^{R,N}I_n = R\,\Delta\, {}^{N}I_{n,n+1}$$

3. By Postulate 13, Corollary 1,

$${}^{R,N}\bar{E}_{n,n+1} = {}^{R,N}E_{n,n+1} - {}^{R,N}I_n$$

4. Substituting (1) and (2) in (3), we have,

$${}^{R,N}\bar{E}_{n,n+1} = R\,\Delta\, {}^{N}E_{n,n+1} - R\,\Delta\, {}^{N}I_n$$

i.e., $${}^{R,N}\bar{E}_{n,n+1} = R\,(\Delta\,{}^{N}E_{n,n+1} - \Delta\,{}^{N}I_{n})$$

5. But, by assumption,

$$\Delta\,{}^{N}\bar{E}_{n,n+1} = \Delta\,{}^{N}E_{n,n+1} - \Delta\,{}^{N}I_{n}$$

6. Substituting (5) in (4) we have,

$${}^{R,N}\bar{E}_{n,n+1} = R\,\Delta\,{}^{N}E_{n,n+1}$$

from which the corollary follows.

POSTULATE 14

A. *In the learning of any rote series, if at any syllable presentation, $r-1$, the effective excitatory potential ($\bar{E}_{r-1,r}$) is greater than the reaction threshold ($\ell_{r-1,r}$) then the correct reaction whose reaction number is r will take place.*

A'. $n \leq N'P$. $(r-1)$ slpnnm (a,P) . r slpnnm (b,P) . rnth'$(r-1,P)$ < $\bar{E}_{r-1,r}$ (bg'a,P) . ⊃ . $\hat{c}$ (c ctrn a) ε 1

B. One of the most universal observations in all kinds of learning is that one, and usually more, repetitions are necessary before the forming associative tendencies become strong enough to evoke the newly associated reaction. These observations have given rise to the concept of the reaction threshold. This concept seems first to have been introduced by Hull (*27*) in 1917. Despite the seemingly central importance of this concept it has received remarkably little attention from psychologists; we have found only one study which has employed it - that of Simley (*59*). Because of the analogy of the stimulus limen or sensory threshold, the term *reaction* threshold has been applied to it. While doubtless related in some sense, the two concepts must be clearly distinguished: the stimulus limen is the stimulus value which divides those stimuli which will just evoke a reaction from those which will not, whereas the reaction threshold is the value of the effective excitatory potential which divides those which will just mediate reaction from those which will not. The above differentiation should probably be supplemented by the implicit assumption that in the determination of the sensory limen there should be a standard effective excitatory potential for the stimulus to evoke the reaction employed and that in the determination of the reaction threshold a standard strength of stimulus should be employed.

The concept, *reaction threshold*, is elucidated as U9 (p. 25). The sign used for this concept in the symbolic logic is rnth. In other portions of the present work, however, the concept is represented by the sign ℓ, e.g., $\ell_{n,n+1}$, or $\ell_{r-1,r}$. Since subscripts are here used in the

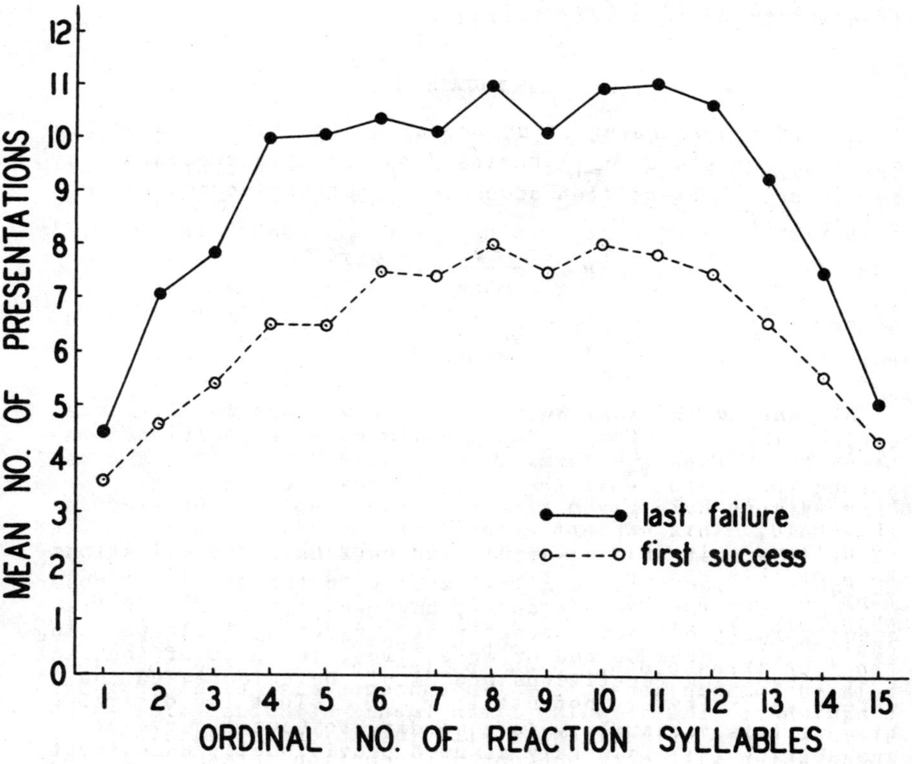

Figure 9. Graphs showing the wide gap separating, on the average, the first appearance of a learned reaction above the reaction threshold in a rote series from the last failure to give the reaction before learning is complete in the sense of having satisfied the criterion of two successive perfect repetitions. Derived from data published by Hull (29).

same manner as in the representation of excitatory potentials, the two sets of subscripts employed above are identical in meaning. In contrast to the usage with ordinary excitatory potentials, a single left-hand superscript is used with ℓ to indicate the repetition at which the threshold value in question obtains at the particular syllable presentation indicated by the first subscript. Thus, ${}^{R}\ell_{r-1,r}$, e.g., ${}^{7}\ell_{4,5}$ means the excitatory potential at the seventh repetition separating those excitatory potentials which, under suitable stimulus conditions, would mediate reaction, from those which would not.

POSTULATE 15

A. *In the learning of any rote series, the value of the reaction threshold* ${}^{R}\ell_{r-1,r}$ *varies from syllable presentation to syllable presentation according to Gauss' normal probability curve,*

$$y = \frac{N}{\sigma\sqrt{2\pi}} e^{-\frac{x^2}{2\sigma^2}}$$

where

$$x = {}^{R}\ell_{r-1,r} - L.$$

A'. $(n, v, u) (\exists j) (\alpha) : \mathrm{Nc}'\alpha > j \,.\, n > 0 \,.\, u > 0 \,.\, \supset \,.$

$$\mathrm{fq}'(v, u, \alpha) - n < \int_{v-u}^{v+u} (\mathrm{Nc}'\alpha \times e^{-(x^2 \div 2\sigma^2)} \div \sigma\sqrt{2\pi})\, dx$$

$$< \mathrm{fq}'(v, u, \alpha) + n$$

B. The basis for assuming fluctuations, or oscillations, at the reaction threshold lies mainly in the universal observation that, for reasons as yet unknown, subjects very frequently fail to react correctly at a given syllable position after having done so one or more times, in spite of the fact that increasing repetitions presumably have increased the effective excitatory potential ($\bar{E}_{n,n+1}$). That the distribution of this variability is according to Gauss' curve of probability (*52*) is merely assumed provisionally as a first approximation. The magnitude of the mean range between the first success and the last failure at the various places throughout a 16-syllable rote series is shown in Figure 9.

It is probable that the phenomenon here under consideration is closely related to that investigated extensively by the early experimental psychologists as "fluctuations of attention." It seems likely that as time goes on, this functional oscillation will find an explicit place not only in the theory of rote learning, but also in that of the most diverse forms of mammalian adaptive behavior.

POSTULATE 16

A. *In any given rote practice there exists one, and only one, limit* L *approached by the mean reaction thresholds of all successively larger and larger classes of syllable presentations, each such class including all syllable presentations beginning over some stretch of time, i.e.,*

$$
L = \begin{cases} {}^{R,N}L_{0,1} = {}^{R,N}L_{1,2} = \cdots\cdots = {}^{R,N}L_{N-1,N} \\ {}^{1,N}L_{r-1,r} = {}^{2,N}L_{r-1,r} = {}^{3,N}L_{r-1,r} = \cdots\cdots \\ {}^{R,6}L_{r-1,r} = {}^{R,7}L_{r-1,r} = {}^{R,8}L_{r-1,r} = \cdots\cdots \end{cases}
$$

where both subscripts and superscripts are used in a sense analogous to that employed in the representation of excitatory potentials.

A'. $\exists ! L$

B. This postulate means in substance that on the average, in any particular rote practice, the reaction thresholds of all syllable presentations of all repetitions tend to be equal whatever the length of the rote series involved. Some indirect indication of the approximate correctness of this postulate is furnished by the fact shown by Hovland (*20*) that when learning is performed by distributed practice, i.e., where the ΔI_n in the equation,

$$
{}_{L}R_{n,n+1} = \frac{L_n}{\Delta\, {}^{N}E_{n,n+1} - \Delta\, {}^{N}I_n}
$$

presumably is considerably reduced (Postulate 13), the relative difference between the number of repetitions required for complete learning at the point of maximum difficulty and that at the two ends of the series, is greatly reduced. According to Postulate 16, if ΔI_n could be reduced to zero, all points in a rote series would be learned with equal facility. Further evidence pointing in the same direction is the fact, reported by Hull (*29*), that recall some 23 hours after the termination of the first massed practice of a rote practice, at which time the inhibitory potential (${}^{N}I_n$) presumably is largely dissipated, was nearly the same at all points in a 15-syllable series.

It will be noted that this postulate makes no statement regarding the value of L in successive rote practices by the same subject. The fact that successive rote series are, on the average, learned with progressively fewer repetitions (see Figure 10) suggests that L may be reduced by preceding practice. This problem presents serious theoretical

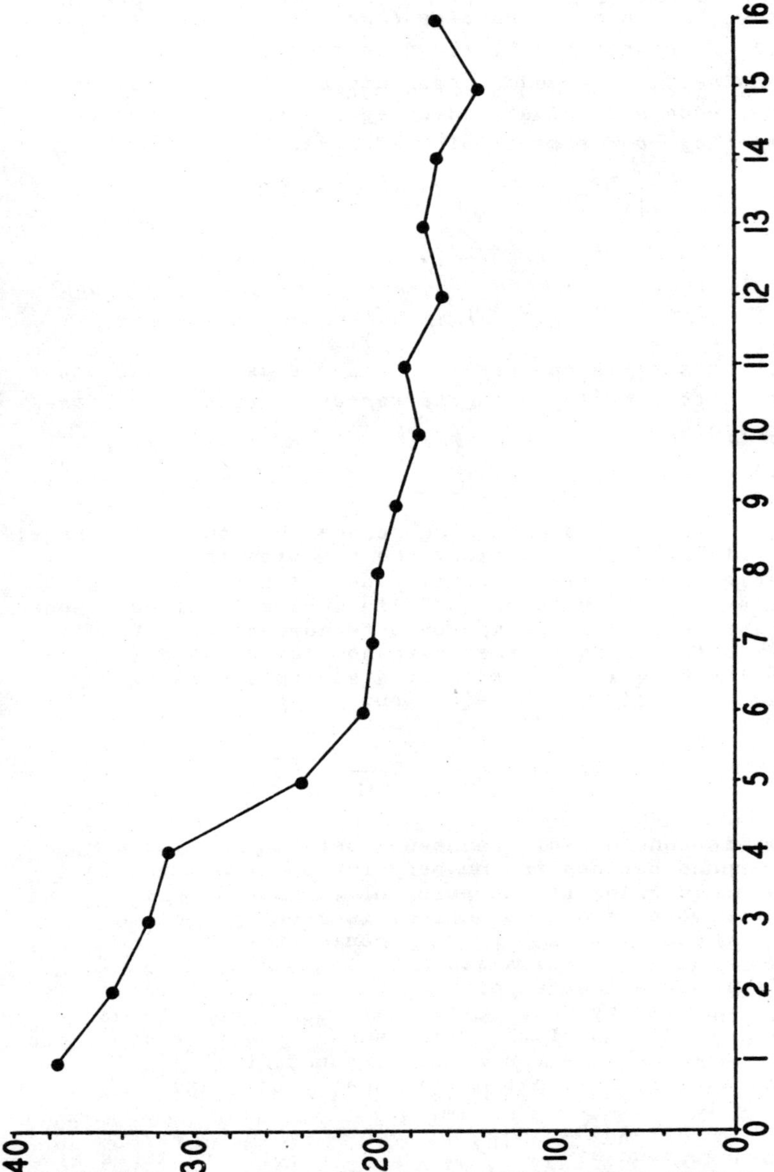

Figure 10. Graph showing the progressively lessening number of repetitions required by twelve subjects to learn successive 16-syllable rote series. After Ward (*63*).

difficulties which ultimately must be met by any satisfactory theory of rote learning. It is not treated in the present system.

POSTULATE 16, COROLLARY 1

A. *In the first massed practice of any rote practice,*

$${}_{L}R_{n,n+1} = \frac{{}_{L}R_{n,n+1}}{\Delta\, {}^{N}E_{n,n+1} - \Delta\, {}^{N}I_{n}}$$

where ${}_{L}R$ *is the mean number of repetitions just necessary to raise the effective excitatory potential, under given conditions, to the reaction threshold.*

Proof:

1. By (4) of the proof of Postulate 13, Corollary 2,

$${}^{R,N}\bar{E}_{n,n+1} = R(\Delta\, {}^{N}E_{n,n+1} - \Delta\, {}^{N}I_{n}) .$$

2. If, now, we let ${}^{R,N}E_{n,n+1} = L$, (1) becomes,

$$L = {}_{L}R\ (\Delta\, {}^{N}E_{n,n+1} - \Delta\, {}^{N}I_{n}) .$$

3. But from (2) we have,

$${}_{L}R_{n,n+1} = \frac{{}_{L}R_{n,n+1}}{\Delta\, {}^{N}E_{n,n+1} - \Delta\, {}^{N}I_{n}}$$

from which the theorem follows.

B. The meaning of this corollary may be illustrated by the following example. Suppose that under given conditions it would require three repetitions to develop enough excitatory potential, if uncomplicated by any associated inhibition, to enable a subject to react successfully at a given syllable position upon reaching it after the delay incidental to the presentation of the other syllables of the series and the brief interval between repetitions. This is equivalent to saying that $L = 3\Delta\, {}^{N}E_{n,n+1}$. Suppose, however, that actually, coincidental with each ΔE, there should also be generated inhibition to the amount of $.5\,\Delta E$, or enough inhibition temporarily to half neutralize the excitation developed by each repetition. This is equivalent to saying that $\Delta I = .5\,\Delta E$. Substituting these values in the equation of Postulate 16, Corollary 1, we have,

$${}_{L}R = \frac{3\,\Delta E}{\Delta E - .5\,\Delta E} = \frac{3}{1 - .5} = 6$$

i.e., under the assumed conditions it would require 6 repetitions to reach the mean reaction threshold.

POSTULATE 17

A. *In any given rote practice there exists one and only one limit, σ, approached by the standard deviations of reaction thresholds of successively larger and larger classes of syllable presentations, each such class including all syllable presentations beginning over some stretch of time, i.e.,*

$$\sigma = \begin{cases} {}^{R,N}\sigma_{0,1} = {}^{R,N}\sigma_{1,2} = \ldots\ldots = {}^{R,N}\sigma_{N-1,N} \\ {}^{1,N}\sigma_{r-1,r} = {}^{2,N}\sigma_{r-1,r} = {}^{3,N}\sigma_{r-1,r} = \ldots\ldots \\ {}^{R,6}\sigma_{r-1,r} = {}^{R,7}\sigma_{r-1,r} = {}^{R,8}\sigma_{r-1,r} = \ldots\ldots \end{cases}$$

where both subscripts and superscripts are used in a sense analogous to that employed in the representation of excitatory potentials.

A'. $E!\sigma$

B. This postulate means in substance that on the average, in any particular rote practice, the standard deviations of the reaction thresholds of all syllable presentations of all repetitions tend to be equal, whatever the length of the rote series involved. This postulate makes no statement regarding the value of σ in successive rote practices by the same subject.

POSTULATE 18

A. *In the learning of any rote series, if a correct reaction occurs,*

$$T_{n,n+1} = v + (h-v)\,e^{-k\bar{E}}$$

where $T_{n,n+1}$ *is the latency of correct reaction* $n+1$, v *is the minimum physiological limit in reaction latency,* h *is the reaction latency computed by extrapolation to where* $\bar{E} = 0$, e *is the Napierian base, a mathematical constant equal to approximately* 2.718, k *is a constant, and* $\bar{E}$ *is the effective excitatory potential for reaction* $n+1$ *to occur at* n *at the instant of its occurrence.*

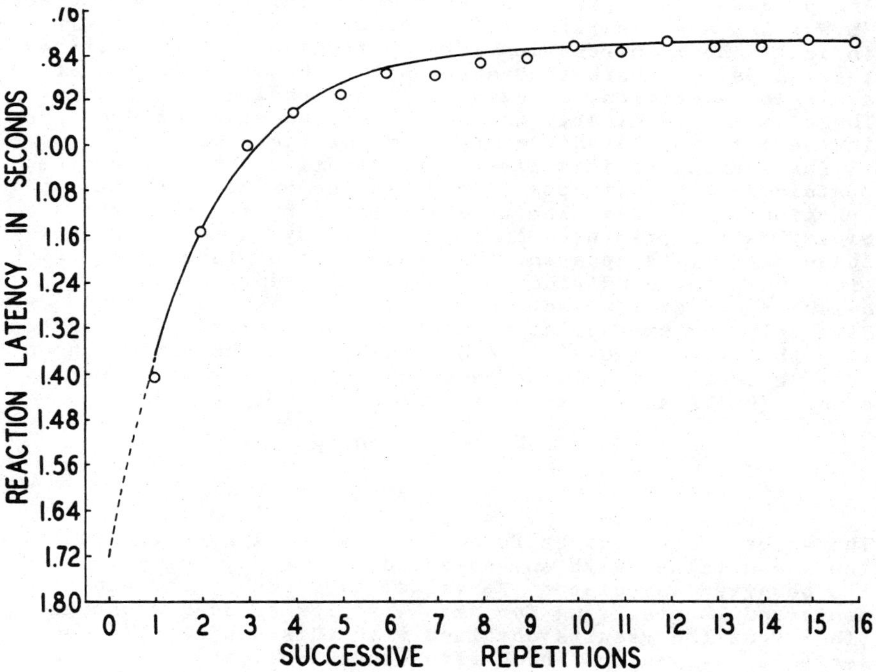

Figure 11. Graphic representation of the relationship of reaction latency during the process of over-learning of nonsense material to the number of repetitions. The small circles represent the mean latencies derived from a table published by Simley (59), reporting the results from his subject M.W. The smooth curve here shown is a graphic representation of an equation kindly fitted to these data by A.P. Weinbach.

A'. $\bar{\text{E}}$'[bg'a, slpnnm'(a,P),P] > rnth'a . ⊃ . T'a

$= \text{v} + (\text{h}-\text{v})\ e^{-\text{k} \times \bar{\text{E}}'[\text{bg}'a,\,\text{slpnnm}'(a,P),P]}$

B. The meaning of this postulate may be illustrated with the aid of experimental results published by Simley (*59*). This investigator gave 20 repetitions to paired associates regardless of whether the associations had yet risen above the threshold of recall, and measured the association times at each successful reaction without interrupting the learning process. A typical set of these results for one subject (M.W.) is shown in Figure 11. The circles in this figure indicate the mean reaction latency from about 125 separate learned pairs, where the reaction threshold was passed after a single repetition, at each of 17 successive repetitions. These data show clearly the negatively accelerated decrease in the reaction latency as repetitions increase.

The meaning of this first-approximation equation of the postulate which attempts to express the value of T as a function of $\bar{\text{E}}$ may also be illustrated by means of the Simley data represented in Figure 11. By an inspection of these data it is apparent that the limit of latency reduction under the conditions of that experiment, is in the neighborhood of .82 sec., which is taken as the value of v. Similarly, by extrapolating backward on the graph, it may be seen that the value at $R = 0$ would be in the neighborhood of 1.72 sec., which is taken as the value of h. If we assume (Problem 1) that $\Delta \text{I} = .09\ \Delta \text{E}$, it follows that

$$\Delta\bar{\text{E}} = \Delta \text{E} - .09\ \Delta \text{E}$$

$$= .91\ \Delta \text{E}$$

The value of k may be found by substituting in the equation the above values with any observed R value, together with its observed associated T value, as given by the experiment, and then solving for k. A series of such substitutions from the results obtained with this subject yielded a series of k values which were appreciably constant, the mean being .55. Accordingly, k is taken as .55.

Suppose, now, that we wish to know the reaction latency (T) which corresponds to a given number of repetitions, e.g. where $R = 3$. Substituting in the equation of Postulate 18, we have,

$$T = .82 + \frac{1.72 - .82}{2.718^{.55 \times 3 \times .91}}$$

$$= .82 + \frac{.90}{2.718^{1.5}}$$

$$= .82 + \frac{.90}{4.49}$$

$$= 1.02$$

Reference to Figure 11 shows that subject M.W. actually averaged 1.00 sec. on the reaction at the third repetition, a reasonably good agreement.

POSTULATE 18, COROLLARY 1

A. *In the first massed practice of any rote series the latency of each correct reaction decreases, and the rate of decrease constantly diminishes, as the number of repetitions increases.*

Proof:

1. From Postulate 18, we have,

$$T_{n,n+1} = v + (h-v)\, e^{-k\bar{E}}$$

2. But by Postulate 13, Corollary 2,

$${}^{R,N}\bar{E}_{n,n+1} = R\Delta\, {}^{N}\bar{E}_{n,n+1}$$

3. Substituting (2) in (1) we have,

$$T_{n,n+1} = v + (h-v)\, e^{-kR\Delta\bar{E}}$$

4. Also by Postulate 13, Corollary 2,

$$\Delta\, {}^{N}\bar{E}_{n,n+1} = \Delta\, {}^{N}E_{n,n+1} - \Delta\, {}^{N}I_{n}$$

5. And by Postulate 3, $\Delta\, {}^{N}E_{n,n+1}$ is constant for all values of R.

6. And by (3) of the proof of Postulate 9, Corollary 2, $\Delta\, {}^{N}I_{n}$ is constant for all values of R.

7. From (4), (5), and (6) it follows that $\Delta\, {}^{N}\bar{E}_{n,n+1}$ must be constant for all values of R.

8. From (3) and (7) it follows that we can regard T as a function of R alone. The corollary follows when we show that the derivative of T with respect to R is negative and increasing. But,

$$\frac{dT}{dR} = -(h-v)\, k\, \Delta\bar{E}\, e^{-kR\Delta\bar{E}}$$

and (h-v), k, $\Delta\bar{E}$, and $e^{-kR\Delta\bar{E}}$ are all positive. Hence, $\frac{dT}{dR}$ is negative. Furthermore,

$$\frac{d^2T}{dR^2} = (h-v)\, k^2\, (\Delta E)^2\, e^{-kR\Delta\bar{E}}$$

and this is positive. Hence, the corollary follows.

B. This corollary is nicely illustrated by Simley's results represented graphically as Figure 11.

Theorems and Problems

THEOREM I

A. *In any syllable-presentation cycle (repetition) of the learning of any rote series, concurrently with the generation of any increment of excitatory potential* ($\Delta E_{r-1,r}$) *there will be generated in association with the segment* ($r-1$) *of any stimulus trace* (${}_s tr$), *the increment of inhibitory potential,* $\frac{\Delta K}{F^{r-s-1}}$, *i.e.,*

$$\Delta\, {}_sI_{r-1,r} = \frac{\Delta K}{F^{r-s-1}}$$

where $s \leq r-1$

Proof:

1. By Postulate 2, after each syllable-presentation cycle there exists a finite increment of excitatory potential ($\Delta E_{r-1,r}$).

2. By (1) and Postulate 4, concurrently with the generation of a finite increment of inhibitory potential ($\Delta E_{r-1,r}$) there will be generated a finite increment of inhibitory potential $\Delta\, {}_{r-1}I_{r-1,r}$ which is associated with stimulus trace segment (${}_{r-1}tr_{r-1}$).

3. Moreover, by D72 and Postulate 5,

$$\Delta\, {}_{r-1}I_{r-1,r} = \Delta K$$

4. By (2), (3), and Postulate 6, the increment of inhibitory potential ($\Delta\, {}_sI_{r-1,r}$) when $r-s=1$, will be ΔK; where $r-s=2$, it will be $\frac{\Delta K}{F}$; where $r-s=3$, it will be $\frac{\Delta K}{F^2}$ and so on.

5. But since the exponent of F in the values of ($\Delta\, {}_sI_{r-1,r}$) in (4) is always one less than the value of $r-s$, the general expression for this exponent must be, $r-s-1$.

6. It follows from (4) and (5) that

$$\Delta\, {}_sI_{r-1,r} = \frac{\Delta K}{F^{r-s-1}}$$

from which the theorem follows.

B. This theorem presents a generalized mathematical statement of the magnitude of the inhibitory potential associated with the segment of any stimulus trace concurrent

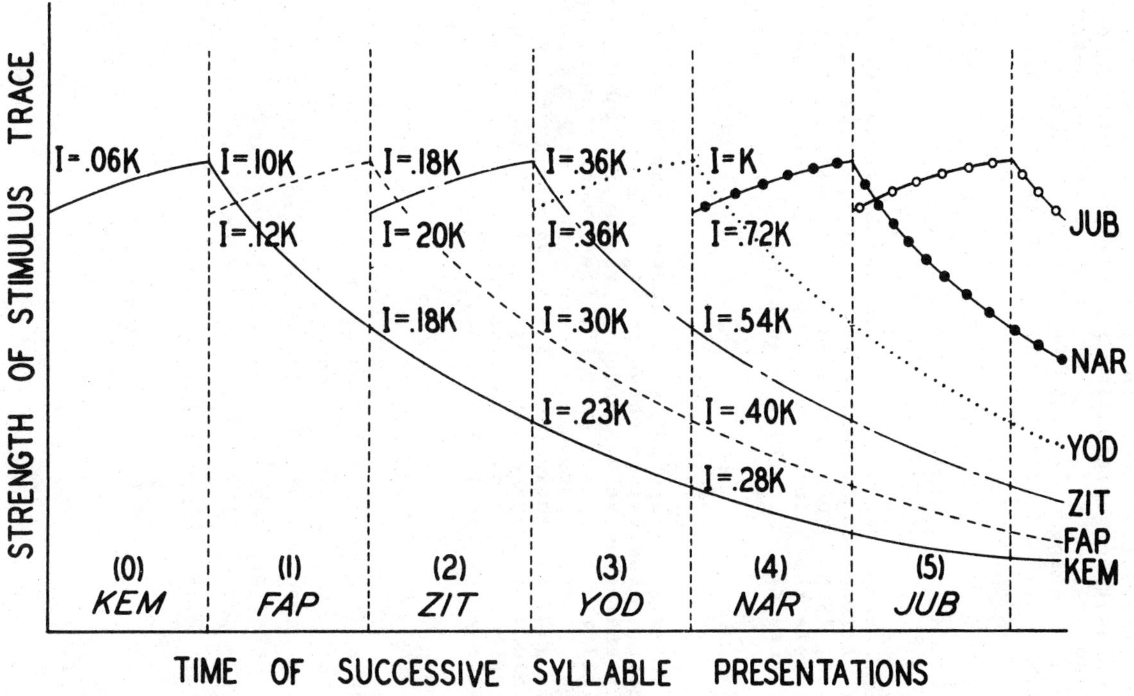

Figure 12. Diagram showing all the stimulus traces of a short rote series. Written in, close to each segment of each trace involved, is the theoretical amount of inhibitory potential associated with the segment in question, which results from the learning, or conditioning, of the fifth or last reaction of the series, "JUB".

with the point of conditioning $(r-1)$ of any syllable reaction (r), as stated verbally in Postulate 5. The meaning of the expression, $\frac{\Delta K}{F^{r-s-1}}$, may be illustrated by its application to one of the values shown in Figure 4, that associated with the stimulus trace originating in syllable FAP. Since the reaction is "NAR", $r = 4$. Also, since 1 is the number of *FAP*, $s = 1$. Taking $F = 1.37$ (Problem 1) as in our previous illustrations, and substituting in the general expression $\frac{\Delta K}{F^{r-s-1}}$, wé have,

$$\frac{\Delta K}{1.37^{4-1-1}} = \frac{\Delta K}{1.37^2} = \frac{\Delta K}{1.87} = .54\ \Delta K$$

As a second example we will take the inhibition generated at the point of final conditioning $(r-1)$ of the same syllable reaction, "NAR", but associated with the trace originating in syllable KEM. In this case $r = 4$ as before, and $s = 0$. (see Figure 4). Accordingly we have,

$$\frac{\Delta K}{1.37^{4-0-1}} = \frac{\Delta K}{1.37^3} = \frac{\Delta K}{2.56} = .40\ \Delta K$$

Both the above values computed by the formula may be verified by inspecting the values shown in Figure 4.

Completely analogous values which represent the inhibitory potentialities resulting from the final conditioning of reaction "JUB" to the compound trace concurrent with stimulus syllable *NAR*, are shown above *NAR* in Figure 12. Corresponding values from the conditioning of "YOD" are shown above *ZIT* in Figure 13, and those from the conditioning of "ZIT" are shown above *FAP* in Figure 14.

THEOREM II

A. *In any syllable-presentation cycle (repetition) of the learning of any rote series, concurrently with the generation of any increment of excitatory potential* $(\Delta E_{r-1,r})$ *there will be generated in association with any segment* (n) *of any stimulus trace* $({}_s\mathrm{tr})$ *the finite increment of inhibitory potential* $\frac{n'-s}{r-s}\,\frac{\Delta K}{F^{r-s-1}}$ *i.e.,*

$$\Delta\,{}_s^{\#}\mathrm{I}_{n,r} = \frac{n'-s}{r-s}\,\frac{\Delta K}{F^{r-s-1}}$$

where $n' = n+1$.

Proof:

1. Replacing in any of the equations of Postulate 7 the value representing the number of the syllable presentation at which the inhibitory potential is active, by n,

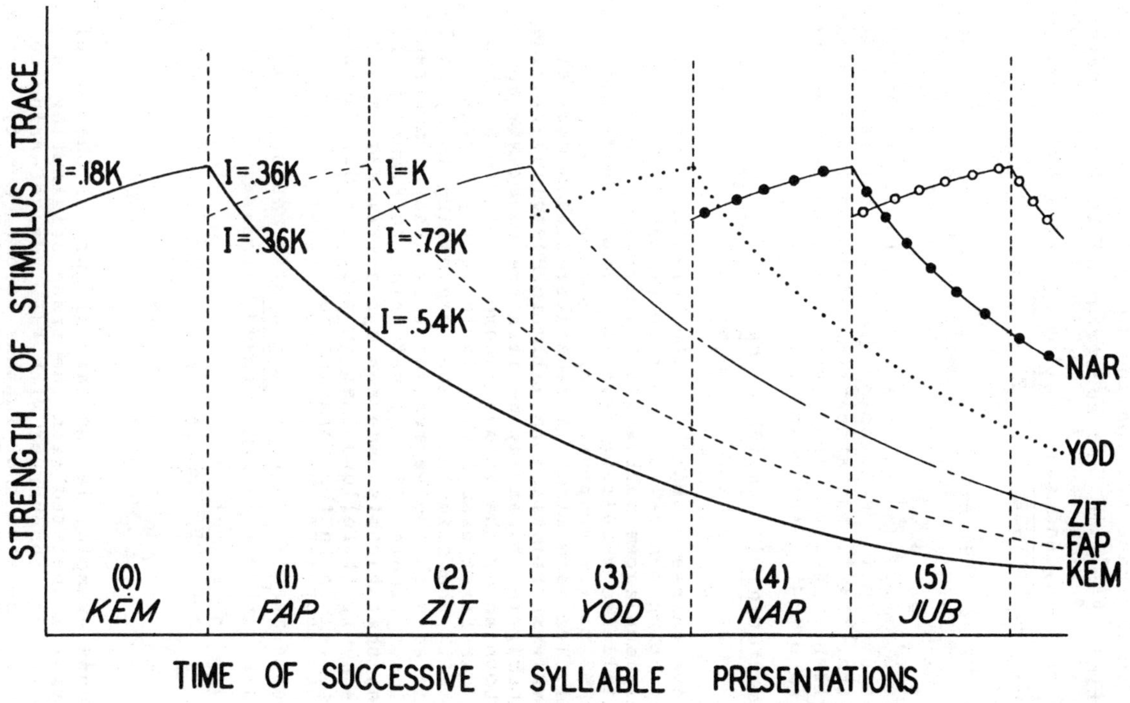

Figure 13. Diagram showing all the stimulus traces of a short rote series. Written in, close to each segment of each stimulus trace involved, is the theoretical amount of inhibitory potential associated with the segment in question, which results from the learning, or conditioning, of the third reaction of the series, "YOD". F is taken as 1.37.

we have the expression

$$\Delta\, {}_{s}^{N}I_{n,r} = \frac{n - (s-1)}{(r-1) - (s-1)} \times \frac{\Delta K}{F^{x}} = \frac{n-s+1}{r-s} \times \frac{\Delta K}{F^{x}} .$$

2. Letting $n = n' - 1$ and substituting in (1) we have,

$$\Delta\, {}_{s}^{N}I_{n,r} = \frac{(n'-1) - s+1}{r-s} \times \frac{\Delta K}{F^{x}} = \frac{n'-s}{r-s} \times \frac{\Delta K}{F^{x}} .$$

3. But, by Theorem I,

$$\Delta\, {}_{s}I_{r-1,r} = \frac{\Delta K}{F^{r-s-1}}$$

from which it follows that in (2)

$$x = r - s - 1.$$

4. Substituting (3) in (2) we have,

$$\Delta\, {}_{s}^{N}I_{n,r} = \frac{n'-s}{r-s} \quad \frac{\Delta K}{F^{r-s-1}}$$

from which the theorem follows.

B. This theorem presents a generalized mathematical statement of what is stated verbally in Postulate 7. The meaning of the expression may be illustrated by its application to some of the values shown in Figure 5. Let us take as our first example the inhibitory potential associated with the stimulus trace originating in syllable FAP, associated with the segment of this trace concurrent with stimulus syllable *ZIT*. By referring to Figure 5, it will be seen that in this case $r = 4$, $s = 1$, and $n = 2$. Since $n' = n + 1$, it follows that $n' = 2 + 1 = 3$. Substituting these values in the formula, we have,

$$ {}_{1}I_{2,4} = \frac{3-1}{4-1} \quad \frac{\Delta K}{1.37^{4-1-1}}$$

$$= \frac{2}{3}\ .54\ \Delta K$$

$$= .36\ \Delta K .$$

As a second example, let us take (Figure 5) the inhibitory potential associated with the trace originating in KEM at the segment concurrent with *FAP*. By referring to Figure 5 it may be seen that in this case , $r = 4$, $s = 0$, and $n = 1$, i.e., $n' = 2$. Substituting in the formula, we have,

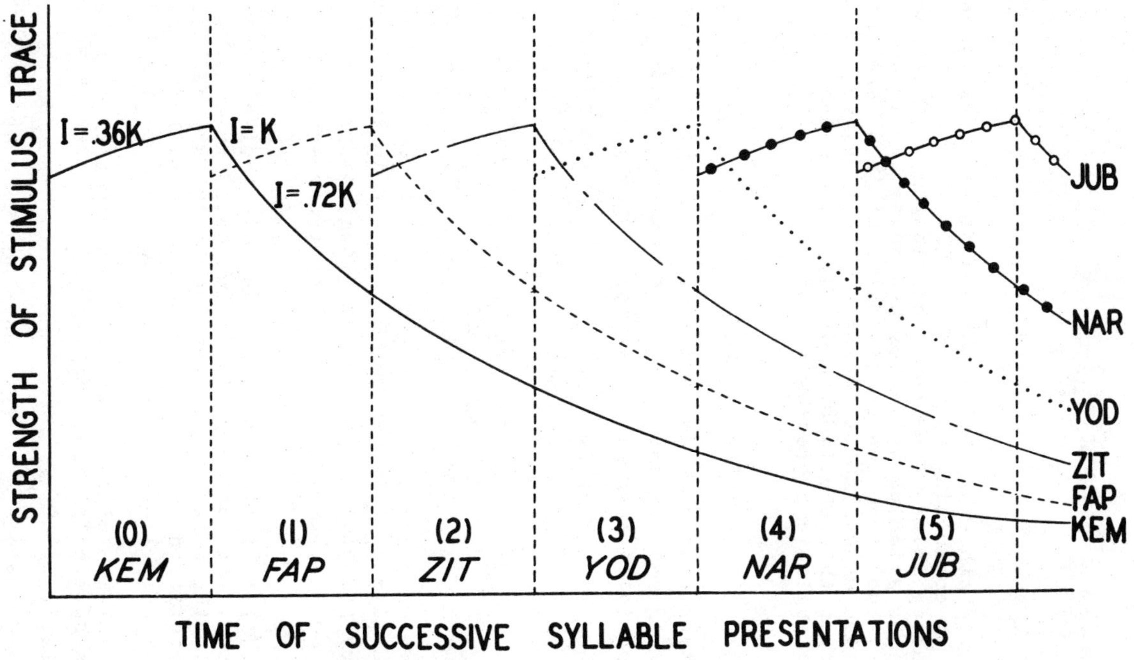

Figure 14. Diagram showing all the stimulus traces of a short rote series. Written in, close to each segment of each trace involved, is the theoretical amount of inhibitory potential associated with the segment in question, which results from the learning, or conditioning, of the second reaction of the series, "ZIT". F is taken as 1.37.

$$ {}_{0}^{N}I_{1,4} = \frac{2-0}{4-0} \frac{\Delta K}{1.37^{4-0-1}} $$

$$ = \frac{2}{4} \frac{\Delta K}{1.37^{3}} $$

$$ = \frac{1}{2} .40 \ \Delta K $$

$$ = .20 \ \Delta K . $$

The results of both examples of the use of the formula may be verified by consulting Figure 5.

Corresponding values from the conditioning of "JUB" are shown above the stimulus syllables preceding *NAR* in Figure 12; values from the conditioning of "YOD" are shown above the stimulus syllables preceding *FAP* in Figure 13; and values from the conditioning of "ZIT" are shown above the stimulus syllable preceding *FAP* in Figure 14.

THEOREM III

A. *In any syllable-presentation cycle (repetition) of the learning of any rote series, the functional sum of all the increments of inhibitory potential associated with a single segment (n) of any stimulus trace* (${}_{s}tr$) *is* $\sum_{C=n-s}^{N-s} \frac{(n^{\text{L}}-s)\ \Delta K}{CF^{C-1}}$ *i.e. (D74),*

$$ \Delta\, {}_{s}^{N}I_{n} = \sum_{C=n^{\text{l}}-s}^{N-s} \frac{(n^{\text{L}}-s)\ \Delta K}{CF^{C-1}} $$

Proof:

1. By Postulate 4, for each $\Delta E_{r-1,r}$ there is an increment of inhibitory potential at every trace segment preceding r in the rote series, i.e., at all of the trace segments, $\| {}_{0\ldots r-1}tr_{s\ldots r-1} \|$.

2. From (1) it follows that for any particular trace segment, ${}_{s}tr_{n}$, at each repetition in the learning of a rote series, there is an increment of inhibitory potential for every reaction number of the rote series following such trace segment (n), i.e., for reaction numbers $n+1$ to N inclusive.

3. From (2) it follows that for each repetition there will be at a particular trace segment, ${}_{s}tr_{n}$, the inhibitory potentials, $\| \Delta\, {}_{s}^{N}I_{n,n+1\ldots N} \|$.

4. Now, by Postulate 9, the functional sum of

$$ \| \Delta\, {}_{s}^{N}I_{n,n+1\ldots N} \| = \sum_{i=n+1}^{N} \Delta\, {}_{s}^{N}I_{n,i} $$

5. Also, from Theorem II,

$$\Delta\, {}_s^N I_{n,i} = \frac{n' - s}{i - s} \; \frac{\Delta K}{F^{i-s-1}}$$

where

$$n + 1 = n' .$$

6. Substituting (5) in (4), we have,

$$\| \Delta\, {}_s^N I_{n,n' \ldots N} \| = \sum_{i=n'}^{N} \frac{n' - s}{i - s} \; \frac{\Delta K}{F^{i-s-1}}$$

7. Letting $C = i - s$, we have,

$$\| \Delta\, {}_s^N I_{n,n+1 \ldots N} \| = \sum_{C=n'-s}^{N-s} \frac{(n' - s)\, \Delta K}{C F^{C-1}}$$

8. But, by D74,

$$\| \Delta\, {}_s^N I_{n,n+1 \ldots N} \| = \Delta\, {}_s^N I_n .$$

9. Substituting (8) in (7), we have,

$$\Delta\, {}_s^N I_n = \sum_{C=n'-s}^{N-s} \frac{(n' - s)\, \Delta K}{C F^{C-1}}$$

from which the theorem follows.

B. The common-sense meaning of this theorem may be understood by a joint consideration of Figures 5 and 12. Figure 5, as we have already seen, shows the theoretical distribution or pattern of the increments of inhibitory potential resulting from the conditioning of reaction "NAR". Figure 12 shows, in a parallel manner, the distribution or pattern of the increments of inhibitory potential resulting from the conditioning of reaction "JUB", which may for present purposes be regarded as the last syllable in the series, i.e., $N = 5$. Figures 13 and 14 show in a similar manner the theoretical distribution or pattern of increments of inhibitory potential resulting from the conditioning of the second and third syllable reactions respectively. The problem of Theorem III is to derive a general mathematical expression which will state the functional sum of the several finite amounts of inhibition resulting from a single repetition and associated with any single segment of any single stimulus trace; e.g., the sum of the inhibition associated with that segment of the stimulus trace originating in the cue syllable, KEM, which is concurrent with stimulus syllable *YOD*.

Since the ordinal number of *YOD* is 3 (i.e., $n = 3$), it follows that since $n' = n + 1$, $n' = 4$. As developed in the proof of the theorem, increments to inhibitory potential at the point in question can come only from the conditioning of reactions whose ordinal numbers (r) are equal to or

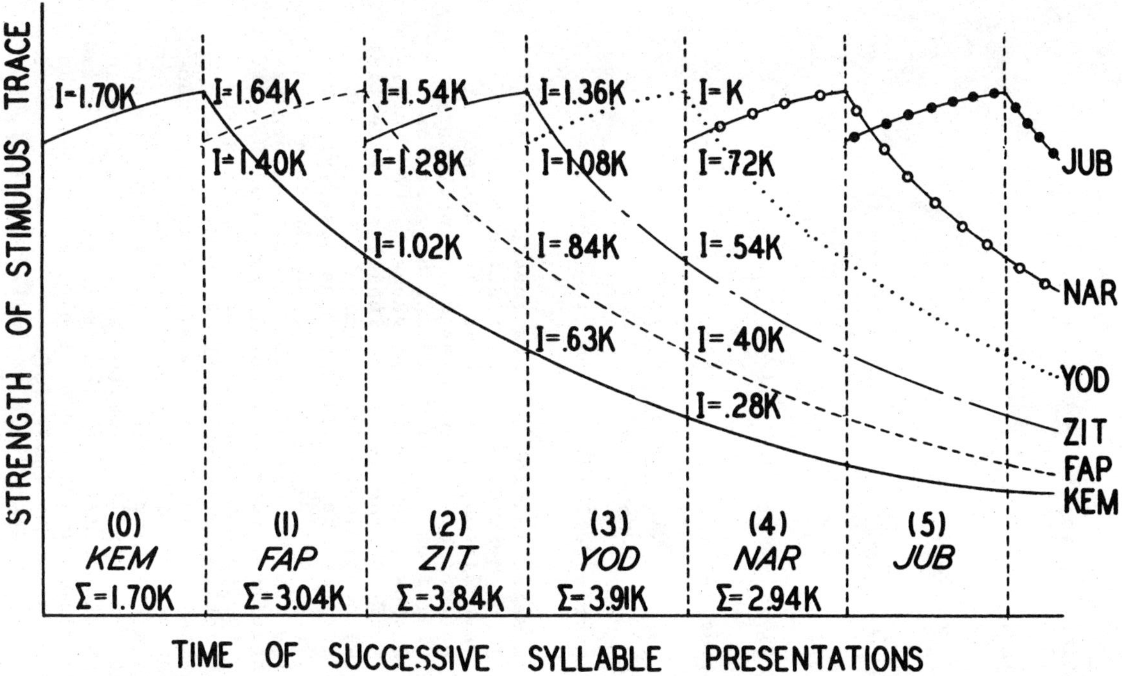

Figure 15. Diagram showing all of the stimulus traces in a 5-syllable rote series. Written in, close to each segment of each trace involved, is the sum of the theoretical amounts of inhibitory potential associated with the segment in question, which resulted from the learning, or conditioning, of *all the syllables in the series*. Diagrams showing separately the theoretical inhibitory-potential values resulting from the conditioning of syllables 2, 3, 4, and 5 are shown in detail in Figures 14, 13, 5, and 12 respectively. Only one amount of inhibitory potential, that of 1.00 K, is associated with the initial segment of the cue syllable (*KEM*) resulting from the learning of "FAP", so no separate diagram is presented for the learning of the first syllable reaction. With this exception, all the values in Figure 15 may be obtained by merely adding together the values appearing at the respective trace segments in Figures 14, 13, 5, and 12. Alternatively, the same values may be computed by substituting in the equation of Theorem III.

greater than n' up to the limit of N. Since $n' = 4$ and $N = 5$, it follows that the increments of inhibitory potential associated with the trace segment in question can arise only from the conditioning of reactions "NAR" and "JUB" represented in detail diagrammatically in Figures 5 and 12 respectively. By consulting Figure 5 it may be seen that the increment of inhibitory potential from a single repetition at the *YOD* segment of the KEM trace resulting from the conditioning of reaction "NAR" is .40 ΔK; and by consulting Figure 12 it may be seen that at the same segment of the same trace there results from the conditioning of reaction "JUB" an inhibitory potential of .23 ΔK, the two making a total of .63 ΔK. This value is entered at $_0tr_3$ in Figure 15.

By consulting the same figures it may be seen that the parallel values for the *YOD* segment of the FAP trace are .54 ΔK and .30 ΔK, making a total of .84 ΔK. By consulting Figures 13, 12, and 5, it may be seen that there are three theoretical inhibitory potential values associated with the *ZIT* segment of the FAP trace, viz., .72 ΔK, .36 ΔK, and .20 ΔK, which make a total of 1.28 ΔK. In a similar manner it may be seen by consulting Figures 14, 13, 12, and 5, that at the *FAP* (initial) segment of the FAP trace there are four theoretical inhibitory potential values, viz.: 1.00 ΔK, .36 ΔK, .18 ΔK, and .10 ΔK, a total of 1.64 ΔK. The above totals, together with all the other total inhibitory values associated with the learning of a 5-syllable rote series, are shown in Figure 15.

The meaning of the general mathematical expression,

$$\sum_{C=n'-s}^{N-s} \frac{(n'-s)\,\Delta K}{CF^{C-1}}$$

may best be understood by using it to derive the value of $^5_0I_3$ which we have already seen amounts to .63 ΔK. In ordinary language the expression means, "the sum of the values yielded by the fraction $\frac{(n'-s)\,\Delta K}{CF^{C-1}}$ as C takes successively the various possible values ranging from $n' - s$ to $N - s$ inclusive." By referring to Figure 12 it will be seen that $N = 5$ and that, since the trace originates in KEM, $s = 0$. Accordingly, $n' = 3 + 1 = 4$; $n' - s = 4 - 0 = 4$; and $N - s = 5 - 0 = 5$. Thus $\sum_{C=n'-s}^{N-s}$ becomes $\sum_{C=4}^{5}$. Accordingly we substitute in the expression

$$\frac{(n'-s)\,\Delta K}{CF^{C-1}}$$

twice: the first time we let $C = 4$, and the second time we let $C = 5$. We then add the results of the two substitutions.

Substituting with $C = 4$ we have,

$$\frac{(4-0)\,\Delta K}{4 \times 1.37^{4-1}} = \frac{4\,\Delta K}{4 \times 1.37^{3}} = \frac{\Delta K}{1.37^{3}} = .40\,\Delta K.$$

Substituting again, this time letting $C = 5$, we have,

$$\frac{(4-0)\,\Delta K}{5 \times 1.37^{5-1}} = \frac{4\,\Delta K}{5 \times 1.37^{4}} = .23\,\Delta K.$$

Summating the results of the two substitutions we have,

$$.40\ \Delta K + .23\ \Delta K = .63\ \Delta K$$

exactly as we found by the procedure employed just above.

In an analogous manner, the equation of Theorem III yields all of the values shown in Figure 15.

THEOREM IV

A. *In the learning of any rote series, the functional sum of all the increments of inhibitory potential associated with a particular compound trace,* $\|{}_{0\ldots n}tr_n\|$, *which results from a single syllable-presentation cycle of a rote practice, is the sum of the totals of inhibitory potential associated with the segments of the separate stimulus traces* (${}_s tr_n$) *making up the compound trace, i.e.,*

$$\Delta I_n = \sum_{s=0}^{n} \|\Delta\, {}_s^{N}I_{n,r\ldots N}\|$$

where $r = n + 1$.

Proof:

1. By Postulate 1, Corollary 1, the stimulus traces having segments in any compound trace, $\|{}_{0\ldots n}tr_n\|$, are those originating in syllable presentations 0 to n inclusive.
2. And, by Postulate 9, the functional sum of the inhibitory potentials of these separate segments of a compound trace must be the sum obtained by simple addition.
3. It follows from (1), (2), and D75 that,

$$\Delta I_n = \|\Delta\, {}_{0\ldots n}^{N}I_{n,r\ldots N}\|$$

$$= \sum_{s=0}^{n} \|\Delta\, {}_s^{N}I_{n,r\ldots N}\|$$

from which the theorem follows.

B. This theorem states, in effect, that the functional sum of all the inhibitory increments resulting from a single repetition and active at a given syllable presentation interval (D24) is the sum of the totals of inhibitory potential associated with the several components of the compound trace in question. This is illustrated nicely by Figure 15 which shows the totals of inhibitory potential for all of the component-trace segments of all the compound traces with which inhibitory potentials are associated. Let us take for detailed consideration the compound trace at syllable *ZIT*, where $n = 2$. This compound trace includes trace segments:

$$_{0}tr_{2}, \quad _{1}tr_{2}, \quad \text{and} \quad _{2}tr_{2},$$

the formula for the compound trace as a whole accordingly being $_{0\ldots 2}tr_{2}$.

Theorem III has shown that the total inhibitory potential associated with $_{0}tr_{2}$ is $\Delta\,^{5}_{0}I_{2}$, that with $_{1}tr_{2}$ is $\Delta\,^{5}_{1}I_{2}$, and that with $_{2}tr_{2}$ is $\Delta\,^{5}_{2}I_{2}$. The formula for all these totals taken together is accordingly $\| \Delta\,_{0\ldots 2}^{5}I_{2} \|$ or, still more explicitly, $\| \Delta\,_{0\ldots 2}^{5}I_{2,3\ldots 5} \|$. The equation,

$$\Delta I_{n} = \sum_{s=0}^{n} \| \Delta\,^{N}_{s}I_{n,r\ldots N} \|$$

i.e.,

$$\Delta I_{2} = \sum_{s=0}^{2} \| \Delta\,^{5}_{s}I_{2,3\ldots 5} \|$$

merely represents the summation of the three totals ranging from where $s = 0$ to where $s = 2$ inclusive. These totals may be seen in Figure 15 to be respectively,

$$1.02\ \Delta K, \quad 1.28\ \Delta K, \quad \text{and } 1.54\ \Delta K.$$

THEOREM V

A. *In the learning of any rote series by massed practice, the functional sum of all the increments of inhibitory potential associated with a particular compound trace, resulting from a single syllable-presentation cycle of a rote practice* (ΔI_{n}) *is,*

$$\Delta\,^{N}I_{n} = \dot{n}' \sum_{C=1}^{N} \frac{\Delta K}{CF^{C-1}} - \sum_{C=1}^{n'-1} \frac{(\dot{n}' - \dot{C})\ \Delta K}{CF^{C-1}} - \sum_{C=N-n'+2}^{N} \frac{\overline{C - 1 - (N - n')}^{\,\bullet}\ \Delta K}{CF^{C-1}}$$

where

$$n' = n + 1,$$

where

$$\dot{n}' = 1 + 2 + 3 \ldots n' = \frac{n'(n+1)}{2},$$

and where

$$C = n' - s$$

Proof:

1. By Theorem IV,

$$\Delta I_n = \sum_{s=0}^{n} \| \Delta\, {}_sI_{n,n'\ldots N} \| .$$

2. But, by Theorem III, (1) may be written,

$$\Delta I_n = \sum_{s=0}^{n} \left(\sum_{C=1}^{N-s} \frac{(n'-s)\,\Delta K}{CF^{C-1}} - \sum_{C=1}^{n'-s-1} \frac{(n'-s)\,\Delta K}{CF^{C-1}} \right)$$

3. And (2) may be written,

$$\Delta I_n = \sum_{s=0}^{n} \sum_{C=1}^{N} \frac{(n'-s)\,\Delta K}{CF^{C-1}} - \sum_{s=0}^{n} \sum_{C=1}^{n'-s-1} \frac{(n'-s)\,\Delta K}{CF^{C-1}}$$

$$- \sum_{s=0}^{n} \sum_{C=N-s+1}^{N} \frac{(n'-s)\,\Delta K}{CF^{C-1}}$$

$$= T_1 - T_2 - T_3$$

where T_1, T_2, and T_3 are the indicated double sums which we shall evaluate separately.

4. The first of these sums may be simplified immediately by interchanging the order of summation.

$$T_1 = \sum_{s=0}^{n} \sum_{C=1}^{N} \frac{(n'-s)\,\Delta K}{CF^{C-1}} = \sum_{C=1}^{N} \sum_{s=0}^{n} \frac{(n'-s)\,\Delta K}{CF^{C-1}}$$

$$= \sum_{C=1}^{N} \frac{\dot{n}'\,\Delta K}{CF^{C-1}} = \dot{n}' \sum_{C=1}^{N} \frac{\Delta K}{CF^{C-1}} .$$

5. To evaluate T_2, we note that it is the sum of the expression $\frac{(n'-s)\,\Delta K}{CF^{C-1}}$ where C and s take on certain pairs of values, namely:

$$\begin{array}{l}
s = 0, \quad C = 1, 2, \ldots, n \\
s = 1 \quad C = 1, 2, \ldots, n-1 \\
\ldots\ldots\ldots\ldots\ldots\ldots\ldots\ldots \\
s = i \quad C = 1, 2, \ldots, n-i \\
\ldots\ldots\ldots\ldots\ldots\ldots\ldots\ldots \\
s = n-1, \quad C = 1 \\
s = n, \quad \text{no value for } C.
\end{array}$$

We may collect the values of s corresponding to a particular C, thus:

$$\begin{array}{ll} C = 1, & s = 0, 1, \ldots, n-1 \\ C = 2, & s = 0, 1, \ldots, n-2 \\ \ldots & \ldots \\ C = j, & s = 0, 1, \ldots, n-j \\ \ldots & \ldots \\ C = n, & s = 0. \end{array}$$

Summing the expression $\frac{(n'-s)\,\Delta K}{CF^{C-1}}$ for a particular value of C, we obtain,

$$\sum_{s=0}^{n-C} \frac{(n'-s)\,\Delta K}{CF^{C-1}} = \frac{\Delta K}{CF^{C-1}} \sum_{s=0}^{n-C} (n'-s) = \frac{\Delta K}{CF^{C-1}} (\dot{n}' - \dot{C})$$

Hence, since $n = n' - 1$,

$$T_2 = \sum_{C=1}^{n'-1} \frac{(\dot{n}'-\dot{C})\,\Delta K}{CF^{C-1}}$$

6. Similarly for T_3, we sum $\frac{(n'-s)\,\Delta K}{CF^{C-1}}$ over the values:

$$\begin{array}{ll} s = 0, & \text{no value for } C \\ s = 1, & C = N \\ s = 2, & C = N-1,\ N \\ \ldots & \ldots \\ s = s, & C = N-s+1, \ldots, N \\ \ldots & \ldots \\ s = n, & C = N-n+1, \ldots, N \end{array}$$

Hence, rearranging by values of C,

$$\begin{array}{ll} C = N-n+1, & s = n \\ C = N-n+2, & s = n, n-1 \\ \ldots & \ldots \\ C = C, & s = n, n-1, \ldots, N-C+1 \\ \ldots & \ldots \\ C = N, & s = n, n-1, \ldots, 1 \end{array}$$

Here

$$\sum_{s=N-C+1}^{n} \frac{(n'-s)\,\Delta K}{CF^{C-1}} = \frac{\Delta K}{CF^{C-1}} \sum_{s=N-C+1}^{n} (n'-s)$$

$$= \frac{\Delta K}{CF^{C-1}} \; \overline{\dot{c - (N-n)}}$$

$$= \frac{\Delta K}{CF^{C-1}} \; \overline{\dot{(C-1)-(N-n')}} \; .$$

Hence,

$$T_3 = \sum_{C=N-n+1}^{N} \frac{\overline{\dot{(C-1)-(N-n')}}\,\Delta K}{CF^{C-1}}$$

But since $n = n' - 1$, we have,

$$T_3 = \sum_{C=N-n'+2}^{N} \frac{\overline{(C-1)-(N-n')}^{\,\bullet}\,\Delta K}{CF^{C-1}} .$$

This yields as a formula for $\Delta^N I_n$,

$$\Delta^N I_n = T_1 - T_2 - T_3 .$$

i.e.,

$$\Delta^N I_n = \dot{n}' \sum_{C=1}^{N} \frac{\Delta K}{CF^{C-1}} - \sum_{C=1}^{n'-1} \frac{(\dot{n}'-\dot{C})\,\Delta K}{CF^{C-1}} - \sum_{C=N-n'+2}^{N} \frac{\overline{(C-1)-(N-n')}^{\,\bullet}\,\Delta K}{CF^{C-1}} .$$

B. The problem of this theorem is to derive a general statement of the theoretical amount of inhibition resulting from a single repetition of any rote series as a whole which will be active during the presentation of any stimulus syllable the ordinal number of which is taken as n. As indicated in Theorem IV B, this inhibition is merely the sums of the several vertical entries above the several stimulus syllables as shown in Figure 15. For example, at syllable *KEM* this is simply 1.70 ΔK; at *FAP* it is 1.64 ΔK + 1.40 ΔK = 3.04 ΔK; at *ZIT* it is 1.54 ΔK + 1.28 ΔK + 1.02 ΔK = 3.84 ΔK; and at *YOD*, so frequently referred to above, it is 1.36 ΔK + 1.08 ΔK + .84 ΔK + .63 ΔK = 3.91 ΔK; and so on. These sums for all the points of final conditioning are shown at the bottom of Figure 15.

Theorem V presents a completely general statement of the inhibition active at any point, n, of any rote series, just as in Theorem IV, except that here it has been done in detail in terms of the basic K and F values. Such a detailed equation is necessary if a rigorous mathematical analysis of the characteristics of the system as a whole is to be made. Much use of this equation will be found in the derivation of subsequent theorems.

The meaning of the several portions of the right-hand member of this equation may best be understood by observing its use in determining the $\Delta^N I_n$ in a particular situation, e.g., that at syllable *ZIT* in the series shown in Figure 15, which is already familiar from the discussion of Theorem IV B. In this case, inspection of Figure 15 shows that $N = 5$ and that the ordinal number of *ZIT* is 2, i.e., $n = 2$. Since $n' = n + 1$, $n' = 2 + 1 = 3$. As F we will take, as hitherto, the value 1.37.

In substituting the above values in this equation it will be convenient to do so with each of the three terms of the right-hand member separately. Taking the first term we have,

$$\dot{n}' \sum_{C=1}^{N} \frac{\Delta K}{CF^{C-1}} = \dot{3} \sum_{C=1}^{C=5} \frac{\Delta K}{C\,1.37^{C-1}} .$$

Now,

$$\dot{3} = 1 + 2 + 3 = 6 .$$

Also, the sum of the series of $\frac{\Delta K}{C\ 1.37^{C-1}}$ from $C = 1$ to $C = 5$ is,

$$\sum_{C=1}^{C=5} \frac{\Delta K}{C\,1.37^{0-1}} = \frac{\Delta K}{1 \times 1.37^{1-1}} + \frac{\Delta K}{2 \times 1.37^{2-1}} + \frac{\Delta K}{3 \times 1.37^{3-1}} + \frac{\Delta K}{4 \times 1.37^{4-1}} + \frac{\Delta K}{5 \times 1.37^{5-1}}$$

$$= \frac{\Delta K}{1 \times 1.37^{0}} + \frac{\Delta K}{2 \times 1.37^{1}} + \frac{\Delta K}{3 \times 1.37^{2}} + \frac{\Delta K}{4 \times 1.37^{3}} + \frac{\Delta K}{5 \times 1.37^{4}}$$

$$= \frac{\Delta K}{1 \times 1} + \frac{\Delta K}{2 \times 1.37} + \frac{\Delta K}{3 \times 1.88} + \frac{\Delta K}{4 \times 2.52} + \frac{\Delta K}{5 \times 3.52}$$

$$= \frac{\Delta K}{1} + \frac{\Delta K}{2.74} + \frac{\Delta K}{5.64} + \frac{\Delta K}{10.08} + \frac{\Delta K}{17.60}$$

$$= \Delta K + .36\ \Delta K + .18\ \Delta K + .10\ \Delta K + .06\ \Delta K$$

$$= 1.7\ \Delta K\ .$$

Multiplying the $1.7\ \Delta K$ by the value of $\dot{3}$, found above, we have,

$$6 \times 1.7\ \Delta K = 10.2\ \Delta K$$

as the value of the first term in the right-hand member of the equation.

Turning to the second term of the right-hand member of the equation and substituting, we have,

$$\sum_{C=1}^{\dot{n}-1} \frac{(\dot{n}-\dot{C})\ \Delta K}{CF^{C-1}} = \sum_{C=1}^{2} \frac{(\dot{3}-\dot{C})\ \Delta K}{CF^{C-1}}$$

Writing out this series from $C = 1$ to $C = 3 - 1$, or 2, we have as the entire series,

$$= \frac{(6-1)\ \Delta K}{1 \times 1.37^{1-1}} + \frac{(6-3)\ \Delta K}{2 \times 1.37^{2-1}}$$

$$= \frac{5\,\Delta K}{1.37^{0}} + \frac{3\,\Delta K}{2 \times 1.37^{1}}$$

$$= \frac{5\,\Delta K}{1} + \frac{3\,\Delta K}{2.74}$$

$$= 5\ \Delta K + 1.1\ \Delta K$$

$$= 6.1\ \Delta K$$

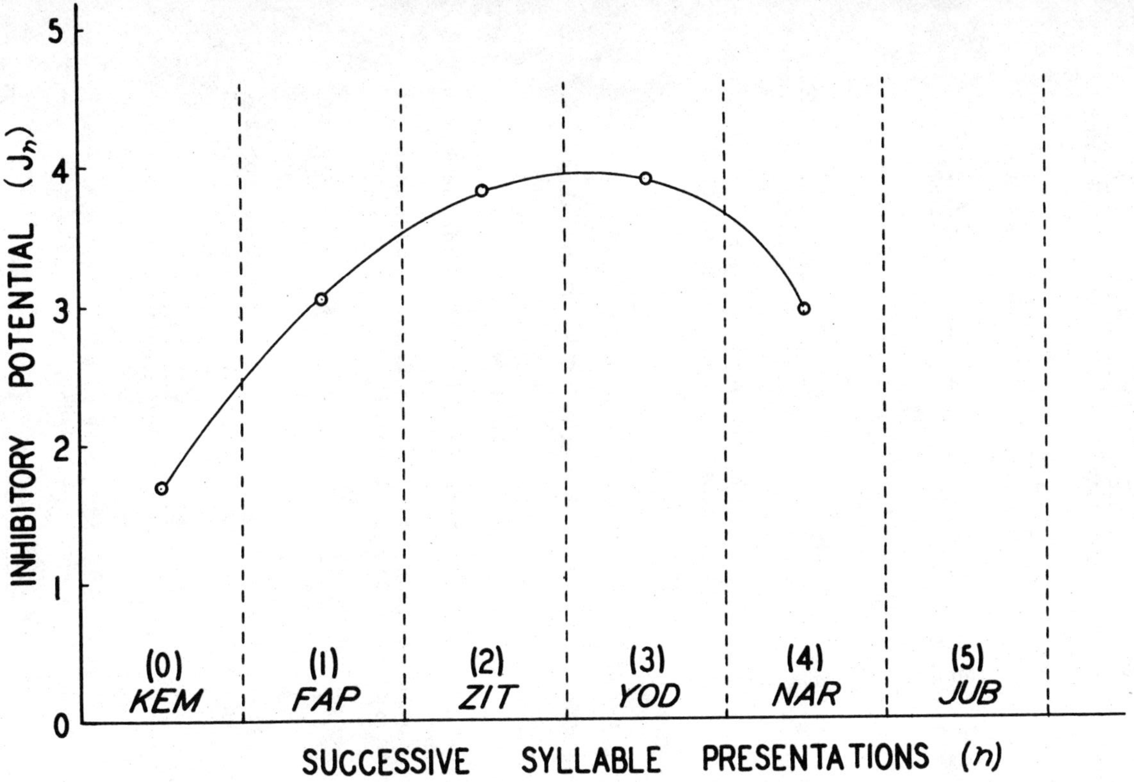

Figure 16. Theoretical distribution of inhibitory potential in terms of ${}^{\prime\prime}J_n$ throughout the learning of a 5-syllable rote series. It will be remembered that the actual occurrence of each correctly learned reaction (r) is concurrent with the presentation of the next preceding syllable (n), i.e., $n = r - 1$.

Turning to the third term of the right-hand member of the equation and substituting, we have,

$$\sum_{C=N-n'+2}^{N} \frac{\overset{\bullet}{\overline{C-1-(N-n')}}\,\Delta K}{CF^{C-1}} = \sum_{C=5-3+2}^{5} \frac{\overset{\bullet}{\overline{C-1-(5-3)}}\,\Delta K}{CF^{C-1}}$$

Writing out the series where $C = 5 - 3 + 2 = 4$ to $C = 5$, we have,

$$= \frac{\overset{\bullet}{\overline{4-1-(5-3)}}\,\Delta K}{4\,F^{4-1}} + \frac{\overset{\bullet}{\overline{5-1-(5-3)}}\,\Delta K}{5\,F^{5-1}}$$

$$= \frac{\dot{1}\,\Delta K}{4 \times 1.37^{3}} + \frac{\dot{2}\,\Delta K}{5 \times 1.37^{4}}$$

$$= \frac{\Delta K}{10.28} + \frac{3\,\Delta K}{17.62}$$

$$= .097\ \Delta K + .17\ \Delta K$$

$$= .26\ \Delta K\,.$$

Now, combining all three terms of the right-hand member of the equation, we have,

$$\Delta\,{}^{5}I_{n'-1} = \Delta\,{}^{5}I_{3-1} = \Delta\,{}^{5}I_{2} = 10.2\,\Delta K - 6.1\,\Delta K - .26\,\Delta K$$

i.e.,

$$\Delta\,{}^{5}I_{2} = 3.84\,\Delta K$$

which will be observed to agree with the value at the bottom of Figure 15.

Proceeding in the same manner to compute the inhibition of the remaining syllable positions of the short rote series presented in Figure 15, we have,

$$\Delta\,{}^{5}I_{0} = 1.7\,\Delta K,\quad \Delta\,{}^{5}I_{1} = 3.04\,\Delta K,\quad \Delta\,{}^{5}I_{3} = 3.91\,\Delta K,\quad \Delta\,{}^{5}I_{4} = 2.94\,\Delta K.$$

These values are shown graphically in Figure 16.

In this connection it may be pointed out that n' is the number of the correct reaction which is opposed by any increment of inhibitory potential, $\Delta\,{}^{N}I_{n}$. This becomes at once apparent when it is recalled that the formula for the increment of a correct excitatory potential (D88) is $\Delta E_{n,n+1}$ and also that $n' = n + 1$.

At this point it may also be well for the reader to become familiar with the concept represented by the symbol, ${}^{N}J_{n}$. By definition (D76),

$${}^{N}J_{n} \quad \frac{\Delta\,{}^{N}I_{n}}{\Delta K}$$

i.e.,

$$\Delta\, {}^{N}I_n = {}^{N}J_n \Delta K.$$

This means that in the example worked out above where it was found that

$$\Delta\, {}^{5}I_2 = 3.84\, \Delta K,$$

the value 3.84 is ${}^{5}J_2$. Similarly with the other $\Delta\, {}^{N}I_n$ values shown at the bottom of Figure 15:

$${}^{5}J_0 = 1.70,\quad {}^{5}J_1 = 3.04,\quad {}^{5}J_3 = 3.91,\quad \text{and}\quad {}^{5}J_4 = 2.94.$$

The significance of ${}^{N}J_n$ is further revealed by the following considerations: We have already seen (D76) that,

$$\Delta\, {}^{N}I_n = {}^{N}J_n \Delta K \qquad (1)$$

Multiplying (1) by R, we have,

$$R \Delta\, {}^{N}I_n = {}^{N}J_n R \Delta K \qquad (2)$$

But by Postulate 9, Corollary 2,

$$R \Delta\, {}^{N}I_n = {}^{R,N}I_n \qquad (3)$$

and by Postulate 9, Corollary 1,

$$R \Delta\, {}^{N}K = {}^{R,N}K \qquad (4)$$

Substituting (3) and (4) in (2), we have,

$${}^{R,N}I_n = {}^{N}J_n\, {}^{R,N}K \qquad (5)$$

By comparing equations (1) and (5) it is seen that ${}^{N}J_n$ occupies the same position whether we are dealing with $\Delta\, {}^{N}I_n$ or with ${}^{R,N}I_n$. Accordingly, ${}^{N}J_n$ is a very useful index of the amount of inhibitory potential at any point in a rote series. For this reason much of the subsequent discussion of inhibitory potential will be in terms of ${}^{N}J_n$. Graphic representations of this index are shown in Figures 16 and 17.

THEOREM V, COROLLARY I

A. *If*

$$\Delta' J_n = J_{n+1} - J_n,$$

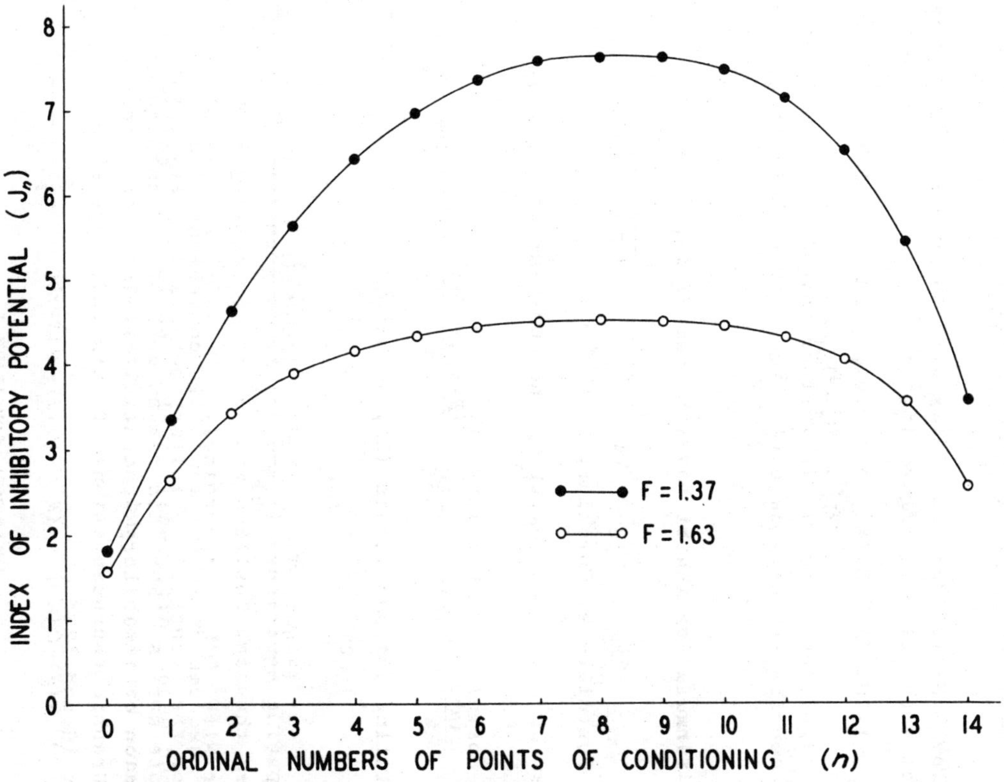

Figure 17. A graphic representation of the index of theoretical inhibitory potential (${}^{N}J_n$) at the several points of conditioning (n) of a 15-syllable rote series by Theorem V for two different values of the factor of reduction (F) as shown in Table 1.

and

$$G\,(n') = \sum_{C=1}^{n} \frac{1}{CF^{C-1}}$$

then

$$\Delta' J_n = (n'+1)\,[G\,(N) - G\,(n')] + (N-n')[G\,(N) - G\,(N-n')] - \frac{F^{n'}-1}{F^{N-1}(F-1)}$$

where (D76),

$$^{N}J_n = \frac{\Delta\, ^{N}I_n}{\Delta K},$$

where

$$\Delta'\, ^{N}J_n = \, ^{N}J_n - \, ^{N}J_{n+1},$$

and where

$$G\,(n') = \sum_{C=1}^{n'} \frac{1}{CF^{C-1}} \,.$$

Proof:

The formula for inhibition (Theorem V) is,

$$(1)\quad \Delta I_n = \dot{n}' \sum_{C=1}^{y} \frac{\Delta K}{CF^{C-1}} - \sum_{C=1}^{n'-1} \frac{(\dot{n}'-\dot{C})\,\Delta K}{CF^{C-1}} - \sum_{C=N-n'+2}^{N} \frac{\overline{C-1-(N-n')}\,\Delta K}{CF^{C-1}}$$

This can be rewritten, omitting the constant factor ΔK, in the form,

$$(2)\quad J_n = \frac{n'(n'+1)}{2}\,[G\,(N) - G\,(n'-1)] - \frac{(N-n')\,(N-n'+1)}{2}\,[G\,(N) - G(N-n'+1)]$$

$$+ \frac{2F^{n'} - F^{n'-1} - n'\,(F-1) - F}{2F^{n'-2}\,(F-1)^2}$$

$$+ \frac{(N-n'-1)\,F^{n'} - (N-n')\,F^{n'-1} + (2n'-N)\,(F-1) + 1}{2F^{N-1}\,(F-1)^2}$$

In the theorems on inhibition it is convenient to use the first and second differences of J_n. We let $\Delta' J_n$ be the difference between J_n and J_{n+1}, i.e.,

$$\Delta' J_n = (n'+1)[G\,(N) - G\,(n')] + (N-n')\,[G\,(N) - G\,(N-n')] - \frac{F^{n'}-1}{F^{N-1}\,(F-1)}$$

from which the corollary follows.

THEOREM V, COROLLARY 2

A. *If,*

$$\Delta_2 J_n = \Delta J_{n+1} - \Delta J_n ,$$

and

$$G(n') = \sum_{C=1}^{n'} \frac{1}{CF^{C-1}} ,$$

then,

$$\Delta_2 J_n = \frac{-1}{F^{n'}} + G(N-n'-1) - G(n'+1).$$

Proof:

This follows directly from Theorem V, Corollary 1.

PROBLEM I

A. *Given the number of syllables in a rote series and the mean number of repetitions required by a group of subjects to just reach the reaction threshold at the several syllable positions marking the several points of reinforcement,*[7] *it is required to find the value of the factor of reduction (F), the value of the basic amount of inhibitory potential (ΔK) in terms of the increment of excitatory potential (ΔE), and the magnitude of the mean reaction threshold (L), also in units of the increment of excitatory potential (ΔE).*

Proof:

In rote series learned by massed practice, the mean number of repetitions required for learning the *n*'th syllable just to the point of reaction is given (Postulate 16, Corollary 1, and D76) by

$$ {}_{L}R_{n,r} = \frac{L}{\Delta\, {}^{N}E_{n,r} - {}^{N}J_n\, \Delta K} \qquad (1)$$

where the right-hand subscripts of ${}_{L}R_{n,r}$ are used in the

7. The procedure for computing the mean number of repetitions required to just bring E to the reaction threshold where the available empirical data are from many different subjects with different learning rates and each subject in various stages of practice, has not yet been worked out with precision. It is evident, however, that this value closely approximates one less than one-half the sum of (1) the mean of the ordinal number of presentations at the first success, and (2) the mean of the ordinal number of presentations at the last failure. (See Problem 3 and Figure 22.) This is the method employed in the determination of the empirical values which appear in Table 2.

same sense as in ${}^{R,N}E_{n,r}$ (see U6). This means that the first right-hand subscript (n) represents the point in the rote series at which the reaction involved will occur when correct, and the second right-hand subscript (r) represents the number of the reaction in question, where, accordingly, $n + 1 = r = n'$.

Given the observed values of ${}_{L}R$ for a series of 15 syllables, we shall calculate the values of $\frac{L}{\Delta E}$, $\frac{\Delta K}{\Delta E}$, and F by solving equation (1), for $n' = 1$, $n' = 8$, and $n' = 15$, where n' is the ordinal number of the reaction syllable being conditioned. To do this we have first, from (1):

$$
\begin{aligned}
{}_{L}R_{0,1}\left(1 - \frac{\Delta K}{\Delta E} \cdot J_0\right) &= \frac{L}{\Delta E} \\
{}_{L}R_{7,8}\left(1 - \frac{\Delta K}{\Delta E} \cdot J_7\right) &= \frac{L}{\Delta E} \qquad (2) \\
{}_{L}R_{14,15}\left(1 - \frac{\Delta K}{\Delta E} \cdot J_{14}\right) &= \frac{L}{\Delta E}
\end{aligned}
$$

Subtracting the first of these equations from the second, and the first from the third, and letting $\delta_1 = R_8 - R_1$ and $\delta_2 = R_{15} - R_1$, we have:

$$
\begin{aligned}
{}_{L}R_{7,8}\,J_7 - {}_{L}R_{0,1}\,J_0 &= \delta_1 \cdot \frac{\Delta E}{\Delta K} \\
{}_{L}R_{14,15}\,J_{14} - {}_{L}R_{0,1}\,J_0 &= \delta_2 \cdot \frac{\Delta E}{\Delta K}
\end{aligned} \qquad (3)
$$

Multiplying the first of these equations by δ_2 and the second by δ_1, and subtracting, we have,

$$
\delta_1\,{}_{L}R_{14,15}\,J_{14} - \delta_2\,{}_{L}R_{7,8}\,J_7 + \delta_3\,{}_{L}R_{0,1}\,J_0 = 0 \qquad (4)
$$

where $\delta_3 = {}_{L}R_{14,15} - {}_{L}R_{7,8}$. In this equation, everything is known except F. Hence we can calculate F from (4), then calculate $\frac{\Delta K}{\Delta E}$ from either of equations (3), and then calculate $\frac{L}{\Delta E}$ from any one of equations (2).

From the formula of Theorem V, we can express J_0, J_7, and J_{14} in terms of F as follows:

$$J_0 = 1 + \frac{1}{2}f + \frac{1}{3}f^2 + \frac{1}{4}f^3 + \frac{1}{5}f^4 + \frac{1}{6}f^5 + \frac{1}{7}f^6 + \frac{1}{8}f^7 + \frac{1}{9}f^8 + \frac{1}{10}f^9 + \frac{1}{11}f^{10} + \frac{1}{12}f^{11} + \frac{1}{13}f^{12} + \frac{1}{14}f^{13} + \frac{1}{15}f^{14}$$

$$J_7 = 1 + \frac{3}{2}f + \frac{6}{3}f^2 + \frac{10}{4}f^3 + \frac{15}{5}f^4 + \frac{21}{6}f^5 + \frac{28}{7}f^6 + \frac{36}{8}f^7 + \frac{35}{9}f^8 + \frac{33}{10}f^9 + \frac{30}{11}f^{10} + \frac{26}{12}f^{11} + \frac{21}{13}f^{12} + \frac{15}{14}f^{13} + \frac{8}{15}f^{14} \qquad (5)$$

$$J_{14} = 1 + f + f^2 + f^3 + f^4 + f^5 + f^6 + f^7 + f^8 + f^9 + f^{10} + f^{11} + f^{12} + f^{13} + f^{14}$$

where we have written $f = \frac{1}{F}$.

Substituting the values of ${}_L R_{0,1}$, ${}_L R_{7,8}$, and ${}_L R_{14,15}$ given by (5), into equation (4), we have, letting $q_1 = \delta_1 \, {}_L R_{14,15}$, $q_2 \doteq \delta_2 \, {}_L R_{7,8}$, $q_3 = \delta_3 \, {}_L R_{0,1}$:

$$(q_1 + q_2 + q_3) + \frac{1}{2}(2q_1 + 3q_2 + q_3)f + \frac{1}{3}(3q_1 + 6q_2 + q_3)f^2 + \frac{1}{4}(4q_1 + 10q_2 + q_3)f^3 + \frac{1}{5}(5q_1 + 15q_2 + q_3)f^4 + \frac{1}{6}(6q_1 + 21q_2 + q_3)f^5$$
$$+ \frac{1}{7}(7q_1 + 28q_2 + q_3)f^6 + \frac{1}{8}(8q_1 + 36q_2 + q_3)f^7 + \frac{1}{9}(9q_1 + 35q_2 + q_3)f^8 + \frac{1}{10}(10q_1 + 33q_2 + q_3)f^9 + \frac{1}{11}(11q_1 + 30q_2 + q_3)f^{10} \qquad (6)$$
$$+ \frac{1}{12}(12q_1 + 26q_2 + q_3)f^{11} + \frac{1}{13}(13q_1 + 21q_2 + q_3)f^{12} + \frac{1}{14}(14q_1 + 15q_2 + q_3)f^{13} + \frac{1}{15}(15q_1 + 8q_2 + q_3)f^{14} = 0.$$

The roots of this equation will be the desired values of f. However, F was assumed to be real and > 1. Hence, the only significant roots of (6) are those between 0 and 1.

B. *Example A.*

The above method of calculation will be applied to the empirical data given in Table 2, in which ${}_L R_{0,1} = 3.1$, ${}_L R_{7,8} = 8.6$, ${}_L R_{14,15} = 4.0$. Hence we have,

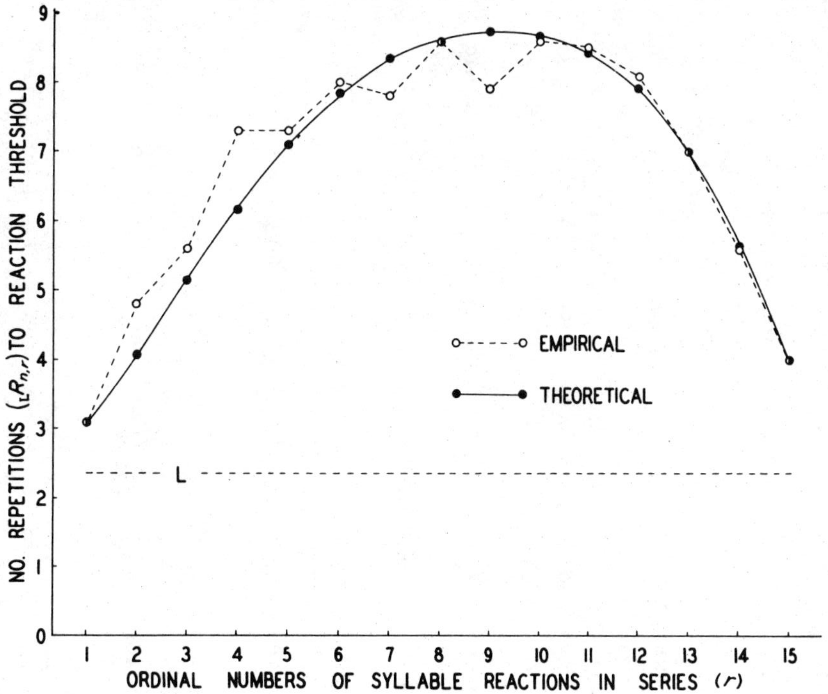

Figure 18. A graphic comparison between (1) the number of repetitions ($_LR_{n,r}$) *empirically* just necessary for each syllable of a 15-syllable rote series to reach the central tendency of the reaction threshold, and (2) the number of repetitions *theoretically* necessary according to Theorem V and Postulate 16, Corollary 1, when the value of F = 1.63, that of ΔK = .163, and that of L = 2.32, Problem I, Example A. It will be remembered that the occurrence of each correctly learned reaction (r) is concurrent with the presentation of the next preceding syllable, as indicated in the expression $_LR_{n,r}$, where $n = r - 1$.

$$
\begin{aligned}
\delta_1 &= {}_L R_{7,8} - {}_L R_{0,1} = 5.5 \\
\delta_2 &= {}_L R_{14,15} - {}_L R_{0,1} = .9 \\
\delta_3 &= {}_L R_{14,15} - {}_L R_{7,8} = -4.6 \\
q_1 &= \delta_1 \, {}_L R_{14,15} = 22.0 \\
q_2 &= -\delta_2 \, {}_L R_{7,8} = -7.74 \\
q_3 &= \delta_3 \, {}_L R_{0,1} = -14.26
\end{aligned} \qquad (7)
$$

Substituting these values of q_1, q_2, q_3 in equation (6), we have:

$$
\begin{aligned}
0 &+ 3.26\,f + 1.77\,f^2 - .92\,f^3 - 4.07\,f^4 - 7.47\,f^5 - 11.00\,f^6 \\
&- 14.61\,f^7 - 9.68\,f^8 - 4.97\,f^9 - .41\,f^{10} + 4.04\,f^{11} + 8.40\,f^{12} \\
&+ 12.69\,f^{13} + 16.92\,f^{14} = 0.
\end{aligned} \qquad (8)
$$

Equation (8) has a significant root at $f = .6128$, i.e., $F = 1.63$.

Inspection shows that the only root of (8) between zero and 1 lies between .5 and .6. By finding the remainders on substituting these quantities and linearly interpolating, we obtain a better approximation of the root. In two such steps we obtain .6128 as a root correct to three places. Then, from the first equation of (3):

$$
\frac{\Delta K}{\Delta E} = \frac{\delta_1}{{}_L R_{7,8}\,J_7 - {}_L R_{0,1}\,J_0} = \frac{5.5}{38.56} = .163 .
$$

Hence,

$$\Delta K = .163\ \Delta E .$$

Then, from the first equation of (2), we have,

$$
\frac{L}{\Delta E} = {}_L R_{0,1}\left(1 - \frac{\Delta K}{\Delta E} J_0\right) = 3.1\ (1 - .163 \cdot 1.548) = 2.32 .
$$

Hence,

$$L = 2.32$$

Example B.

As a second example of the procedure outlined above we shall take a set of data which, while not threshold data (as presumably are those of Example A) will yield values which will serve to illustrate the mathematical characteristics of the system, particularly where F has a value less than the critical number 1.55. (See Theorem XXV A.) Accordingly, in the present series we have ${}_{L}R_{0,1}$ = 2.65, ${}_{L}R_{7,8}$ = 6.95, and ${}_{L}R_{14,15}$ = 3.3. The solution of this problem yields: F = 1.37, ΔK = .09, and L = 2.25.

C. It is noteworthy that all of the above constants come out as positive real numbers, which the theory requires. This fact tends to support to a certain extent the soundness of the analysis. The value of L, in relationship to the empirical learning data from which it was computed, is shown for Example A in Figure 18. This figure also shows that learning scores (${}_{L}R_{n,n+1}$) calculated from the constants found above, agree rather well with the corresponding observed values. This also supports, to a certain extent, the soundness of the postulates involved in the systematic development thus far.

PROBLEM II

A. *Given the indices of inhibitory potentiality* (J's) *at the several points of final conditioning in a rote series, together with the values of the constants* ΔK *and* L, *it is required to find the number of repetitions required to raise the effective excitatory potential* ($\bar{E}$) *to the central tendency of the reaction threshold* (L).

Proof:

The formula for the determinations here required is given in Postulate 16, Corollary 1,

$$ {}_{L}R_{n,r} = \frac{L}{\Delta E - \Delta I} \qquad (1)$$

By definition, D76,

$$ \Delta\, {}^{N}I_{n} = {}^{N}J_{n}\, \Delta K \qquad (2)$$

Substituting (2) in (1), we have,

$$ {}_{L}R_{n,r} = \frac{L}{\Delta E - {}^{N}J_{n}\, \Delta K} \qquad (3)$$

by means of which the number of repetitions required to pass the reaction threshold at each point of final conditioning may be calculated.

B. The use of the above equation for the calculation of the number of repetitions required for the threshold learning of each syllable reaction in a rote series by massed practice may be illustrated by the computation of this value for the second syllable of a 15-syllable series, taking L = 2.25 ΔE, ΔK = .09 ΔE, ${}^{15}J_{2}$ = 3.365 (as shown by Table 1).

TABLE 1

The indices of inhibitory potentiality ($^{N}J_{n}$) resulting at the various points of final conditioning from the learning of rote series of two different lengths and for two different values of F. These J-values have been computed by successive substitutions in the equation of Theorem V. The detailed method of one such computation is shown in Theorem V B. It will be recalled (D76) that $\Delta {}^{N}I_{n} = J_{n} \Delta K$.

Ordinal number of syllable at point of conditioning (n) where $n+1 = r = n'$	Indices of inhibitory potential ($^{N}J_{n}$) at the several points of conditioning (D66) of rote series			
	Where F = 1.63 (Problem I Example A)		Where F = 1.37 (Problem I Example B)	
	15-syllable series	12-syllable series	15-syllable series	12-syllable series
0	1.548	1.548	1.792	1.788
1	2.645	2.643	3.365	3.361
2	3.370	3.365	4.651	4.620
3	3.835	3.827	5.641	5.580
4	4.130	4.113	6.387	6.278
5	4.313	4.279	6.934	6.747
6	4.423	4.359	7.318	7.010
7	4.484	4.363	7.565	7.065
8	4.508	4.282	7.688	6.883
9	4.499	4.067	7.685	6.387
10	4.448	3.598	7.539	5.417
11	4.332	2.575	7.205	3.618
12	4.094		6.594	
13	3.613		5.536	
14	2.581		3.670	

Substituting these values appropriately in the above equation, we have,

$$ {}_{L}R_{1,2} = \frac{2.25\ \Delta E}{\Delta E - 3.365 \times .09\ \Delta E} $$

$$ = \frac{2.25}{1 - .303} $$

$$ = \frac{2.25}{.697} $$

$$ = 3.23 $$

By successive application of this procedure all of the R values for the various syllable reactions in a 12-syllable and a 15-syllable rote series have been computed. They are shown in Table 2, parallel with the corresponding J values which are shown in Table 1. Frequent reference will be made to this table in the B-portions of later theorems.

THEOREM VI

A. *If the increment of inhibition at the m'th syllable in a given rote series is greater than the increment of inhibition at the n'th syllable, then the number of repetitions required to raise the effective excitatory potential to the mean reaction threshold (L) at the m'th syllable is greater than the number of repetitions required at the n'th syllable.*

Proof:

By D76,

$$ {}^{N}J_{m} = \frac{\Delta\, {}^{N}I_{m}}{\Delta K} $$

Also we have, by Postulate 16, Corollary I:

$$ {}_{L}R_{m,m+1} = \frac{L}{\Delta E - \Delta K\, {}^{N}J_{m}} $$

Therefore,

$$ {}_{L}R_{m,m+1} - {}_{L}R_{n,n+1} = \frac{L}{\Delta E - \Delta K\, {}^{N}J_{m}} - \frac{L}{\Delta E - \Delta K\, {}^{N}J_{n}} = \frac{L\, \Delta K\, ({}^{N}J_{m} - {}^{N}J_{n})}{(\Delta E - \Delta K\, {}^{N}J_{m})\,(\Delta E - \Delta K\, {}^{N}J_{n})} $$

This may be written in the form:

$$ {}_{L}R_{m,m+1} - {}_{L}R_{n,n+1} = \frac{\Delta K}{L}\, R_{m} R_{n}\, ({}^{N}J_{m} \quad {}^{N}J_{n}) $$

TABLE 2

The number of repetitions (${}_{L}R_{n,r}$) (a) theoretically and (b) empirically necessary to raise the effective excitatory potential ($\bar{E}_{n,r} = E_{n,r} - {}^{N}I_{n}$) to the mean reaction threshold (L) for each of the reactions of a 12-syllable and a 15-syllable rote series (exclusive of the cue syllable) and for two different sets of F, ΔK, and L values. In these computations the J-values were taken from the corresponding columns of Table 1. The detailed procedure for the computation of ${}_{L}R_{1,2}$ of the 15-syllable series is presented in Problem II B.

Ordinal number of reaction conditioned (r) where $n+1 = r = n'$	Number of repetitions (${}_{L}R_{n,r}$) empirically found necessary to raise effective excitatory potential ($\bar{E}$) to reaction threshold (L)	Number of repetitions (${}_{L}R_{n,r}$) theoretically necessary to raise the effective excitatory potential ($\bar{E}$) to the reaction threshold (L)			
		Where F = 1.63, ΔK = .163, and L = 2.32 (Problem I, Example A)		Where F = 1.37, ΔK = .09, and L = 2.25 (Problem I, Example B)	
		15-syllable series	12-syllable series	15-syllable series	12-syllable series
1	3.1	3.10	3.10	2.68	2.68
2	4.8	4.08	4.08	3.23	3.23
3	5.6	5.14	5.13	3.87	3.85
4	7.3	6.19	6.17	4.57	4.52
5	7.3	7.09	7.03	5.30	5.17
6	8.0	7.81	7.66	5.98	5.73
7	7.8	8.32	8.03	6.59	6.10
8	8.6	8.62	8.03	7.06	6.18
9	7.9	8.75	7.68	7.31	5.91
10	8.6	8.69	6.88	7.29	5.29
11	8.5	8.44	5.60	7.00	4.39
12	8.1	7.89	4.00	6.41	3.33
13	7.0	6.97		5.53	
14	5.6	5.64		4.49	
15	4.0	4.01		3.36	

Now by assumption,

$$J_m > J_n$$

$$\therefore \qquad J_m - J_n > 0.$$

But,

$$\frac{\Delta K}{L} \; {}_{L}R_{m,m+1} \; {}_{L}R_{n,n+1} > 0$$

for all of these factors are positive. Hence,

$${}_{L}R_{m,m+1} - {}_{L}R_{n,n+1} > 0$$

i.e.,

$${}_{L}R_{m,m+1} > {}_{L}R_{n,n+1}\,.$$

B. This theorem states in effect that in any rote series if one syllable has a greater loading of inhibition (${}^{N}I_n$) than another, the first will require more repetitions for complete learning than the second. For example, where $N = 15$ and $F = 1.37$ in Table 1 (and Figure 17), syllable #1 has a loading of 1.79 ΔK of inhibition (ΔI_n), whereas syllable #2 has a loading of 3.37 ΔK of inhibition. The theorem implies that syllable #2 should require more repetitions for complete learning than will syllable #1.

The method of determining the number of repetitions theoretically required to raise the effective excitatory potential ($\bar{E}$) above the reaction threshold (L) has been shown in detail in Problem II B, and the results of the application of this procedure to the theoretical inhibitory potential values of Table 1 (and Figure 17) are shown in detail in Table 2 (and Figure 18). By a comparison of the two tables and remembering that n is always the ordinal number of the syllable marking the action of the inhibitory potential, that the point of action of this inhibitory potential is the same as the point of conditioning, and that the ordinal number of the reaction syllable whose excitatory potential ($E_{n,r}$) is opposed to this inhibitory potential is always one greater than its point of conditioning, i.e., that the ordinal number (r or n') of the reaction syllable is always $n + 1$, we find that,

when $n = 2,\ r = 2 + 1 = 3,\ \Delta\, {}^{N}I_n = \Delta\, {}^{15}I_2 = 4.65$, and

$${}_{L}R_{n,r} = {}_{L}R_{2,3} = 3.87,$$

when $n = 3,\ r = 3 + 1 = 4,\ \Delta\, {}^{N}I_{n-1} = \Delta\, {}^{15}I_3 = 5.64$, and

$${}_{L}R_{n,r} = {}_{L}R_{3,4} = 4.57,$$

i.e., $\qquad 4.57 > 3.87,$

exactly as the theorem states.

THEOREM VII

A. *In the first massed practice of the learning of any rote series, if* a, b, c, d *are four syllable positions in a rote series, and if* $J_a - J_b > J_c - J_d > 0$ *and* $J_b > J_d$, *then:*

$$ {}_LR_{a,a+1} - {}_LR_{b,b+1} > {}_LR_{c,c+1} - {}_LR_{d,d+1} . $$

Proof:

From the proof of Theorem VI we have,

$$ {}_LR_{a,a+1} - {}_LR_{b,b+1} = \frac{\Delta K}{L}\, {}_LR_{a,a+1}\, {}_LR_{b,b+1} (J_a - J_b) $$

$$ {}_LR_{c,c+1} - {}_LR_{d,d+1} = \frac{\Delta K}{L}\, {}_LR_{c,c+1}\, {}_LR_{d,d+1} (J_c - J_d) . $$

But, by assumption, $J_a - J_b > J_c - J_d > 0$. Furthermore, since $J_a > J_c$ and $J_c > J_d$, we have, by Theorem VI, ${}_LR_{a,a+1} \geq {}_LR_{c,c+1}$, and ${}_LR_{b,b+1} > {}_LR_{d,d+1}$. Hence,

$$ {}_LR_{a,a+1} - {}_LR_{b,b+1} = \frac{\Delta K}{L}\, {}_LR_{a,a+1}\, {}_LR_{b,b+1} (J_a - J_b) > \frac{\Delta K}{L}\, R_{a,a+1}\, {}_LR_{b,b+1} (J_c - J_d) $$

$$ \geq \frac{\Delta K}{L}\, {}_LR_{c,c+1}\, {}_LR_{d,d+1} (J_c - J_d) = {}_LR_{c,c+1} - {}_LR_{d,d+1} . $$

B. The meaning of this theorem may be illustrated by the theoretical results assembled for the 15-syllable series shown in Tables 1 and 2, which are based on the analyses carried out in Problems I B and II B, respectively. By referring to Table 1 it will be seen that the assumed inequalities are satisfied if syllable a is #13, b is #14, c is #2, and d is #1. Then,

$$ J_a = {}^{15}J_{13} = 5.54 $$

$$ J_b = {}^{15}J_{14} = 3.67 $$

$$ J_c = {}^{15}J_{2} = 4.65 $$

$$ J_d = {}^{15}J_{1} = 3.37 $$

Then the inequalities,

$$ J_a - J_b > J_c - J_d > 0 $$

and

$$ J_b > J_d $$

become respectively,

$$5.54 - 3.67 > 4.63 - 3.37 > 0$$

i.e.,

$$1.87 > 1.28 > 0$$

and

$$3.6 > 3.37.$$

Turning, now, to the number of repetitions theoretically required to reach the reaction threshold at each of these positions, we find, by Table 2,

$${}_{L}R_{a,a+1} = {}_{L}R_{13,14} = 4.49$$

$${}_{L}R_{b,b+1} = {}_{L}R_{14,15} = 3.36$$

$${}_{L}R_{c,c+1} = {}_{L}R_{2,3} = 3.87$$

$${}_{L}R_{d,d+1} = {}_{L}R_{1,2} = 3.23$$

Substituting these values in the inequality proved by Theorem VII we have,

$$R_a - R_b > R_c - R_d$$

$$4.49 - 3.36 > 3.87 - 3.23$$

i.e.,

$$1.13 > .64$$

which is in conformity with the theorem.

C. Turning to the corresponding experimental results, we have (Table 2),

$$5.6 - 4.0 > 5.6 - 4.8 > 0$$

i.e.,

$$1.6 > .8 > 0$$

which also satisfies the inequality.

THEOREM VIII

A. *In the learning of any rote series, the greater the mean inhibitory potential at a given point of conditioning, the greater, on the average, will be the time elapsing between the stimulus and the reaction.*

Proof:

By Postulates 3, 8, and 12, the excitatory potential (E) is the same at all points of conditioning in a rote series.

Moreover, by Postulate 10,

$$\bar{E}_{n,r} = \mathcal{E}_{n,r} - \mathcal{I}_n$$

where $n + 1 = r$.

It follows from the foregoing considerations that the greater the inhibition at a given point of conditioning, the less will be the effective excitatory potential at that point and hence (by Postulate 18) the greater will be the reaction time.

B. Let us suppose that in a rote series of 15 syllables (Table 1) 7 massed repetitions have just occurred which has brought the most of the syllable reactions above the reaction threshold, and that the J-values at ${}^{15}J_0$ and ${}^{15}J_1$ are 1.79 and 3.36 respectively. Since $\Delta\,{}^{N}I_n = {}^{N}J_n\,\Delta K$, and since $\Delta K = .09\ \Delta E$, the $\Delta\,{}^{15}I_0$ at the cue syllable must be,

$$\Delta\,{}^{15}I_0 = 1.79 \times .09\ \Delta E$$

$$= .16\ \Delta E$$

$$\therefore \qquad \Delta\bar{E}_{0,1} = \Delta E - .16\,\Delta E = .84\ \Delta E.$$

In the case of the second point of conditioning we have,

$$\Delta I_1 = 3.36 \times .09\ \Delta E$$

$$= .3\ \Delta E.$$

$$\therefore \qquad \Delta\bar{E}_{1,2} = \Delta E - .3\ \Delta E = .70\ \Delta E.$$

Substituting these values in the equation of Postulate 18,

$$T_{n,n+1} = v + (h - v)\,e^{-k\bar{E}},$$

using the values for v, h, and k derived in Postulate 18 B, and recalling that the first massed practice (Postulate 13, Corollary 2),

$$\bar{E}_{n,n+1} = R\,\Delta\bar{E}_{n,n+1}$$

the latency of the first point of conditioning will be,

$$T_{0,1} = .82 + \frac{1.72 - .82}{2.718^{.55 \times 7 \times .84}}$$

$$= .82 + \frac{.90}{2.718^{3.23}}$$

$$= .82 + \frac{.90}{25.3}$$

$$= .86 \text{ sec.}$$

In a similar manner the reaction latency at the second point of conditioning will be,

$$T_{1,2} = .82 + \frac{1.72 - .82}{2.718^{.55 \times 7 \times .7}}$$

$$= .82 + \frac{.90}{2.718^{2.7}}$$

$$= .82 + \frac{.90}{14.9}$$

$$= .88 \text{ sec.}$$

i.e., .88 sec. > .86 sec., as the theorem states.

THEOREM IX

A. *In the first massed practice of a rote learning, if a, b, c, and d are four syllable presentation numbers, and if*

$$J_a - J_b > J_c - J_d > 0$$

and

$$J_a > J_c, \quad J_b > J_d$$

then the corresponding reaction latencies satisfy the inequality:

$$T_a - T_b > T_c - T_d .$$

Proof:

By Postulate 18 and Postulate 13, Corollary 1,

$$T_{n,n+1} = v + (h-v)\, e^{-kR(\Delta E_{n,n+1} - \Delta {}^{N}I_n)} .$$

But

$$\Delta {}^{N}I_n = \Delta K\, {}^{N}J_n .$$

Hence,

$$T_n = v + (h-v)\, e^{-kR\Delta E}\, e^{kR\Delta K J_n}$$

$$\therefore \quad T_a - T_b = (h-v)\, e^{-kR\Delta E}\left(e^{kR\Delta K J_a} - e^{kR\Delta K J_b}\right)$$

$$= (h-v)\, e^{-kR\Delta E}\, e^{kRJ_b}\left(e^{kR\Delta K (J_a - J_b)} - 1\right).$$

Likewise,

$$T_c - T_d = (h-v)\, e^{-kR\Delta E}\, e^{kRJ_d}\, (e^{kR\Delta K\,(J_c - J_d)} - 1)$$

But, since

$$J_a - J_b > J_c - J_d > 0$$

then

$$e^{kR\Delta K\,(J_a - J_b)} - 1 > e^{kR\Delta K\,(J_c - J_d)} - 1 > 0.$$

And since

$$J_b > J_d$$

then

$$e^{kRJ_b} > e^{kRJ_d} > 0.$$

$$\therefore \qquad T_a - T_b > T_c - T_d .$$

B. The meaning of this theorem may be illustrated by means of the formulation of the results of Problem I (p. 103) which involved the learning of a 15-syllable series (exclusive of the cue syllable). The results of the solution of this problem are given in Table 1. The inequalities with respect to J-values assumed in the postulate will be satisfied if *a* is conditioning point #13, *b* is conditioning point #14, *c* is conditioning point #2, and *d* is conditioning point #1. According to Table 1, the J-values at the specified points of conditioning will be,

$$J_a = 5.54, \quad J_b = 3.67, \quad J_c = 4.65, \text{ and } J_d = 3.37.$$

Substituting these values in the first inequality supposed in the theorem we have,

$$5.54 - 3.67 > 4.63 - 3.37 > 0$$

i.e.,

$$1.87 > 1.28 > 0 .$$

Substituting in the following pair of inequalities supposed in the theorem we have,

$$5.54 > 4.65, \quad 3.67 > 3.37.$$

Having shown that the reactions chosen satisfy the conditions of the theorem so far as J-values are concerned, it remains to show that the inequality stated in the theorem holds of the reaction latencies. This will be done by computing the reaction latency on the basis of the equation fitted to Simley's data exactly as in the case of Theorem VIII, again assuming that seven learning repetitions had occurred just before the reaction latencies under consideration were taken. The results of these computations are:

$$T_a = T_{13,14} = .953 \text{ sec.}, \quad T_b = T_{14,15} = .888 \text{ sec.},$$

$$T_c = T_{2,3} = .998 \text{ sec.}, \quad T_d = T_{1,2} = .881 \text{ sec.}$$

$$\therefore \qquad .953 - .888 > .916 - .881$$

i.e.,

$$.065 > .035$$

just as the theorem states.

THEOREM X

A. *In a rote learning consisting of a single massed practice, of two points equidistant (in time) from the maximum point of effective excitatory potential ($\bar{\text{E}}$) after the termination of learning, the rate of increase of $\bar{\text{E}}$ at the point preceding the maximum is greater than the rate of decrease at the point following the maximum.*

Proof:

The effective excitatory strength is, by Postulate 10 and Definition D79,

$$\bar{\text{E}}_{n,r}\,(t) = \mathcal{E}_{n,r}\,(t) - \mathcal{I}_n\,(t).$$

Hence, by Postulates 12 and 13,

$$\bar{\text{E}}\,(t) = ae^{-bt} - ce^{-dt}\,.$$

Let

$$t = t' + \frac{1}{d-b}\,\log\frac{c}{a}\,.$$

Then

$$\text{E}(t') = a\left(\frac{c}{a}\right)^{-\frac{b}{d-b}} e^{-bt'} - c\left(\frac{a}{c}\right)^{-\frac{d}{d-b}} e^{-dt'}$$

which reduces to

$$\bar{\text{E}}\,(t') = a^{\frac{d}{d-b}}\, c^{\frac{-b}{d-b}}\,(e^{-bt'} - e^{-dt'})$$

$$\therefore \qquad \frac{d\bar{\text{E}}\,(t)}{dt} = \frac{d\bar{\text{E}}\,(t')}{d\,t'} = a^{\frac{d}{d-b}}\, c^{\frac{-b}{d-b}}\,(de^{-dt'} - be^{-bt'})\,.$$

The point of maximum effective excitatory potential occurs where the derivative vanishes, namely:

$$t'_{max} = \frac{\log \frac{d}{b}}{d - b}.$$

Hence,

$$\frac{d\bar{E}\ (t'_{max} - \Delta t')}{d\ t'} = a^{\frac{d}{d-b}}\ c^{\frac{-b}{d-b}}\ (\frac{d}{b})^{\frac{-d}{d-b}}\ d\,[e^{d\Delta t'} - e^{b\Delta t'}].$$

Likewise,

$$\frac{-d\bar{E}\ (t'_{max} + \Delta t')}{d\,t'} = a^{\frac{d}{d-b}}\ c^{\frac{-b}{d-b}}\ (\frac{d}{b})^{\frac{-d}{d-b}}\ d\,[e^{-b\Delta t'} - e^{-d\Delta t'}].$$

To prove the theorem we must show that:

$$e^{d\Delta t'} - e^{b\Delta t'} > e^{-b\Delta t'} - e^{-d\Delta t'}.$$

That is,

$$e^{b\Delta t'}\ (e^{(d-b)\Delta t'} - 1) > e^{-d\Delta t'}\ (e^{(d-b)\Delta t'} - 1).$$

That is,

$$e^{b\Delta t'} > e^{-d\Delta t'}.$$

But this is obviously satisfied for all positive values of $\Delta t'$. Hence the theorem follows.

B. This theorem may be illustrated by means of the same supposititious materials as those employed in Postulates 12 B and 13 B. To this end the values shown in Table 3 have been calculated. This shows the values of E, I, and $\bar{E}$ at intervals of only .2 t in the region of the reminiscence effect. All three sets of values are shown in Figure 19.

An inspection of the middle curve in Figure 19 and of the final column of values in Table 3 shows that the effective excitatory potential ($\bar{E}$) begins spontaneously to rise at once following the termination of active learning, reaching its maximum in the neighborhood of $t = 2$. A simple inspection of Figure 19 shows that this rise is steeper than its fall. This difference between the steepness of the rise and that of the fall of this reminiscence effect may be shown numerically by means of the final column of Table 3. Suppose we lay off a distance of 1.8 time units on both sides of the maximum reminiscence effect and note the change in $\bar{E}$ at those points during .2 units of time. On the anterior side of the maximum reminiscence effect, $t_{.2}$ yields an $\bar{E}$ of 4.075 and $t_{.4}$ yields an $\bar{E}$ of 4.138, the difference being .063. On the posterior side of the reminiscence effect, $t_{3.8}$ yields an $\bar{E}$ of 4.191 and $t_{3.6}$ yields an $\bar{E}$ of 4.223, the difference being in this case .032. But since .063 > .032, it follows that

TABLE 3

Theoretical values of excitatory potential (E), of inhibitory potential (I), and of effective excitatory potential ($\bar{E}$) for various lengths of time (t) following the termination of a first massed practice, computed on the assumption that the absolute excitatory potential diminishes (Postulate 12 B) according to the equation $E(t) = \frac{8}{2.718^{.1t}}$, that the inhibitory potential diminishes (Postulate 13 B) according to the equation $I(t) = \frac{4}{2.718^{.3t}}$, and (Postulate 10) that $\bar{E} = E - I$.

Time since the termination of the first massed practice (t)	Absolute excitatory potential (E)	Inhibitory potential (I)	Effective excitatory potential ($\bar{E}$)
.0	8.000	4.000	4.000
.2	7.842	3.767	4.075
.4	7.686	3.548	4.138
.6	7.534	3.341	4.193
.8	7.385	3.147	4.238
1.0	7.239	2.963	4.276
1.2	7.095	2.791	4.304
1.4	6.955	2.629	4.326
1.6	6.817	2.475	4.342
1.8	6.682	2.331	4.351
2.0	6.550	2.197	4.353
2.2	6.421	2.068	4.353
2.4	6.293	1.947	4.346
2.6	6.169	1.834	4.335
2.8	6.046	1.727	4.319
3.0	5.927	1.626	4.301
3.2	5.809	1.532	4.277
3.4	5.694	1.442	4.252
3.6	5.582	1.359	4.223
3.8	5.471	1.280	4.191
4.0	5.363	1.205	4.158
4.2	5.257	1.135	4.122
4.4	5.153	1.069	4.084
4.6	5.051	1.006	4.045
4.8	4.950	.948	4.002
5.0	4.853	.893	3.960
6.0	4.391	.661	3.730
7.0	3.973	.490	3.483
8.0	3.595	.363	3.232
9.0	3.253	.269	2.984
10.0	2.943	.199	2.744
12.0	2.410	.109	2.301
14.0	1.972	.060	1.912
16.0	1.615	.033	1.582
18.0	1.322	.018	1.304
20.0	1.082	.010	1.072

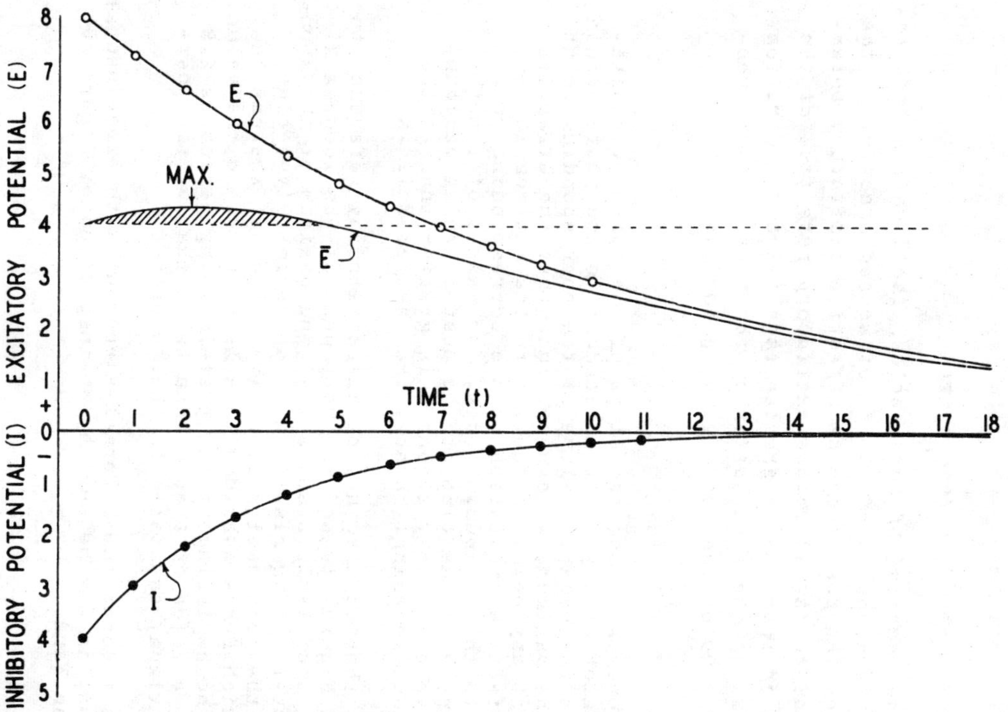

Figure 19. Diagram showing the theoretical relationship of Ward's reminiscence phenomenon to the differential rate of decay of excitatory and inhibitory potentials at a given point of final conditioning. The curves representing the excitatory potential and the inhibitory potential are from the same formulae as those in Figure 6. The effective excitatory potential is simply the excitatory potential less the inhibitory potential, i.e., $\bar{E} = E - I$. The area representing the reminiscence effect is hatched.

the rate of increase in the reminiscence effect is greater than the rate of its decrease at the comparable points examined.

C. This theorem is evidently in formal agreement with observed fact, as may be seen by comparing the curve of $\bar{E}$ in Figure 19 with Ward's experimental results shown in Figure 37.

THEOREM XI

A. *In the learning of rote series, the duration from the time of the termination of the first massed practice to the time at which the post-learning effective excitatory potential is equal to the effective excitatory potential at the termination of learning is greater than twice the duration from the time of the termination of the first massed practice to the time of the post-learning maximum.*

Proof:

From Theorem X, since the slope of the curve of excitation is greater at any point antecedent to the post-learning maximum than the negative slope at the corresponding point following the maximum, we can conclude that the effective excitation at any point antecedent to the maximum is less than the effective excitation at the corresponding point subsequent to the maximum. Hence, for values of t not greater than twice the time at the post-learning maximum, the effective excitatory strength is greater than at the termination of learning. Hence, the theorem follows.

B. This theorem also may be illustrated by reference to Table 3 and Figures 19 and 37, very much as was Theorem X. An inspection of Table 3 shows that the maximum of the spontaneous post-learning rise in $\bar{E}$ is approximately 2 time units from the termination of active learning, whereas the point at which $\bar{E}$ has subsided to a level equal to the value it had at the beginning of the post-learning period is 4.8 time units, a value of t more than twice that of the post-learning maximum, as stated in the theorem.

C. This theorem is in formal agreement with experimental observations, as may be seen by an inspection of Figure 39 adapted from Ward.

THEOREM XII

A. *In the learning of rote series, the optimum constant time interval for* n *evenly distributed repetitions is a decreasing function of* n.

Proof:

Let t_1 be the optimum time interval for n repetitions. Then the effective excitatory tendency at the termination of the n'th repetition is:

$$\bar{E}(t_1, n) = \sum_{i=0}^{n-1} (ae^{-bit_1} - ce^{-dit_1}).$$

Since t_1 is the optimum time interval, we know that

$$\frac{d}{dt_1} \bar{E}(t_1, n) = 0.$$

Now, in general, $\bar{E}(t,n)$ increases for values of t less than t_1 and decreases for values of t greater than t_1. Hence the theorem is proved when we show that

$$\frac{d}{dt_1} \bar{E}(t_1, n+1) < 0.$$

But,

$$\bar{E}(t_1, n+1) = \bar{E}(t_1, n) + ae^{-bnt_1} - ce^{-dnt_1}.$$

Hence,

$$\frac{d\bar{E}(t_1, n+1)}{dt_1} = \frac{d}{dt_1}(ae^{-bnt_1} - ce^{-dnt_1}).$$

But nt_1 is larger than the time $(\bar{t})$ at which $\bar{E}(t)$ is maximum; for otherwise we would have:

$$\frac{d}{dt_1} \bar{E}(t_1, n) > 0$$

which is contradictory to assumption. Therefore,

$$\frac{d}{dt_1}(ae^{-bnt_1} - ce^{-dnt_1}) < 0$$

and the theorem follows.

B. It will be recalled that the method of distributed repetitions or practice (D34) in rote learning is contrasted with the method of massed repetitions or practice (D32). By the method of *massed* practice there is a pause of only a few seconds between successive repetitions of the rote series; because of its small magnitude this "empty" time is neglected in the present systematization. By *distributed* practice, on the other hand, "empty" time intervals of no

TABLE 4

This table shows some of the critical steps leading to the determination of the theoretically optimal duration of evenly distributed rest pauses in rote learning with 12 repetitions for series of repetitions involving one and two such pauses. For other assumed characteristics of the theoretical situation, see text.

Length of interpolated time interval (t)	Effective excitatory potential where single interval is interpolated in middle of series		Effective excitatory potential where two equal intervals are interpolated, one after the 4th repetition and the second after the 8th repetition		
	$\bar{E}$ at end of time interval	Total $\bar{E}$ at once after 12th repetition, i.e., 1.849 added to column (1)	$\bar{E}$ from repetitions 1, 2, 3, and 4 at end of second interpolated interval	$\bar{E}$ from repetitions 5, 6, 7, and 8 at end of second interpolated interval	Total $\bar{E}$ at once after 12th repetition, i.e. after adding 1.232 to the sum of columns (3) and (4)
	(1)	(2)	(3)	(4)	(5)
.0	1.849	3.698	1.232	1.232	3.696
.2	1.971	3.820	1.388	1.314	3.934
.4	2.083	3.932	1.515	1.388	4.135
.6	2.183	4.032	1.617	1.455	4.304
.8	2.273	4.122	1.696	1.516	4.444
1.0	2.354	4.203	1.755	1.569	4.556
1.2	2.426	4.275	1.800	1.617	4.649
1.4	2.488	4.337	1.828	1.659	4.719
1.6	2.544	4.393	1.845	1.696	4.773
1.8	2.593	4.442	1.851	1.728	4.811
2.0	2.633	4.482	1.848	1.755	4.835
2.2	2.669	4.518	1.837	1.781	4.850
2.4	2.699	4.548	1.819	1.799	4.850
2.6	2.719	4.568	1.796	1.816	4.844
2.8	2.743	4.592	1.769	1.828	4.829
3.0	2.757	4.606	1.737	1.838	4.807
3.2	2.767	4.616	1.693	1.845	4.770
3.4	2.774	4.623	1.667	1.849	4.748
3.6	2.777	4.626	1.628	1.851	4.711
3.8	2.775	4.624	1.588	1.850	4.670
4.0	2.771	4.620	1.546	1.848	4.626
4.2	2.764	4.613	1.504	1.844	4.580
4.4	2.755	4.604	1.461	1.837	4.530
4.6	2.743	4.592	1.419	1.829	4.480
4.8	2.729	4.578	1.377	1.819	4.428
5.0	2.713	4.562	1.334	1.809	4.375
6.0	2.607	4.456	1.163	1.738	4.133
7.0	2.472	4.321	.945	1.648	3.825
8.0	2.319	4.168	.785	1.547	3.564
9.0	2.161	4.010	.649	1.441	3.322
10.0	1.922	3.771	.534	1.334	3.100

practice of greater or less duration are interposed at one or more points throughout the learning. By the method of *evenly distributed practice* (D35) the "empty" intervals are of equal duration and are so introduced that the number of repetitions in the groups of repetitions remaining without such interruption is equal. For example, an even distribution of 12 repetitions might be accomplished by introducing one no-practice interval in the middle of the series (thus dividing the 12 repetitions into two massed practices), or by introducing two no-practice periods, the first between the fourth and fifth repetitions and the second between the eighth and ninth repetitions (thus dividing the 12 repetitions into three massed practices) and so on. One of the most common methods of distributing the repetitions is to interpose a no-practice period after every single repetition; in this case each individual syllable-presentation cycle becomes a massed practice. By the method of *well distributed* practice the time interval used in evenly distributed practice is so chosen that it will yield the maximum effective excitatory potential ($\bar{E}$) at the termination of the last repetition.

The rote-learning situation chosen for the concrete illustration of this theorem is that of learning the ninth syllable of the 15-syllable series shown with theoretical analysis in Table 2, the learning to be carried to 12 repetitions, a point well beyond the reaction threshold. Table 1 shows that where $F = 1.37$, J_9 of a 15-syllable series is 7.685. Taking, as in previous illustrations, $\Delta K = .09\ \Delta E$, it follows that $\Delta I_9 = .09 \times 7.685 = .69165$.

We shall make a comparison of the optimum time interval where two equal intervals are inserted after the fourth and eighth repetitions of a 12-repetition learning, as contrasted with the optimal time interval where only one interval is inserted after the sixth repetition. The method of determining the optimal time interval is like that employed in Theorem X B and illustrated in detail by Table 3, where the maximum $\bar{E}$ was seen (in the last column) to fall at close to $t = 2.0$ with an $\bar{E}$ value of 4.353.

In the present case the procedure was to compute E and I at this point ($n = 9$) after 6 repetitions. Since ΔE is taken as the unit of measurement, $E = 6$. Since $I = 6 \times .69165\ \Delta E$, $I_n = 4.15152$. Accordingly, at any time t after the termination of learning,

$$E(t) = \frac{6}{2.718^{.1t}}$$

and

$$I(t) = \frac{4.15152}{2.718^{.3t}}.$$

Throughout the critical region the values of E and I were computed at intervals of $.2\ t$, and from these two columns of results the difference, or $\bar{E}$, was found; this is shown in column (1) of Table 4. But after the rest interval,

6 more repetitions are given. By Postulates 3 and 5, these give the same increase of E and I as the first 6 repetitions did, i.e., E = 6 and I = 4.15152, or a net increase in effective excitatory potential of 6 - 4.1515, or 1.8485. By the present set of postulates this increase in $\bar{E}$ from repetitions 7 to 12 is the same regardless of the amounts of E and I at n = 9 during the period when these repetitions are occurring. It is also to be noticed that the status of the E and I resulting from the first 6 repetitions ceases to change the moment practice is resumed. Accordingly the value 1.849 when added to the values in column (1) yields the theoretical amount of E at the termination of the learning process, as shown in column (2). By glancing down column (2) it will be observed that the $\bar{E}$ becomes maximal at about t = 3.6. This means that under the assumed conditions the optimum time interval where only one rest pause is employed is 3.6 time units.

We proceed now to determine the optimal length of the rest pause separating practice where two such intervals are employed. The first step in the method of determining the final effective excitatory potential when two rest pauses are introduced is to determine the E and I values where R = 4 (repetitions 1, 2, 3, and 4) and determine the $\bar{E}$ from this at the end of a *double* rest interval. This is shown in column (3) of Table 4. The double interval is taken on the ground that the amount of both E and I at the end of the first pause merely remains stationary during repetitions 5, 6, 7, and 8, but at the beginning of the second pause the loss in E and I is resumed as if never interrupted. This process gives the status of the portions of E and I acquired during repetitions 1, 2, 3, and 4 at the end of the second time interval, i.e., at the beginning of the ninth repetition.

We must now take up the E and I acquired during repetitions 5, 6, 7, and 8. The procedure for determining the $\bar{E}$ from these repetitions after a single rest interval (the second) is the same as that employed in columns (1) and (2); the results are shown in column (4).

If, now, we add parallel $\bar{E}$-entries in columns (3) and (4) we shall have the sum of the $\bar{E}$ from all the practice up to the beginning of the ninth repetition. And if to this is added the $\bar{E}$ resulting from repetitions 9, 10, 11, and 12 (4 - 2.768 = 1.232), we shall have the total $\bar{E}$ for the learning of the series as a whole where two rest pauses are introduced. This is shown in column (5) of Table 4.

A glance down column (5) shows that the maximum $\bar{E}$ falls in the neighborhood of T = 2.2. We have seen above that a single interpolated interval in the same kind of repetition series comes out with an optimal value of 3.6. But since 2.2 < 3.6, it follows that $t_{n=2} < t_{n=1}$, exactly as Theorem XII states.

THEOREM XIII

A. *In the learning of rote series, for any given number of repetitions, the advantage of distributed practice over massed practice, in terms of effective excitatory strength, is an increasing linear function of the inhibition induced by one repetition.*

Proof:

Let n repetitions be distributed with constant intervals (t_0). The effective excitatory strength at the termination of this distributed practice is:

$$\bar{E}_d = \sum_{i=0}^{n-1} (ae^{-bit_0} - ce^{-dit_0}).$$

The effective excitatory strength at the termination of n repetitions by massed practice is:

$$\bar{E}_m = n \cdot (a - c).$$

The advantage of distributed practice over massed practice for n repetitions, in terms of effective excitatory strength, is:

$$A = \bar{E}_d - \bar{E}_m$$

$$= \sum_{i=0}^{n-1} (ae^{-bit_0} - ce^{-dit_0} - a + c).$$

Hence,

$$\frac{dA}{dc} = \sum_{i=0}^{n-1} (1 - e^{-dit_0}).$$

But this is positive, and independent of c. Hence, the theorem follows.

B. This theorem may be illustrated by the following concrete example. Suppose we have four learning situations, A, B, C, and D, in which

$$\Delta I_A = .8, \quad \Delta I_B = .6, \quad \Delta I_C = .4, \text{ and } \Delta I_D = .2 .$$

Now if each situation is learned by distributed practice involving 12 repetitions and two interpolated rest intervals of .5 time units each, Theorem XIII states that the difference between the excitatory potential at the end of the 12 repetitions in each case, and what would have been yielded by massed practice in that case, will form a linear progression, varying with the size of ΔI. By means of computations exactly analogous to those employed in the 2-interval distribution of practice in Theorem XII B (columns 3, 4, and 5 of Table 4) we have, for the effective excitatory potential at the end of the 12 repetitions,

$$\bar{E}_A = 3.100, \quad \bar{E}_B = 5.180, \quad \bar{E}_C = 7.262, \quad \bar{E}_D = 9.344 .$$

But by massed practice the $\bar{E}_A = 12\ (1 - .8) = 2.4$, and so on for the other situations. Accordingly by massed practice,

$$\bar{E}_A = 2.4, \quad \bar{E}_B = 4.8, \quad \bar{E}_C = 7.2, \quad \bar{E}_D = 9.6 .$$

We may now determine the advantage accruing in each case by subtracting the result by massed practice from the outcome by distributed practice. The results of this procedure are as follows:

When $\Delta I = .8$, the advantage of distributed practice is $3.100 - 2.4 = .70$
" $\Delta I = .6$, " " " " " " $5.180 - 4.8 = .38$
" $\Delta I = .4$, " " " " " " $7.262 - 7.2 = .062$
" $\Delta I = .2$, " " " " " " $9.344 - 9.6 = -.256$

Now, Theorem XIII states, first, that these differences should increase as ΔI increases, i.e.,

$$.70 > .38 > .062 > -.256$$

which is evidently true.

Secondly, the theorem states that,

$$.70 - .38 = .38 - .062 = .062 - (-.256)$$

i.e., $$.32 = .318 = .318$$

which is also evidently true except for inaccuracies due to dropped decimals.

THEOREM XIV [8]

A. *In the learning of rote series, the advantage of well distributed practice over massed practice, in terms of repetitions, is an increasing function of the mean number of repetitions required for learning by massed practice.*

8. A "well-distributed practice" is any distributed practice (D34) such that at the termination of learning $\bar{E}_{n,r}$ is greater than would be the case by massed practice (D32). Through an oversight this concept was not included among those formally defined.

Proof:

Let R be the mean number of repetitions required to learn a given syllable by massed practice. Let R_A be the advantage, in terms of repetitions, of distributed practice over massed practice. Then we see that R_A is the largest integer such that:

$$\sum_{i=0}^{R-R_A-1} (ae^{-bit_1} - ce^{-dit_1}) \geq L = R(a-c) \quad (1)$$

(where t_1 is the time interval between repetitions).

Now, if we increase R to $R+1$, there will be a corresponding positive increase Δc in c. Then $(R+1)_A$ will be the largest integer such that:

$$\sum_{i=0}^{R-(R+1)_A} [ae^{-bit_1} - (c+\Delta c)e^{-dit_1}] > L = (R+1)(a-c-\Delta c) \quad (2)$$

We shall show that R_A satisfies (2) automatically. For, replacing $(R+1)_A$ by R_A in the left member of (2), we have:

$$\sum_{i=0}^{R-R_A} [ae^{-bit_1} - (c+\Delta c)e^{-dit_1}] = \sum_{i=0}^{R-1-R_A} [ae^{-bit_1} - ce^{-dit_1}]$$

$$- \Delta c \sum_{i=0}^{R-R_A} e^{-dit_1} + ae^{-b(R-R_A)t_1} - ce^{-d(R-R_A)t_1}.$$

Then, from inequality (1) and from the assumption that the practice is well distributed, we have:

$$\sum_{i=0}^{R-R_A} [ae^{-bit_1} - (c+\Delta c)e^{-dit_1}] > L + (a-c) - \Delta c \sum_{i=0}^{R-R} e^{-dit_1}$$

$$> L + (a-c) - \Delta c(R+1).$$

But,

$$R(a-c) = L = (R+1)(a-c-\Delta c).$$

Hence,

$$(R+1)\Delta c = a-c.$$

Therefore,

$$\sum_{i=0}^{R-R_A} [ae^{-bit_1} - (c+\Delta c)e^{-dit_1}] > L$$

i.e., R_A satisfies inequality (2) automatically. But $(R+1)_A$ is the largest integer satisfying (2). Hence,

$$R_A \leq (R + 1)_A$$

which completes the proof.

B. Suppose that we have three rote learning situations, A, B, and C, such that when a single interval of $t = 1.0$ is interposed in the middle of the series of repetitions the reaction threshold will just be reached in 8 repetitions, 10 repetitions, and 12 repetitions respectively. By reversing the procedure which generated columns (1) and (2) of Table 4, it is found that the above learning outcomes will occur when,

$$\Delta I_A = .80,\ \Delta I_B = .86, \text{ and } \Delta I_C = .90\,.$$

Now by Postulate 16, Corollary 1, when learning is by massed practice,

$$_LR_{n,n+1} = \frac{L}{\Delta E_{n,n+1} - \Delta I_n}\,.$$

Substituting appropriately in this equation we find that the number of repetitions required by massed practice is, for the three assumed conditions with respect to ΔI,

$$_LR_{(A)} = 11.13,\quad _LR_{(B)} = 16.36,\quad _LR_{(C)} = 23.83\,.$$

Subtracting from each of the above values the number of repetitions required to learn by distributed repetitions with a single pause between massed repetitions, we have the advantage due to distributed practice:

$$A = 3.13,\ B = 6.36,\ C = 11.83\,.$$

The theorem in effect states that since by massed practice,

$$_LR_{(A)} < {}_LR_{(B)} < {}_LR_{(C)}$$

it should follow that

$$3.13 < 6.36 < 11.83\,,$$

which is the fact.

C. The study of Lyon (*39*) gives data on the advantage of distributed practice over massed with varying lengths of lists. The rate of presentation was two syllables per second. The criterion of learning difficulty employed was the number of minutes required for learning. This criterion is, of course, perfectly correlated with the number of trials or repetitions required for learning. Massed practice involved continuous uninterrupted learning. With distributed practice only one presentation per day was employed. The results are given in Table 5.

TABLE 5

Mean number of minutes required for learning by massed and distributed practice with varying lengths of lists of nonsense syllables. Data from Lyon (39).

Length of list	Minutes required for learning		Minutes advantage of distributed practice
	Massed practice	Distributed practice	
8	1/4	1/8	1/8
12	6	1 1/2	4 1/2
16	9	3 2/3	5 1/3
24	16	5	11
32	28	6	22
48	43	14	29
72	138	25	113

Table 5 clearly shows that the advantage of distributed practice becomes greater, the greater the number of trials required for learning by massed practice. When one-fourth minute is required for learning by massed practice, the saving of massed over distributed practice is only one-eighth minute, but when 138 minutes are required for learning by massed practice, the saving is 113 minutes.

Additional data confirming these results were obtained in an experiment by Hovland (26). Lists of 8, 11, and 14 syllables were employed. The number of repetitions required with learning by massed and distributed practice are given in Table 6. It will be observed that the same

TABLE 6

Average number of repetitions required to learn three lengths of series by massed and distributed practice. Data from Hovland (26).

Length of list	Massed practice	Distributed practice	Advantage distributed practice
8	7.00	5.93	1.07
11	11.64	9.43	2.21
14	17.14	12.14	5.00

trend noted in Lyon's curves is obtained but that the range of lengths used is much smaller.

On the basis of the preceding experimental results it may be said that the theorem has a satisfactory empirical confirmation.

THEOREM XV

A. *In the learning of rote series the advantage of well-distributed practice over massed practice, in terms of repetitions, is a concave function of the mean number of repetitions required for learning by massed practice.*

Proof:

It will be necessary, in the proof of this theorem, to extend certain functions, defined originally only for positive integers, to functions defined for all real numbers. We shall also interpret the term "concave" in terms of derivatives instead of differences i.e., a function of $f(x)$ is concave if:

$$\frac{d^2 f(x)}{d x^2} > 0 .$$

Consider the sum:

$$\sum_{i=0}^{z} (ae^{-bit_1} - ce^{-dit_1}) .$$

Taking a, b, d, and t_1 to be constant, this is a function of z and c, defined for all positive values of c and all positive integral values of z. Under the conditions of the theorem, this is an increasing, concave function of z. Then we define $g(z, c)$ to be a function defined for all positive, real values of z and c, coinciding with the above sum for all integral values of z, and such that it is an increasing, concave function of z; i.e.,

$$\frac{\partial g(z, c)}{\partial z} > 0, \quad \frac{\partial^2 (z, c)}{\partial z^2} > 0 .^{9}$$

We can assume further, without inconsistency, that

$$\frac{\partial g(z, c)}{\partial z} = a - c$$

when $z = 0$.

9. The symbol ∂ indicates differentiation of $g(z - c)$ considered as a function of z alone.

Likewise, consider the sum:

$$\sum_{i=0}^{z} e^{-dit_1}$$

and define the increasing, convex function of $f(z)$ for all positive, real values, reducing to this sum for all integral values of z. We can assume further, without inconsistency, that $\frac{df(z)}{dz} = 1$ when $z = 0$. Now, $_{L}R_{(A)}$ is (see the proof of Theorem XIV) the largest integer such that:

$$\sum_{i=0}^{R-R_A-1} (ae^{-bit_1} - ce^{-dit_1} \geq L.$$

In terms of the functions defined above, if we let

$$Z = {}_{L}R - {}_{L}R_A - 1$$

then R_A is defined by

$$g(z, c) = L).$$

Now R, and therefore z, is implicitly a function of the initial increment of inhibition, c. Furthermore, L is a constant. Hence, differentiating with respect to c, we have:

$$\frac{\partial g(z,c)}{\partial z} \cdot \frac{dz}{dR} \cdot \frac{dR}{dc} + \frac{\partial g(z,c)}{\partial c} = 0.$$

But, clearly, we can choose $g(z,c)$ and $f(z)$ so that

$$\frac{\partial g(z,c)}{\partial c} = -f(z).$$

Hence,

$$\frac{\partial g(z,c)}{\partial z} \cdot \frac{dz}{dR} \cdot \frac{dR}{dc} = f(z).$$

Differentiating again with respect to c, we have:

$$\frac{\partial^2 g(z,c)}{\partial z^2} \left(\frac{dz}{dR}\right)^2 \left(\frac{dR}{dc}\right)^2 + \frac{\partial g(z,c)}{\partial z} \cdot \frac{d^2 z}{dR^2} \left(\frac{dR}{dc}\right)^2 + \frac{\partial g(z,c)}{\partial z} \cdot \frac{dz}{dR} \frac{d^2 R}{dc^2}$$

$$= \frac{df(z)}{dz} \cdot \frac{dz}{dR} \cdot \frac{dR}{dc}.$$

But, by assumption, $g(z,c)$ is a concave function of z, and hence the first term of the left member of this equation is positive. Therefore,

$$\frac{\partial g(z,c)}{\partial z} \frac{d^2 z}{dR^2} \left(\frac{dR}{dc}\right)^2 < \frac{dz}{dR} \cdot \left(\frac{df(z)}{dz}\right) \cdot \frac{dR}{dc} - \frac{\partial g(z,c)}{\partial z} \left(\frac{d^2 R}{dc^2}\right)$$

Now, since

$$z = R - R_A - 1,$$

$$\frac{d^2 z}{dR^2} = \frac{d}{dR} \left(1 - \frac{dR_A}{dR}\right) = \frac{-d^2 R_A}{dR^2}$$

The theorem which we are to prove is equivalent to the inequality:

$$\frac{d^2 R_A}{dR^2} > 0$$

i.e.,

$$\frac{d^2 z}{dR^2} < 0 .$$

But, from the above inequality, since $\frac{\partial g(z,c)}{\partial z}$ and $\left(\frac{dR}{dc}\right)^2$ are positive and $\frac{dz}{dR} < 1$, it is sufficient to show that:

$$\frac{df(z)}{dz} \frac{dR}{dc} < \frac{\partial g(z,c)}{\partial z} \frac{d^2 R}{dc^2} .$$

Now,

$$R(a - c) = L.$$

Hence, differentiating,

$$(a - c) \frac{dR}{dc} - R = 0 .$$

Differentiating again:

$$(a - c) \frac{d^2 R}{dc^2} - 2 \frac{dR}{dc} = 0.$$

Hence,

$$\frac{dR}{dc} = \frac{1}{2} (a-c) \frac{d^2 R}{dc^2}$$

Furthermore, since $f(z)$ is a convex function,

$$\frac{df(z)}{dz} < \left.\frac{df(z)}{dz}\right]_{z=0} = 1 .$$

And, since $g(z,c)$ is a concave function of z,

$$\frac{\partial g(z,c)}{\partial z} = \frac{\partial g(z,c)}{\partial z}\bigg]_{z=0} = a - c .$$

Therefore,

$$\frac{df(z)}{dz}\ \frac{dR}{dc} < 1 \cdot \frac{1}{2}\ (a-c)\ \frac{d^2 R}{dc^2} < \frac{1}{2}\ \frac{\partial g(z,c)}{\partial z}\ \frac{d^2 R}{dc^2}$$

$$< \frac{\partial g(z,c)}{\partial z}\ \frac{d^2 R}{dc^2} .$$

Hence,

$$\frac{d^2 R_A}{dR^2} > 0 .$$

(Theorem XIV can be derived easily from the generalized functions of this proof. It is easily shown, however, that Theorem XV cannot be proved by the methods of Theorem XIV, and, in fact, is not true in the sense of finite differences for integral values of R and R_A.)

B. This theorem finds exemplification in the theoretical material presented in Theorem XIV B. The present theorem states in effect that since by massed practice,

$${}_LR_C > {}_LR_B > {}_LR_A$$

it follows that,

$$11.83 - 6.36 > 6.36 - 3.13$$

i.e.,

$$5.47 > 3.23$$

which is the fact.

C. The same data of Lyon quoted above in connection with Theorem XIV are employed to give confirmation of Theorem XV. In Figure 20 the difference in the number of trials by massed and distributed practice is plotted against the mean number of repetitions required for learning by massed practice. Inspection of the figure shows that the advantage of distributed practice is a concave function of the number of repetitions required for learning by massed practice, although the concavity is not pronounced.

The differences between the number of trials required for learning by massed and distributed practice and the number of repetitions required for learning by massed

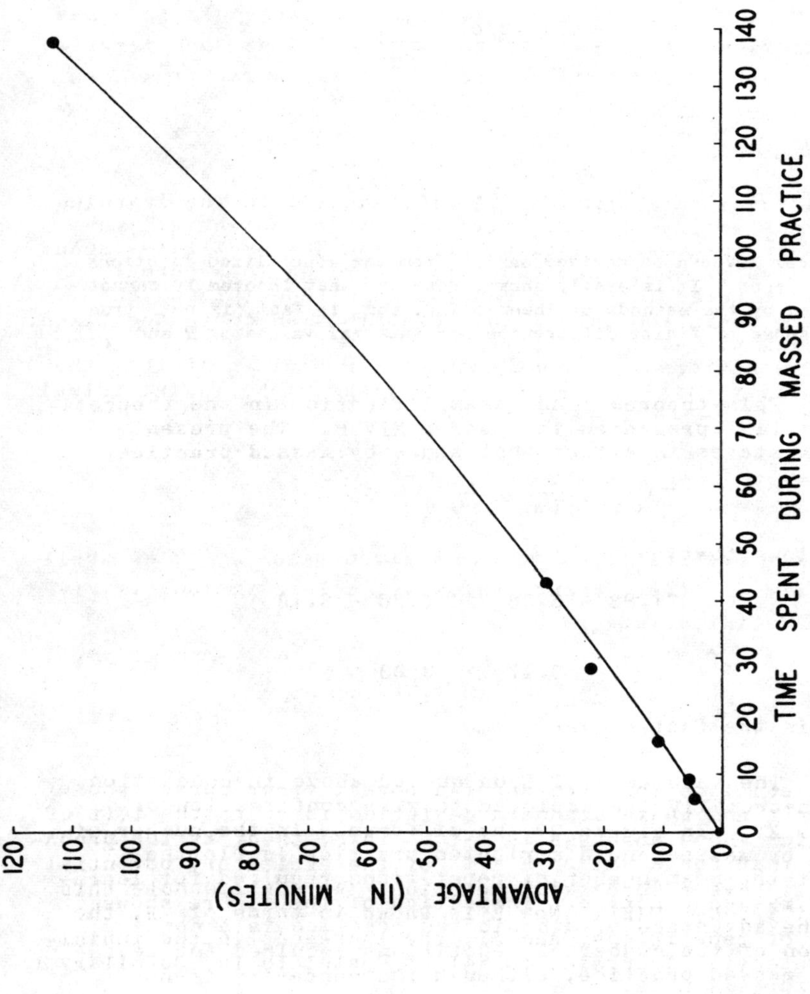

Figure 20. Advantage of distributed practice over massed (in minutes saved) relative to length of time required for learning by massed practice. Data from Lyon (39).

practice are similarly related in Hovland's study (*26*), as shown in the final column of Table 6 (p. 131).

The available evidence appears to confirm the theorem.

PROBLEM III

A. *In the case of the conditioning of any single syllable reaction in a rote series learned by massed practice, given the value of the reaction threshold* (L), *the standard deviation of the variability of the reaction threshold* (σ), *and the increment of inhibitory potential* (ΔI) *at each repetition: to find the mean repetition at which the first successful reaction will occur.*

Proof:

According to Postulates 15, 16, and 17, in the learning of a rote series the effective excitatory potential just necessary to evoke a syllable reaction (ℓ) oscillates about a positive central magnitude (L) according to a normal chance distribution. The central magnitude (L) and the standard deviation (σ) of this distribution curve are fixed, known quantities.

It is an immediate consequence of Postulate 15 that the probability that ℓ has a value lying in the infinitesimal region (ℓ_1, $\ell_1 + \Delta\ell$) is given by,

$$\frac{1}{\sqrt{2\pi}\,\sigma}\, e^{-\frac{(\ell_1 - L)^2}{2\sigma^2}}\, \Delta\ell \, .$$

Hence the probability that for a given value, ℓ_1, we shall have $\ell < \ell_1$, i.e., a successful syllable reaction, is given by the definite integral,

$$\frac{1}{\sqrt{2\pi}\,\sigma} \int_{-\infty}^{\ell_1} e^{\frac{-(\ell - L)^2}{2\sigma^2}}\, d\ell$$

and is equal to the area under a normal error curve, whose mean is L and whose standard deviation is σ, to the left of the point whose abscissa is $-\ell_1$. This, then, is the probability that for a value $\bar{E} = \ell$ of the excitatory potential we shall have a successful reaction. We shall denote this probability by $P(\bar{E})$. Now $\bar{E}$ is known in terms of *R*, the number of repetitions, and ΔI, the increment in the inhibitory potential according to (4) of Postulate 13, Corollary 2. This expression is,

$${}^{R,N}E_{n,n+1} = R(\Delta^{N}E_{n,n+1} - \Delta^{N}I_{n}) \, .$$

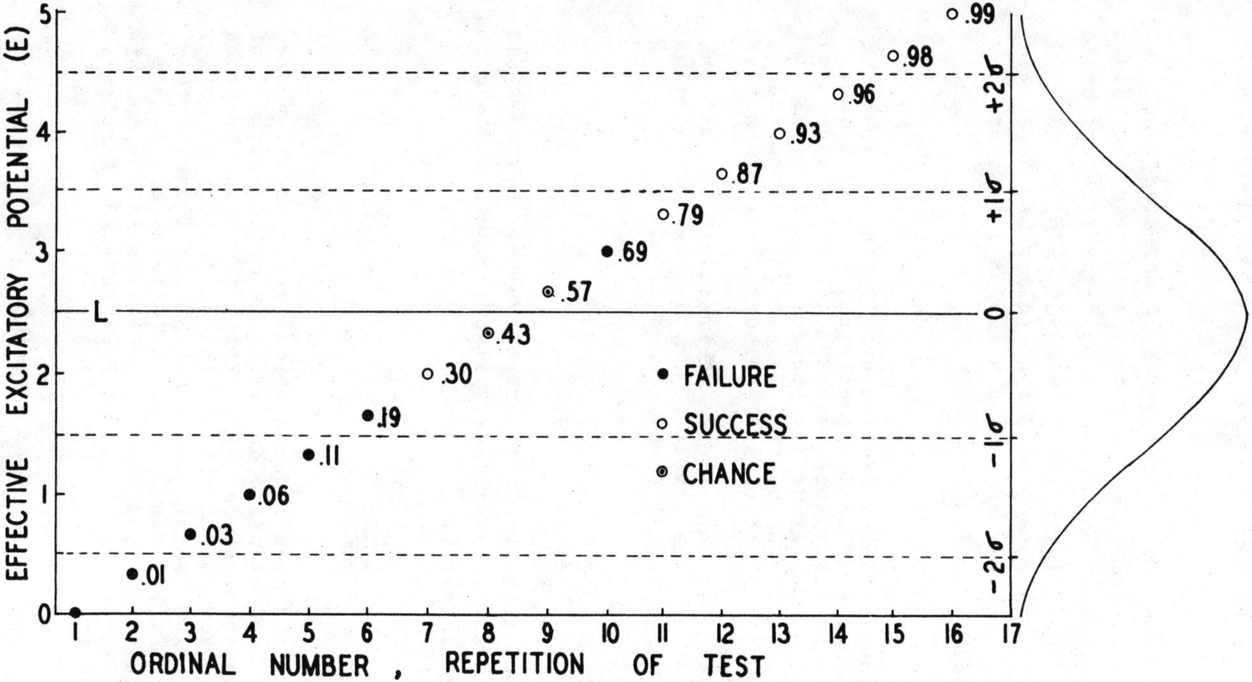

Figure 21. Diagram representing the chances of success at the successive repetitions which lead to the final complete conditioning of a syllable reaction. In this diagram $L = 2.5$, $\sigma = 1.00$, and $\Delta I = .66\frac{2}{3}$. The normal probability integral is plotted at the right of the diagram. The chance of a successful reaction at each repetition is shown in decimals along the diagonal row of circles which represents the rectilinear curve of learning (Postulates 3, 5, and 13, Corollary 2). See text for an account of the method of determining the series of probability values.

TABLE 7

This table shows (1) the relative frequencies (in percentages) of the probability integral over each one-tenth of the standard deviation, and (2) the cumulative percentage values for the same integral. The distribution of values is cut off arbitrarily at 2.5 σ at each side of the central tendency. Note that one half of the probability distribution is shown in columns a, b, c, and d, and that the other half is shown in columns a′, b′, c′, and d′. It will be observed that columns c and d are derived directly from columns a and b respectively. Adapted from Thorndike (62).

Deviations from the central tendency of probability	Percentage of probability lying between each σ entry and the one preceding it in column (a)	Deviations from a point at -2.5 σ from the central tendency of probability	Percentage of probability lying between -2.5 σ and the entry in column (a)	Deviations from the central tendency of probability	Percentage of probability lying between each σ entry and the one preceding it in column (a′)	Deviations from a point at -2.5 σ from the central tendency of probability	Percentage of probability lying between -2.5 σ and the entry in column (a′)
(a)	(b)	(c)	(d)	(a′)	(b′)	(c′)	(d′)
-2.5 σ	.00	.0 σ	.00	.0 σ	3.98	2.5 σ	49.37
-2.4 σ	.20	.1 σ	.20	+.1 σ	3.98	2.6 σ	53.35
-2.3 σ	.25	.2 σ	.45	+.2 σ	3.94	2.7 σ	57.29
-2.2 σ	.32	.3 σ	.77	+.3 σ	3.87	2.8 σ	61.16
-2.1 σ	.40	.4 σ	1.17	+.4 σ	3.75	2.9 σ	64.91
-2.1 σ	.49	.5 σ	1.66	+.5 σ	3.60	3.0 σ	68.51
-1.9 σ	.60	.6 σ	2.26	+.6 σ	3.43	3.1 σ	71.94
-1.8 σ	.72	.7 σ	2.98	+.7 σ	3.23	3.2 σ	75.17
-1.7 σ	.86	.8 σ	3.84	+.8 σ	3.01	3.3 σ	78.18
-1.6 σ	1.02	.9 σ	4.86	+.9 σ	2.78	3.4 σ	80.96
-1.5 σ	1.20	1.0 σ	6.06	+1.0 σ	2.54	3.5 σ	83.50
-1.4 σ	1.39	1.1 σ	7.45	+1.1 σ	2.30	3.6 σ	85.80
-1.3 σ	1.60	1.2 σ	9.05	+1.2 σ	2.06	3.7 σ	87.86
-1.2 σ	1.83	1.3 σ	10.88	+1.3 σ	1.83	3.8 σ	89.69
-1.1 σ	2.06	1.4 σ	12.94	+1.4 σ	1.60	3.9 σ	91.29
-1.0 σ	2.30	1.5 σ	15.24	+1.5 σ	1.39	4.0 σ	92.68
-.9 σ	2.54	1.6 σ	17.78	+1.6 σ	1.20	4.1 σ	93.88
-.8 σ	2.78	1.7 σ	20.56	+1.7 σ	1.02	4.2 σ	94.90
-.7 σ	3.01	1.8 σ	23.57	+1.8 σ	.86	4.3 σ	95.76
-.6 σ	3.23	1.9 σ	26.80	+1.9 σ	.72	4.4 σ	96.48
-.5 σ	3.43	2.0 σ	30.23	+2.0 σ	.60	4.5 σ	97.08
-.4 σ	3.60	2.1 σ	33.83	+2.1 σ	.49	4.6 σ	97.57
-.3 σ	3.75	2.2 σ	37.58	+2.2 σ	.40	4.7 σ	97.97
-.2 σ	3.87	2.3 σ	41.45	+2.3 σ	.32	4.8 σ	98.29
-.1 σ	3.94	2.4 σ	45.39	+2.4 σ	.25	4.9 σ	98.54
.0 σ	3.98	2.5 σ	49.37	+2.5 σ	.20	5.0 σ	98.74
					.00	5.1 σ	100.00

Let us denote by $P(R)$ the probability of a successful reaction on the Rth repetition. Then,

$$P(R) = P(\bar{E}) \text{ when } \bar{E} = R(\Delta\bar{E} - \Delta I).$$

For obvious psychological and practical reasons it is convenient to break off the normal curve for values of $R = 0$ and R equal to a value such that the corresponding value of E is just twice L. Let us call this value of R, R_{max}. This approximation introduces an entirely negligible error into the computation.

The probability that the first success will occur on the Rth repetition is equal to the product of the probabilities that it will fail on each of the $R-1$ previous trials, and succeed on the Rth, since these events are independent. If we denote this probability by $Q(R)$, we have,

$$Q(R) = P(R) \prod_{r=1}^{R-1} [1 - P(r)].$$

Note that $P(r)$ for $r = 0$ is zero, and that $P(r)$ for $r = R_{max}$ is unity. Now the mean value of the number of repetitions for the first success, which we denote by R_1 is,

$$R_1 = \sum_{r=1}^{\infty} rQ(r),$$

by definition of mean.

B. Let us assume for the present purposes that $L = 2.5$, $\sigma = 1.0$, and $\Delta I = .66\frac{2}{3}$. Under these assumptions the process of the final conditioning of a single syllable reaction is represented diagrammatically in Figure 21. In this figure and throughout the present discussion it is assumed that the probability integral passes abruptly to zero at a distance of 2.5 σ from its maximum value. With this restriction the value of the integral is shown in the right-hand portion of the diagram and also in Table 7.

Since by (4) of Postulate 8, Corollary 2,

$$^{R,N}\bar{E}_{n,n+1} = R(\Delta {}^{N}E_{n,n+1} - \Delta {}^{N}I_n)$$

and since $\Delta {}^{N}I_n = .66\frac{2}{3}$, it comes about that,

$$^{R,N}\bar{E}_{n,n+1} = \frac{1}{3} R.$$

By substituting the various values of R in this equation we obtain the amounts of $\bar{E}_{n,n+1}$ at the various stages of the learning process. These are represented by the row of circles running diagonally across the diagram (Figure 21).

In this connection it will be recalled (p. 14) that each syllable-presentation cycle (i.e., repetition or R) serves

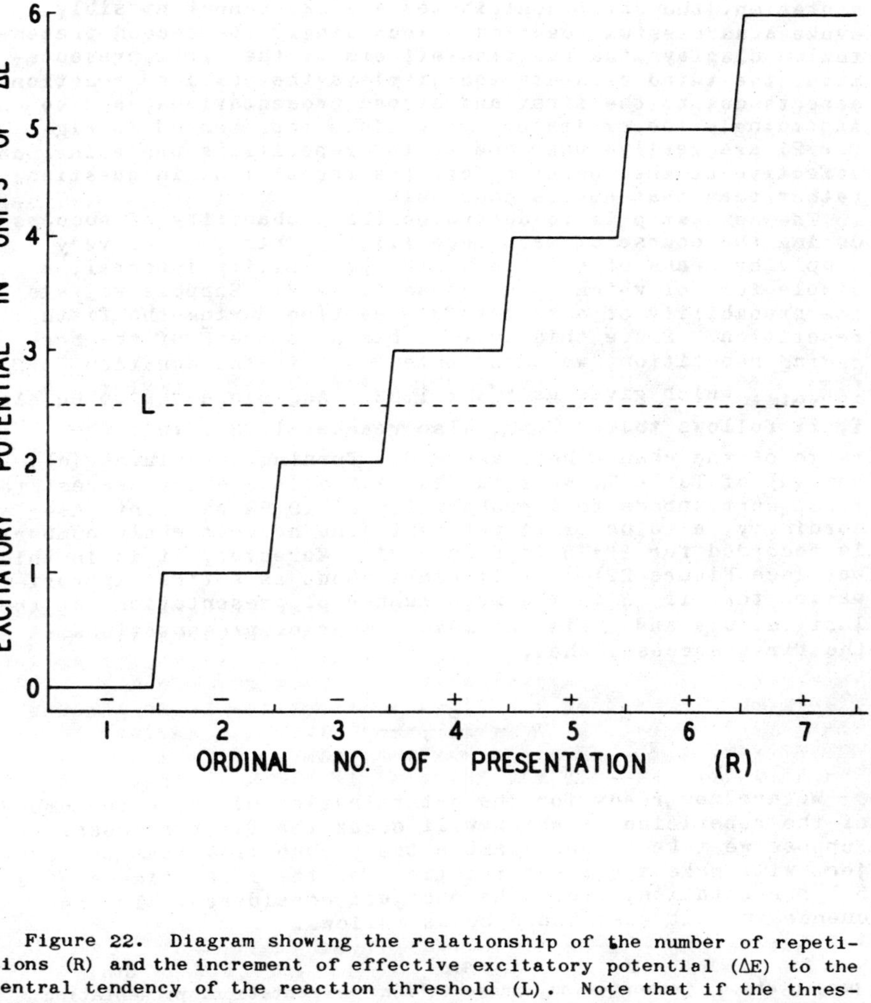

Figure 22. Diagram showing the relationship of the number of repetitions (R) and the increment of effective excitatory potential (ΔE) to the central tendency of the reaction threshold (L). Note that if the threshold be taken as half way between the first success and the last failure it would, in terms of repetitions, be $\frac{4+3}{2}$, or 3.5. It is evident from an inspection of the figure, however, that the threshold in terms of ΔE is 2.5 or one less. It thus comes about that the central tendency of the reaction threshold, in terms of effective excitatory potential, is $\frac{4+3}{2} - 1$, i.e., one less than the mean of the number of presentations at the first success and the last failure.

a double purpose: (A) it serves as a test of the amount of effective excitatory potential ($\bar{E}$) produced by the *preceding* presentations, and (B) it contributes a new increment of effective excitatory potential ($\Delta\bar{E}$) to the sum already accumulated (see Figure 22). For this reason the first presentation, though it contributes its $\Delta\bar{E}$, cannot possibly evoke a successful reaction. Accordingly the second presentation displays the reaction-effects of the first presentation, the third presentation displays the combined reaction-effects due to the first and second presentations, and so on. Accordingly the excitatory potentials represented in Figure 21 are really those due to the repetitions preceding, and effective at the *beginning* of, the repetitions in question rather than that at its conclusion.

The next step is to determine the probability of success during the course of each repetition. This is done very simply by means of a table of the probability integral, a simple form of which is given as Table 7. Suppose we take the probability of a successful reaction during the fifth repetition. Since this must be based on the $\bar{E}$ of the *preceding* repetition, we substitute $R = 4$ in the equation for ${}^{R,N}\bar{E}_{n,n+1}$, which gives us ${}^{4,N}\bar{E} = 1.33$. And since σ also equals 1, it follows that ${}^{4,N}E_{n,n+1}$ also reaches 1.33 σ into the range of the chance here assumed. Turning to columns (c) and (d) of Table 7, we find that 1.3 σ, the entry nearest to 1.33, corresponds to a probability of 10.88 per cent. Accordingly, a value of 11 per cent (the nearest whole number) is recorded for $R = 5$ in Figure 21. Moreover, it is in this way (see Figure 22) that it comes about as a first approximation that if ${}_{F}R$ is the mean number of presentations at the last failure and ${}_{S}R$ is the mean number of presentations at the first success, then,

$${}_{L}R = \frac{{}_{F}R + {}_{S}R}{2} - 1.$$

We are now ready for the determination of the mean number of the repetition at which will occur the first success. Suppose we take as our example the probability that the subject will make a correct reaction for the first time at the 5th presentation, i.e., the one just considered. The sequence in that case would be as follows:

1st presentation,	2nd presentation,	3rd presentation,	4th presentation,	5th presentation
−	−	−	−	+

where a minus sign indicates a failure of correct reaction and a plus sign indicates a success. Now it is a well known mathematical principle that the probability of a compound event is the product of the probabilities of several independent events. By referring to Figure 21 it may be seen that at the second repetition there is a chance of failure of .99 (i.e., 1.00 - .01), at the third repetition, a chance of failure of .97 at the fourth repetition, a chance of failure of .94, and so on. On the other hand, the chance of success at R = 5 is, as we have seen, .11. Accordingly the probability that a subject will fail at repetitions 1, 2, 3, and 4 and succeed at R = 5 must be,

$$1.00 \times .99 \times .97 \times .94 \times .11 = .099.$$

In a similar manner, the probability of a first success at the sixth repetition must be,

$$1.00 \times .99 \times .97 \times .94 \times .89 \times .19 = .169 .$$

By repetitions of this procedure were developed the several entries in column (a) of Table 8, where may be found the two probability values calculated just above.

TABLE 8

The distribution of the probabilities of having (a) a first success and (b) a final failure followed by five uninterrupted successes (arbitrarily taken as the standard of perfect learning) at each of the several repetitions shown in Figure 20. For method of calculation, see text.

R	First success		Last failure	
	Probability	Probability multiplied by R	Probability	Probability multiplied by R
	(a)	(b)	(c)	(d)
1	.000	.000	.000	.000
2	.010	.020	.000	.000
3	.030	.090	.000	.000
4	.058	.232	.002	.008
5	.099	.495	.009	.045
6	.169	1.014	.031	.186
7	.190	1.330	.079	.553
8	.186	1.488	.146	1.168
9	.149	1.323	.182	1.638
10	.077	.770	.186	1.860
11	.027	.297	.168	1.848
12	.0065	.078	.104	1.248
13	.0008	.010	.065	.845
14	.0005	.007	.039	.546
15			.020	.300
16			.010	.160
Σ	1.0008	7.150	1.040	10.400
Mean		7.142		10.000

The mean repetition at which the first success will occur is the quotient of the sum of the products of each probability multiplied by the corresponding number of repetitions, divided by the sum of the probabilities. In this way the mean repetition at which the first correct reaction will occur is 7.14, as shown in Table 8 and as indicated (approximately) by the hollow circle above $R = 7$ in Figure 21. A simpler, but less precise, method of making this determination is to find the R-value corresponding to the maximum probability appearing in column (a).

PROBLEM IV

A. *In the case of the conditioning of any single syllable reaction in a rote series learned by massed practice, given the value of the reaction threshold (L), the standard deviation of the variability of the reaction threshold (σ), and the increment of inhibitory potential (ΔI) at each repetition: to find the mean repetition at which the last failure preceding permanent success occurs.*

Proof:

This differs only slightly from Part A of the previous problem. We are now concerned with the probability that the last failure will fall on the Rth repetition. The probability that a successful reaction will result from the Rth repetition is given, as before, by $P(R)$. If we denote by $q(R)$ the probability that the last failure occurs on the Rth repetition, we have,

$$q(R) = [1 - P(R)] \prod_{r=R+1}^{\infty} P(r)$$

since $1 - P(R)$ is the probability of failure on the Rth repetition and $\prod_{r=R+1}^{\infty} P(r)$ is the probability that the remaining trials will result in successes.

If the mean value of the number of repetitions required to obtain the last failure is R_C, then,

$$R_C = \sum_{r=1}^{\infty} rq(r) .$$

B. In this problem we shall assume that L = 2.5, $\sigma = 1.0$, and $\Delta^{N}I_n = .66\frac{2}{3}$ exactly as in Problem III B. Accordingly, the probability values which appear in Figure 21 may also be employed to exemplify the present problem. Here also we shall need to employ the well-known mathematical principle that the probability of a compound event is the product of the probabilities of the separate independent components of the compound. Actually the criterion of complete learning in rote series ordinarily taken is two successive repetitions of the series *as a whole* without failure. Obviously this is a function not only of the maturity of the learning connection considered, but of the other syllables of the series. *As the nearest simple equivalent of complete learning applicable to a single reaction by itself we have taken 5 successive successful reactions.* Under these conditions the last failure is a compound event consisting of one failure followed by 5 successive correct reactions. The probability of such an event at any repetition is evidently the product of the probability of a failure at this repetition multiplied by the product of success at the next 5 repetitions.

Consider, for example, the probability that the last failure will occur at the twelfth repetition. Reference to Figure 21 shows that the probability of success at $R = 12$ is .88, which means that the probability of failure at this point (1.00 - .88) must be .12. Accordingly the probability of failure at this repetition will be:

$$.11 \times .93 \times .96 \times .98 \times .99 \times 1.00 = .104 .$$

In a similar manner, the probability that the last failure will occur at the eleventh repetition is:

$$.21 \times .87 \times .93 \times .96 \times .98 \times .99 = .167 .$$

By successive repetitions of this procedure were obtained the several items in column (c) of Table 8. The mean value of the syllable at which the last failure would theoretically occur was found from columns (c) and (d) in a manner exactly analogous to that employed for finding the first successful reaction in Problem III B. In this way it is found that the mean last failure is $_{F}R = 10.0$, as shown in Table 8 and as indicated by the solid circle above $R = 10$ in Figure 21.

PROBLEM V

A. *In the case of the conditioning of any single syllable reaction in a rote series learned by massed practice, given the value of the reaction threshold (L), the standard deviation of the variability of the reaction threshold (σ) and the increment of inhibitory potential (ΔI) at each repetition: to find the mean number of failure repetitions before complete learning.*

Proof:

In this problem we are dealing with R_{max} trials in which the probability of success on the Rth trial or repetition is given by the quantity P(R) defined in Part A of Problem III. Now it is well known in the theory of mathematical probability that if the probability of the occurrence of any event on the rth trial is given by P_r, then the mean number of times that this event will occur in n trials is $\sum_{r=1}^{n} P_r$. In our case the probability of failure to identify the syllable on the Rth repetition is $1 - P(R)$. Hence the mean number of failures in R_{max} repetitions is:

$$\sum_{r=1}^{R_{max}} [1 - P(r)] = R_{max} - \sum_{r=1}^{R_{max}} P(r).$$

B. In this problem we shall once more assume that $L = 2.5$, $\sigma = 1.0$, and $\Delta I = .66\frac{2}{3}$, exactly as in Problems III B and IV B.

The substance of the above proof is that the mean number of failure repetitions required for perfect learning is the sum of the probability of failure at each of the possible repetitions. Figure 21, as we have seen, gives the probability of success at each repetition. If each of these values is subtracted from 1.00, we have a series of values giving the probability of failure at each repetition. For example, at $R = 1$, the probability of failure is $1.00 - .00$ or 1.00; that at $R = 4$ is $1.00 - .06$, or .94, and so on. Accordingly we have as the theoretical mean number of repetitions for complete learning of the supposititious series before us:

$$1.00 + .99 + .97 + .94 + .89 + .81 + .70 + .57 + .43$$
$$+ .31 + .21 + .13 + .07 + .04 + .02 + .01 = 8.09 .$$

Thus the mean ordinal number of repetitions-of-test at which failures occur is 8.09, which would be 8.00 except for imperfect decimals.

THEOREM XVI

A. *The mean number of repetitions up to and including the first success, and the mean number of repetitions up to and including the last failure, are increasing functions of the mean number of repetitions required for learning.*

Proof:

Postulate 15 may be regarded as postulating a fluctuation of the threshold such that, on any given trial, the threshold is at ℓ, where the values of ℓ are distributed on a normal curve with mean value at L.

A successful reaction will occur on the rth repetition if and only if $R\Delta\bar{E} > \ell$. The probability that this will occur is equal to the area under the normal curve of ℓ to the left of $R\Delta\bar{E}$. Let us designate this probability by $P_r({}_NR)$ where ${}_NR$ is the mean number of repetitions required for learning. ${}_NR$ is the smallest integer such that:

$$ {}_NR\,\Delta\bar{E} > L . $$

Now let $Q_r({}_NR)$ be the probability that the first success falls on the rth repetition. Then, the following relations follow from basic theorems on probability:

(1) $$ Q_r({}_NR) = P_r({}_NR) \prod_{i=-\infty}^{r-1} [1 - P_i({}_NR)] $$

(2) $$ \sum_{r=-\infty}^{s} Q_r({}_NR) = 1 - \prod_{i=-\infty}^{s} [1 - P_i({}_NR)] $$

(3) $$ \sum_{r=-\infty}^{\infty} Q_r({}_NR) = 1 $$

(4) $$ {}_sR = \sum_{r=-\infty}^{\infty} rQ_r({}_NR) $$

(5) $$ {}_sR = \sum_{s=1}^{\infty} \sum_{r=s}^{\infty} Q_r({}_NR) - \sum_{s=-\infty}^{-1} \sum_{r=-\infty}^{s} Q_r({}_NR) $$

where ${}_sR$ is the mean number of repetitions up to and including the first success.

Now, by definition, $P_r({}_NR)$ is a decreasing function of ${}_NR$, for all values of r. Hence, by (2), $\sum_{r=-\infty}^{s} Q_r({}_NR)$ is a decreasing function of ${}_NR$ for all s. Hence, by (3), $\sum_{r=s}^{\infty} Q_r({}_NR)$ is an increasing function of ${}_NR$, for all s. Hence, by (5), R_1 is an increasing function of ${}_NR$, which proves the first half of the theorem.

Now let $p_r({}_NR)$ be the probability that an unsuccessful reaction will occur on the rth repetition; let $q_r({}_NR)$ be the probability that the last failure falls on the rth repetition; and let ${}_FR$ be the mean number of repetitions up to and including the last failure. Then, corresponding to equations (1) to (4) above, we have:

(6) $$ q_r({}_NR) = p_r({}_NR) \prod_{i=r+1}^{\infty} [1 - p_i({}_NR)] $$

(7) $$ \sum_{r=s}^{\infty} q_r({}_NR) = 1 - \prod_{i=s}^{\infty} 1 - p_i({}_NR) $$

(8) $$ \sum_{r=-\infty}^{\infty} q_r({}_NR) = 1 $$

(9) $$ {}_FR = \sum_{r=-\infty}^{\infty} rq_r({}_NR) . $$

The remainder of the proof follows, *mutatis mutandis*, the proof of the first half of the theorem.

[(1) It is clear that the truth of this theorem, as well as of succeeding theorems, depends in no way upon the normality of the distribution of values of ℓ. Any other reasonable distribution would serve as well.

(2) The appearance of negative values of repetitions is due to the fact that, on the assumption of a normal distribution, the threshold value may at times be negative. This could be avoided by assuming a bounded or asymmetrical distribution, which would not vitiate the theorems stated.]

B. Suppose we take as our examples the learning of two syllables where $L = 2.5$ and $\sigma = 1$, but in the case of one syllable, $\Delta I = .5$, and in the case of the other, $\Delta I = .66\frac{2}{3}$. The case where $\Delta I = .66\frac{2}{3}$ has been worked out in Problems III B, IV B, and V B. Exactly analogous computations have been carried out for the case where $\Delta I = .5$. As a result of these computations we have the following values:

	$\Delta I = .5$	$\Delta I = .66\frac{2}{3}$
Mean number failures preceding complete learning	5.56	8.08
Mean number repetitions to, and including, first success	5.61	7.14
Mean number repetitions to, and including, last failure	6.51	10.00

The theorem states, in effect, that if $8.08 > 5.56$, then $7.14 > 5.61$, and also $10.00 > 6.51$. Since these inequalities are all valid, the theorem finds exemplification.

C. Verification of this theorem is presented by Figure 23, where it may be seen that the larger the mean number of failures preceding final success, the larger the number of repetitions both to the first success and the final failure. Taking Hull's numerical values (*29*) from which Figure 23 was plotted, we may set up a table comparable to that of Part B, above:

	Reaction syllable # 3	Reaction syllable # 5
Mean number of failures preceding complete learning	5.1	7.1
Mean number of repetitions to and including first success	5.4	6.5
Mean number of repetitions to and including last failure	7.8	10.1

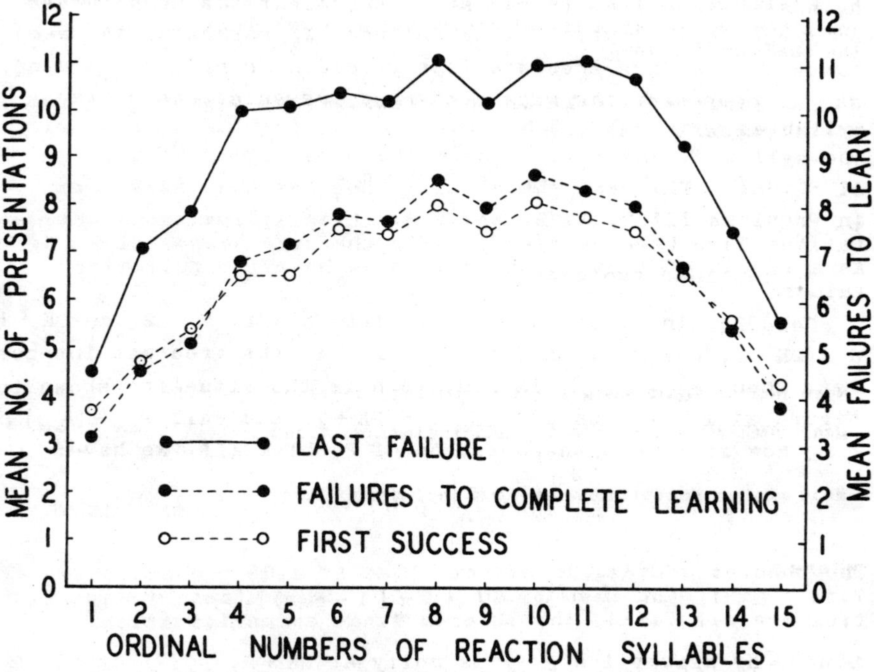

Figure 23. A graphic representation of the relationship at the several syllables of a 15-syllable rote series between the mean number of failures before complete learning, the mean number of repetitions to and including the first success, and the mean number of repetitions to and including the last failure. Adapted from Hull (29).

The theorem states, in effect, that if $7.1 > 5.1$, then $6.5 > 5.4$ and also $10.1 > 7.8$. Accordingly, Theorem XVI finds empirical confirmation.

THEOREM XVI, COROLLARY I

A. *In the learning of a given syllable of a rote series by massed practice, if we let R_a represent the mean number of presentations at the last failure, R_b represent the mean number of failure presentations preceding complete learning, and R_c represent the mean number of presentations at the first success, then:*

(1) $$R_a - R_b \;>\; R_b - R_c\,.$$

Proof:

We consider two cases:

Case I. In Figure 24 I, II, let $a = 1$, $b = 2$, $c = K + 1$, $d = 2K + 1$, $e = R_{max}$, and $f = 2K + 2$. If the greatest integer less than R_{max}, $[R_{max}]$, is odd, we have the situation shown in Figure 24 I. Let us put $[R_{max}] = 2K + 1$, and call $R_{max} - [R_{max}]$, Δ. Now if we use the notation of Problem III, we have:

$$R_a = \sum_{r=1}^{\infty} rQ\,(r)$$

where

$$Q\,(R) = P\,(R) \prod_{r=1}^{R-1} [\,1 - P\,(r)\,]\,.$$

Since, by assumption $P\,(R)$ is unity at $2K + 2$, $Q\,(R)$ at $2K + 2$ is zero and the infinite sum giving R_a becomes finite.

Similarly,

$$R_c = \sum_{r=-\infty}^{\infty} rq\,(r)$$

where

$$q\,(R) = [\,1 - P\,(R)\,] \prod_{r=R+1}^{\infty} P\,(r)\,.$$

Now consider Figure 24 b, in which we have shown the same normal curve shifted Δ units to the left. For this distribution we can calculate a new quantity, R_c', defined similarly to R_c by the equation:

$$R_c' = \sum_{-\infty}^{\infty} rq'\,(r)$$

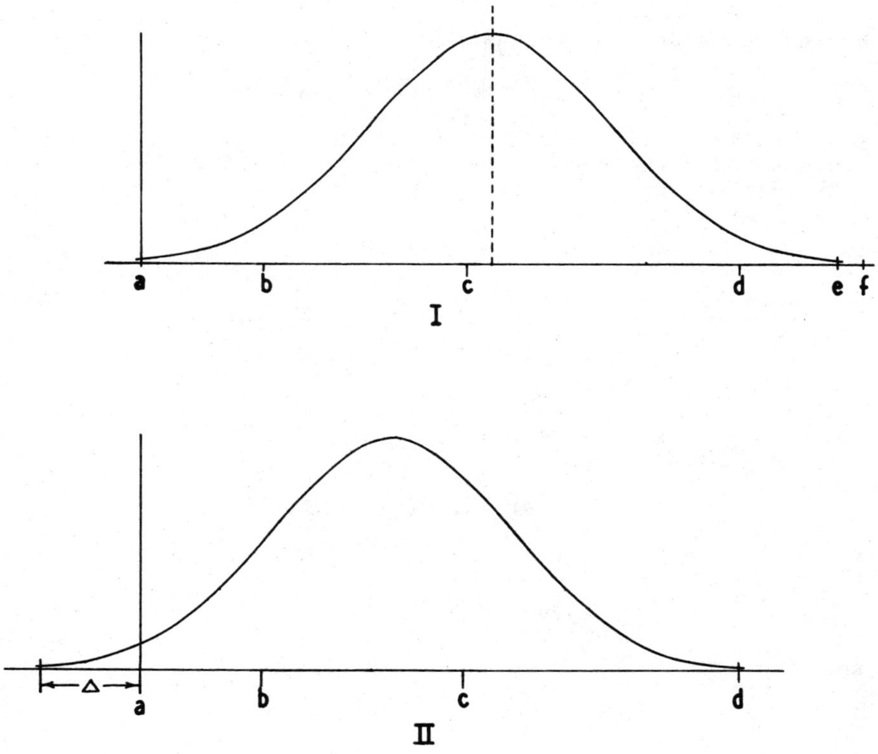

Figure 24. Two diagrams of the normal probability distribution to illustrate the proof of Theorem XVI, Corollary 1.

where

$$q'(R) = [1 - P'(R)] \prod_{r=R+1}^{\infty} P'(r).$$

We note that allowing the lower limit of the summation sign to be minus infinity does not affect the value of R_c'. Of course, the values of $P'(R)$ are calculated as in Problem III. However, it is now obvious that

$$R_c \geq R_c'.$$

We now consider the inequality,

(2) $$R_a - R_b > R_b - R_c'.$$

The corollary follows *a fortiori* from the truth of this, since we have replaced R_c by a smaller quantity. We have only to show that

$$R_a + R_c' > 2R_b.$$

By our choice of $P'(R)$ we now observe that

$$Q(1) = q'([R_{max}] - R)$$

$$Q(2) = q'([R_{max}] - 1)$$

and, in general,

(3) $$Q(R) = q'([R_{max}] - R + 1).$$

Now,

$$R_a = \sum_{r=-\infty}^{\infty} rQ(r)$$

and

$$R_c' = \sum_{r=-\infty}^{\infty} rq'(r).$$

Since the index of summation on the right is a dummy, we replace it by $([R_{max}] - r + 1)$. Then,

$$R_a + R_c' = \sum_{r=-\infty}^{\infty} rQ(r) + \sum_{r=-\infty}^{\infty} ([R_{max}] - r + 1)\, q'(R_{max} - r + 1)$$

$$= \sum_{r=-\infty}^{\infty} [rQ(r) + ([R_{max}] - r + 1)]\, Q(r) \qquad \text{by (3)}$$

$$= \sum_{r=-\infty}^{\infty} ([R_{max}] + 1\, Q(r)$$

$$= ([R_{max}] + 1) \sum_{r=-\infty}^{\infty} Q(r)$$

since $[R_{max}] + 1$ is independent of r. Now since we must have the first reaction success on at least one trial, and since they are mutually exclusive, we have,

$$\sum_{r=-\infty}^{\infty} Q(r) = 1 .$$

Accordingly,

$$R_a + R_c' = [R_{max}] + 1. \tag{4}$$

Notice that this argument is independent of the assumption that $[R_{max}] = 2K + 1$.

Let us now estimate the value of R_b. Consider the areas to the right of b and to the right of d in Figure 24 I. These are $1 - P(2)$ and $1 - P(2K + 1)$ respectively. Obviously,

$$2 - [P(2) + P(2K + 1)] = 1 - \text{Error}.$$

Since there is one K point on each side of the middle line, we can pair all the P's off in this way. Hence,

$$R_b = 1 + K - E.$$

The 1 arises from the fact that we have certain failure on the first trial. The E is the sum of all the errors from the separate pairs. It is obviously less than one half. Now,

$$1 + K = R_b + E,$$

and since,

$$[R_{max}] = 2K + 1,$$

we have from (4),

$$R_a + R_c' = 2(K + 1).$$

Therefore,

$$R_a + R_c' = 2R + 2E,$$

and

$$R_a + R_c' > 2R_b .$$

If $R_{max} = 2K + 2$ the argument still holds, since although E is now zero we can show that $R_c = R_c' + 1$.

In Figure 24 I, II, let $a = 1$, $b = 2$, $c = K + 1$, $d = 2K + 2$, $e = R_{max}$, and $f = 2K + 3$.

Case II. Here $[R_{max}] = 2K + 2$. By exactly the argument as that used before, we have,

$$R_a + R_c' = [R_{max}] + 1 .$$

Our problem is the estimation of R_b. As before, we pair the areas on the ends together and get a sum less than unity. There are more divisions on the right of the middle, L, than on the left. The one nearest the middle is less than one half. Hence,

$$R_b = 1 + K + \frac{1}{2} - E ,$$

where, as before, E is $\leq \frac{1}{2}$. Thus,

$$K + \frac{3}{2} = R_b + E .$$

But,

$$R_a + R_c' = 2K + 3 .$$

$$\therefore \quad R_a + R_c' = 2R_b + 2E .$$

$$\therefore \quad R_a + R_c' > 2R_b .$$

If $R_{max} = 2K + 3$, the argument still holds and by exactly the same reasoning as before.

B. This corollary states, in effect, that the mean number of presentations up to and including that of the last failure exceeds the mean number of failure presentations preceding complete learning by a greater amount than the mean number of failure presentations preceding complete learning exceeds the mean number of presentations up to and including that of the first success. It may be illustrated by the theoretical values presented in Theorem XVI B, where $R_a = 10$, $R_b = 8.08$, and $R_c = 7.14$. Substituting these values in the inequality,

$$R_a - R_b > R_b - R_c ,$$

i.e.,

$${}_FR - {}_NR > {}_NR - {}_SR ,$$

we have,

$$10.0 - 8.08 > 8.08 - 7.14 ,$$

i.e.,

$$1.92 > .94$$

which constitutes a concrete exemplification of the corollary.

C. This corollary is capable of empirical test. Typical empirical evidence is presented graphically in Figure 23. The corollary states, in effect, that the middle curve in Figure 23 is closer to the lower curve than it is to the upper one. That this is true may be seen at a glance. The numerical values from which Figure 23 was plotted are shown in Table 9.

TABLE 9

In this table, n represents the ordinal number of the syllable learned; R_a, the mean ordinal number of the last failure; R_c, the mean ordinal number at the first success; R_b, the mean number of failures preceding "complete" learning; and R_b', the value of the mean number of failures preceding complete learning, computed by the formula,

$$R_b' = \frac{R_a + R_c}{2} - 1.$$

These values are derived from a table published by Hull (29).

n = 1	2	3	4	5	6	7	8	9	10	11	12	13	14	15
R_a = 4.5	7.1	7.8	10.0	10.1	10.4	10.2	11.1	10.2	11.0	11.1	10.7	9.3	7.5	5.6
R_b = 3.1	4.5	5.1	6.8	7.1	7.8	7.7	8.5	8.0	8.7	8.4	8.0	6.8	5.5	3.8
R_c = 3.7	4.7	5.4	6.5	6.5	7.5	7.4	8.0	7.5	8.1	7.8	7.5	6.6	5.6	4.3
R_b' = 3.1	4.9	5.6	7.3	7.3	8.0	7.8	8.6	7.9	8.6	8.5	8.1	7.0	5.6	4.0

Table 9 enables us to substitute a considerable variety of R_a values in the inequality,

$$R_a - R_b > R_b - R_c .$$

If we take the series of R values where $n = 4$, we have,

$$10.0 - 6.8 > 6.8 - 6.5$$

i.e.,

$$3.2 > .3 ,$$

which agrees with the corollary. So far as the present evidence goes, the corollary is confirmed.

While it is clear that this corollary finds empirical confirmation, it is important to note that the same type of inequality as that here observed would also be produced by an asymmetry in the probability distribution of σ such as is suggested in Theorem XIX C and Figures 26 and 27. From the evidence at present available it seems probable that both factors contribute to the inequality observed in Figure 22.

THEOREM XVII

A. *The mean number of repetitions between the first success and the last failure is an increasing function of the mean number of repetitions required for learning.*

Proof:

From the symmetry of the distribution of ℓ about L, we know that:

$$_{L}R - R_m = R_m - {}_{\ell}R .$$

Hence,

$$\frac{_{L}R - {}_{\ell}R}{R_m} = 2 - 2\frac{_{\ell}R}{R_m} .$$

We shall show that $\frac{_{\ell}R}{R_m}$ is a decreasing function of R_m and hence $\frac{_{L}R - {}_{\ell}R}{R_m}$ is an increasing function of R_m, which is a stronger result than that stated in the theorem.

Let R_m be arbitrary, and consider $R'_m = R_m \cdot x$, where $x > 1$. By equation (4) of the proof of Theorem XVI,

$$_{\ell}R = \sum_{r=-\infty}^{\infty} r Q_r (R_m)$$

$$\frac{_{\ell}R'}{x} = \sum_{r=-\infty}^{\infty} \frac{r}{x} Q_r (R_m \cdot x) .$$

But,

$$\sum_{r=-\infty}^{[xs]} Q_r (R_m \cdot x) = 1 - \prod_{i=-\infty}^{[\alpha s]} p_i (R_m \cdot x)$$

$$> 1 - \prod_{i=-\infty}^{s} p_i (R_m) = \sum_{r=-\infty}^{s} Q_r (R_m) .$$

Then, emulating the proof of Theorem XVI, using equation (5) of that proof, we have:

$$\sum_{r=-\infty}^{\infty} \frac{r}{x} Q_r (R_m \cdot x) < \sum_{r=-\infty}^{\infty} r Q_r (R_m) .$$

Hence,

$$\frac{_{\ell}R'}{x} < {}_{\ell}R ,$$

i.e.,

$$\frac{_{\ell}R'}{R'_m} < \frac{_{\ell}R}{R_m} .$$

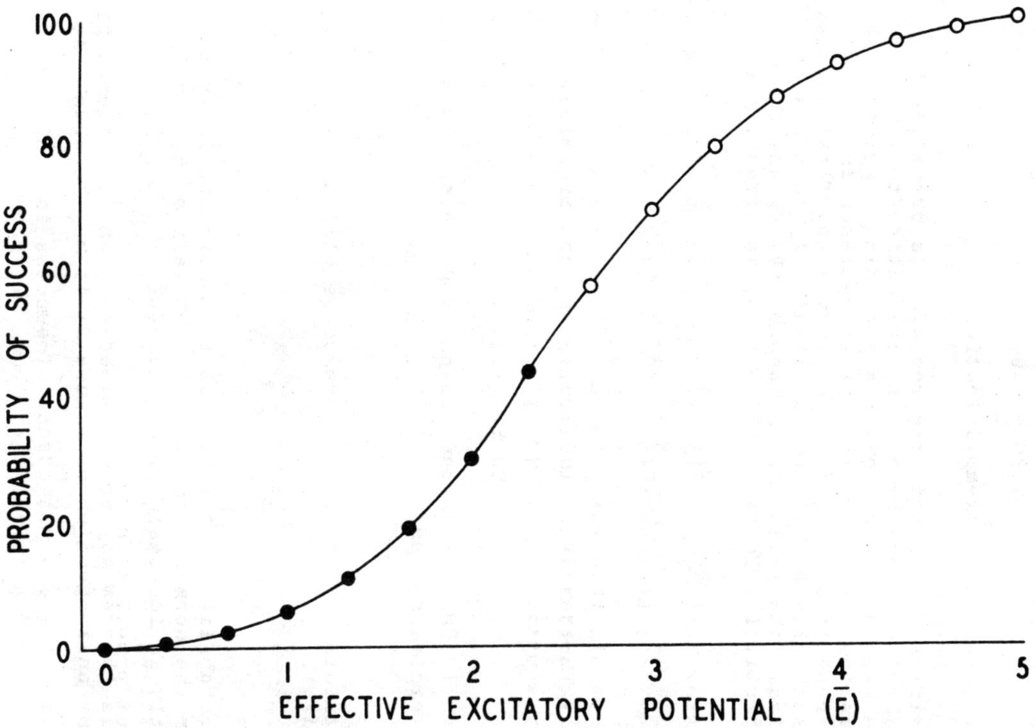

Figure 25. Graphic representation of the curvature of the theoretical probability of recall as a function of effective excitatory potential $\bar{E}$. Each circle represents the probability of a correct recall at each repetition, on the assumption that $\Delta\bar{E} = .33\frac{1}{3}$. Note that this figure is a graphic representation of the probability values which accompany the diagonal row of circles in Figure 21.

B. The theoretical material presented in Theorem XVI B serves here also. This theorem says in effect that if

$$8.08 > 5.56$$

then

$$10.00 - 7.14 > 6.51 - 5.61$$

i.e.,

$$2.86 > .90 .$$

Thus the theorem finds exemplification.

C. Verification of this theorem also is presented by Figure 23, where it may be seen by inspection that the larger the mean number of failures preceding complete learning, the farther apart are the curves representing the number of repetitions required to produce the first success and the last failure. Stated quantitatively in terms of the empirical data presented in Theorem XVI C, where $r = 2$ and 5 respectively, Theorem XVII states, in effect, that if

$$7.1 > 4.5$$

then

$$10.1 - 6.5 > 7.1 - 4.7$$

i.e.,

$$3.6 > 2.4 .$$

Accordingly Theorem XVII finds empirical confirmation.

THEOREM XVIII

A. *Recall (D53) is an increasing function of the mean effective excitatory potential.*

Proof:

By Postulate 14, the correct response will be given if, and only if,

$${}^{R,N}\bar{E}_{n,n+1} > {}^{R}\ell_{n,n+1} .$$

But the probability that ℓ is less than $\bar{E}$ is precisely the area under the normal curve of ℓ to the left of $\bar{E}$, and is consequently an increasing function of $\bar{E}$.

B. This theorem may be explained by means of Figures 21 and 25. By an inspection of Figure 21 it may be seen that

if	$R = 1$,	the	chance	of	success	is	.00
"	$R = 2$,	"	"	"	"	"	.01
"	$R = 3$,	"	"	"	"	"	.03
"	$R = 4$,	"	"	"	"	"	.06
"	$R = 5$,	"	"	"	"	"	.11
"	$R = 6$,	"	"	"	"	"	.19

TABLE 10

This table shows the per cent of successful reactions in the learning of individual syllables at successive presentations. These percentages were derived from all of the syllables which were preceded by 4 failures, 6 failures, and 8 failures respectively before the learning of sixteen 15-syllable series (exclusive of the cue syllable) by each of eight subjects. These data are from unpublished results by Hull (29).

Ordinal number of presentation	Percentage of successful reactions		
	Learning preceded by 4 failures (n = 219)	Learning preceded by 6 failures (n = 205)	Learning preceded by 8 failures (n = 163)
1	0	0	0
2	1	1	0
3	11	5	1
4	38	9	3
5	87	22	7
6	96	42	11
7	93	82	23
8	95	89	47
9	93	88	70
10	99	91	87
11	98	93	86
12	97	95	90
13	98	94	91
14	98	94	96
15	99	94	98
16	99	99	98
17	100	99	98
18	100	99	99
19	99	99	99
20	99	99	100

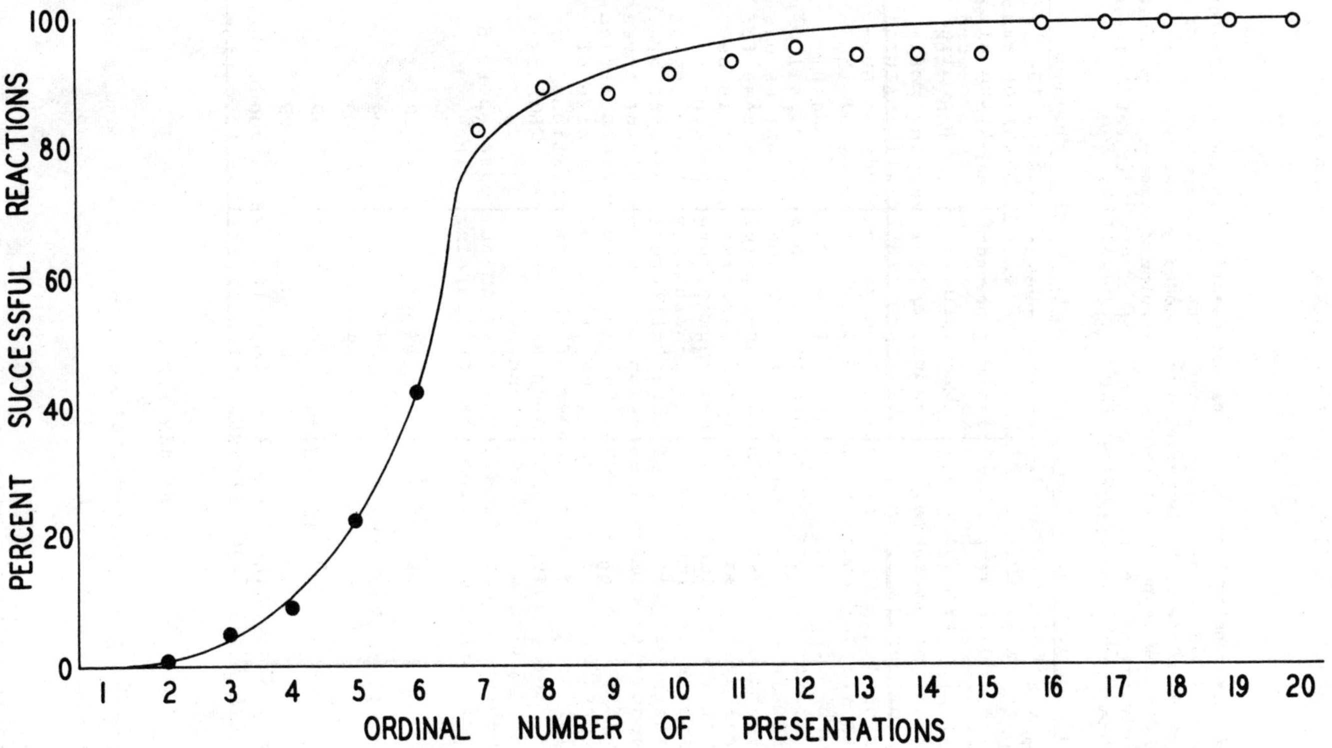

Figure 26. Graphic representation of the empirical per cent of successful reactions at each presentation of 205 separate nonsense syllables at each of which complete learning was preceded by 6 failures of successful reaction. The numerical data from which this figure was constructed are shown in Table 10. Compare with Figure 25.

and so on. Theorem XVIII states, in effect, that since

$$6 > 5 > 4 > 3,$$

and so on, it follows that

$$.19 > .11 > .06 > .03,$$

and so on. Figure 24 shows this increasing relationship graphically. Thus the theorem finds exemplification.

C. A rough empirical test can be made of this theorem. From the original records of the learning of sixteen 15-syllable series by each of eight subjects, a tabulation was made (Table 10) of all the failures of correct reaction at individual syllables where "complete" learning was preceded by a total of 4, 6, and 8 failures respectively. The data from each set of percentages were taken from varying positions in the series, from syllables presumably with naturally differing learning difficulty, from the records of subjects with varying rates of learning for syllables of the same natural difficulty and from records which found the subjects at various stages of practice. In spite of all these variable factors, the rate of learning, except as disturbed by chance, must have been approximately the same for the several syllables in each pool. Accordingly it is believed that the percentage values thus secured present a useful first approximation to the probability of success at various stages of learning of three different rates of effective acquisition. An inspection of any one of the three data columns in Table 10 shows that (despite certain irregularities presumably due to the limitations of the size of the sample) as the ordinal number of the presentations increases, the probability of success increases. This is shown graphically in Figures 26 and 27.

The theorem states, in effect, that in the case where 6 failures of correct reaction precede learning, since

$$6 > 5 > 4 > 3,$$

and so on, it follows that

$$42 > 22 > 9 > 5,$$

and so on. Since this inequality is valid, we conclude that that Theorem XVIII finds presumptive empirical verification.

THEOREM XIX

A. *Recall is a concave function of* $\bar{E}$ *if* $\bar{E} < L$, *and a convex function of* $\bar{E}$ *if* $\bar{E} > L$.

Proof:

The first derivative with respect to $\bar{E}$ of the area under

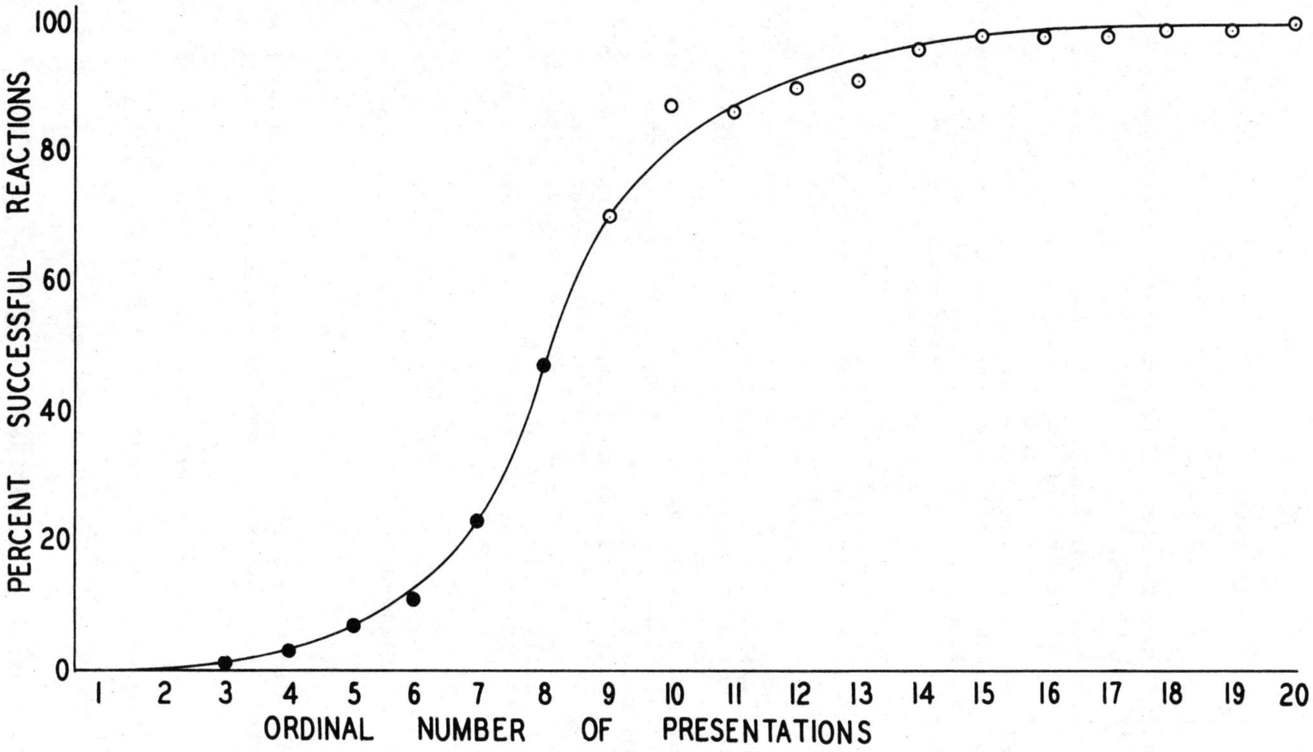

Figure 27. Graphic representation of the empirical per cent of successful reactions at each presentation of 163 separate nonsense syllables at each of which complete learning was preceded by 8 failures of successful reaction. The numerical data from which this figure was constructed are shown in Table 10. Compare with Figure 25.

the normal curve, to the left of $\bar{E}$, is the ordinate of the normal curve at the point, $1 = \bar{E}$. Hence, the second derivative of the area is the slope of the normal curve at the point, $\ell = \bar{E}$. But this is positive if $\bar{E} < L$, and negative if $\bar{E} > L$. Hence, the theorem follows.

B. This theorem may be explained by means of Figures 21 and 25. It will be recalled that in these diagrams, $\bar{E}_L = 2.5$. An inspection of Figure 21 shows the following paired values:

Where	$\bar{E} = 1.33$,	probability	of	success	is	.11
"	$\bar{E} = 1.67$,	"	"	"	"	.19
"	$\bar{E} = 2.00$,	"	"	"	"	.30
"	$\bar{E} = 2.33$,	"	"	"	"	.43
"	$\bar{E} = 2.67$,	"	"	"	"	.57
"	$\bar{E} = 3.00$,	"	"	"	"	.69
"	$\bar{E} = 3.33$,	"	"	"	"	.79
"	$\bar{E} = 3.67$	"	"	"	"	.87

Now let us consider the first three entries in the above table, where the $\bar{E}$ values are all less than 2.5. The theorem states, in effect, that for these values,

$$.42 - .30 > .30 - .19 > .19 - .11$$

i.e.,

$$.13 > .11 > .08$$

which is evidently true, and so the first portion of the theorem finds exemplification.

Turning now to the three entries where $\bar{E}$ is greater than 2.5, the theorem states, in effect, that

$$.87 - .79 < .79 - .69 < .69 - .57$$

i.e.,

$$.08 < .10 < .12$$

which is evidently true and so the second portion of the theorem finds exemplification.

A graphic representation of the opposite nature of the curvature of the series of probability values above and below where $\bar{E}'_L = \bar{E}_L$, i.e., 2.5, is shown in Figure 25, the former being represented as solid circles and the latter as hollow circles.

C. This theorem is susceptible of a rough empirical test, much as was Theorem XVIII. As in Theorem XVIII, the evidence is presented numerically in Table 10 and graphically in Figures 26 and 27. An inspection of the sequence of solid circles in the lower portion of the curve in

Figures 26 and 27 shows clearly the positive acceleration demanded by the theorem. An inspection of the sequence of empty circles in the upper part of the curve shows, in general, a negative acceleration, though here there is considerable irregularity and the negative acceleration is not so marked. The theorem is, however, clearly confirmed to a first approximation in so far as the data cited are relevant.

THEOREM XIX, COROLLARY 1

A. *The rate of change of recall as a curved function of* $\bar{E}$ *is the same at* $\bar{E}'$ *and* $\bar{E}''$ *when* $\bar{E}' - L = L - \bar{E}''$.

Proof:

The recall is the probability that for a given value $\bar{E}_1$ of $\bar{E}$, the reaction will be successful. Hence it is the area under the normal curve to the left of the point whose abscissa is $\bar{E}_1$. The rate of change of the recall as a function of $\bar{E}$ is the derivative of this area with respect to $\bar{E}$. This is, however, for a given value of $\bar{E}_1$ nothing more than the ordinate of the normal curve when the abscissa is $\bar{E}_1$, but points, with the property that $\bar{E}' - L = L - \bar{E}''$, are placed symmetrically about the middle point, L, of the curve. Hence the ordinates of such points will be the same and the corollary follows.

B. This corollary may be illustrated by the theoretical material represented in Figures 21 and 25. It states, in effect, that:

$$ {}^{11,N}\bar{E} - {}^{10,N}\bar{E} = {}^{7,N}\bar{E} - {}^{6,N}\bar{E} $$

i.e., (Figure 21),

$$ .79 - .69 = .30 - .19 $$

$$ .10 = .11 . $$

This is a substantial equality, except for dropped decimals. Accordingly, the corollary finds exemplification.

C. The Hull data presented above in Table 10 and Figures 26 and 27 permit an approximate empirical test of this corollary. The theorem states, in effect, that the upper portions of Figures 26 and 27 should be symmetrical with the lower portions, as is the case with Figure 25. Inspection shows that this is clearly not the case in spite of the fact that the general picture presented by the empirical curves is strikingly similar to that of the theoretical curve. An examination of Table 10 suggests strongly that the asymmetry is more marked in the case of syllables easily learned. We must conclude, then, that this theoretical proposition suffers empirical refutation.

In accordance with recognized scientific practice in cases of the refutation of a theoretical proposition by empirical observation, we must raise the question as to which postulates are indicated as defective by the evidence here cited. It would seem that either the acquisition of $\bar{E}$ has a negative acceleration or the distribution of variability about the central tendency of the reaction threshold (L) is skewed, or both. Accordingly, Postulates 3 and 15 both fall under suspicion and one of the two almost certainly will require recasting. (See Theorem XVI, Corollary 1 C.) If experimental situations could be found in which each of the two factors here under consideration were active singly, a critical solution of this intriguing problem might be expected.

PROBLEM VI

A. *In the learning of rote series, given the central tendency of the variable reaction threshold* (L) *and the probability of recall of a given syllable, of unknown* ΔI, *at the several presentations up to complete learning: it is required to determine the extent of the variability of the reaction threshold* (σ_L) *about its central tendency.*

Proof:

We may assume that the probability of success on a given syllable has been plotted against the ordinal number of the repetition involved. It is noted that the theory has been predicated on the fact that the excitatory potential associated with a repetition of a given syllable is that at the completion of the learning occurring at the repetition. Since the testing at a given repetition discloses the excitatory potential existing at the end of the previous occurrence of the syllable in data obtained experimentally, we must reduce the repetitions used by unity. Let us suppose, therefore, that this has been done. Now the probability of success on a given repetition is the area to the left of the repetitions under a normal curve whose mean is at R_m and whose standard deviation is σ_R. To obtain an approximation of this normal curve from the data given, we use the following procedure. We replace the normal curve by a series of rectangular figures with the property that the area of successive rectangles making up the polygonal curve is equal to the difference between successive ordinates of the original distribution. This scheme has the property that the total area under the polygonal line is unity. Since it is to approximate a normal curve, we shift the abscissa one-half unit to the left. We may now find the mean, R_m, and standard deviation for this distribution by the usual methods. They will both be expressed in units of repetition (R). Their ratio (Z), however, will be a pure number independent of unit. That is, $\frac{\sigma_R}{R_m} = Z$.

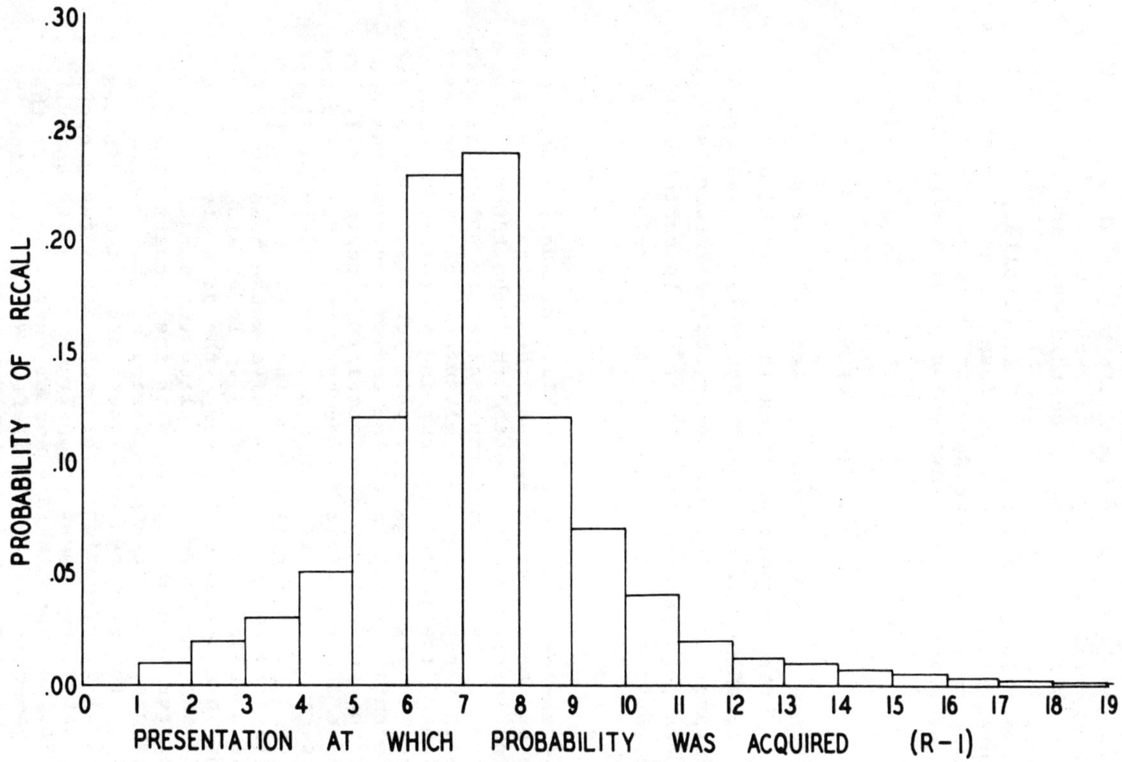

Figure 28. Distribution of probabilities of recall at the several repetitions-of-learning in the learning of 163 nonsense syllables, each showing 8 failures of recall before the criterion of learning was reached. The values shown in this figure were obtained from the smoothed curve shown in Figure 27, by methods explained in the text.

Now,

$$R_m \Delta\bar{E} = E_L$$

and

$$E = \Delta\bar{E}R.$$

The standard deviation, σ_L will be equal to $\sigma_R \Delta\bar{E}$, since it is computed by final differences in E, squaring, averaging, and then subtracting the square root. Hence,

$$\frac{\sigma_R}{R_m} = \frac{\sigma_L}{\bar{E}_L} \doteq Z.$$

Now E_L is known by assumption (Problem I). Hence,

$$\sigma_L \doteq ZE_L .$$

B. We shall take as a concrete example for the illustration of this methodology the data of Problem I, Example A, and the last column of data in Table 10, which are both from the same investigation (*29*). We first plot the data in Table 10 as appears in Figure 27. From the smoothed curve there shown we determine the difference in probability of success at each successive presentation. When plotted, these differences appear as shown in Figure 28.

From the values represented by the rectangular figures of the histogram, both the mean and the standard deviation of the distribution may be computed directly, the standard deviation being $\sigma_L \Delta\bar{E}$, and the mean being $L\Delta\bar{E}$. The values thus computed are:

$$\sigma_L \Delta\bar{E} = 2.53$$

$$L\Delta\bar{E} = 7.33$$

Substituting in the equation,

$$Z = \frac{\sigma_L}{L}$$

we have,

$$Z = \frac{2.53}{7.33} = .35$$

Now, by Problem I, Example A, the value of L for these same data has been determined at 2.32. Substituting this together with the value of Z in the equation,

$$\frac{\sigma_L}{L} = Z$$

we have,

$$\frac{\sigma_L}{2.32} = .35$$

i.e.,

$$\sigma_L = .35 \times 2.32 = .81$$

which is the value sought.

PROBLEM VII

A. *In the learning of rote series, given the mean number of the presentations at the first success and the mean number of the presentations at the last failure, it is required to find the mean number of presentations just necessary to bring the effeotive excitatory potential to the reaction threshold (L).*

Proof:

Suppose we let R_g represent the presentation at which the last increment of effective excitatory potential (ΔE) is added to make up the effective excitatory potential ($\bar{E}$) under consideration, and we also let R_i represent the presentation at which the effective excitatory potential ($\bar{E}$) in question is tested for recall. It is clear from the nature of the process of rote learning that

(1) $$R_i - R_g = 1 .$$

Now by Problems III and IV it is evident that the method of locating the mean repetitions at which the first success occurs is exactly symmetrical with that by which the last failure is located. Moreover, by Postulate 12, the distribution of probability involved is symmetrical. It accordingly follows that

(2) $$\frac{R_a + R_c}{2} = {}_LR_i$$

where ${}_LR_i$ is the central tendency of such probability.

But R_a and R_c are in terms of R_i, whereas the basis for computing $\bar{E}$ must be in terms of R_g. We accordingly substitute (2) in (1), which gives us,

$$\frac{R_a + R_c}{2} - {}_LR_g = 1$$

i.e.,

(3) $${}_LR_g = \frac{R_a + R_c}{2} - 1 .$$

But by Postulate 16, Corollary 1,

(4) $${}_{L}R_{g}\ \Delta\bar{E} = {}_{L}\bar{E}$$

Substituting (3) in (4), we have,

(5) $$\left(\frac{R_a + R_c}{2} - 1\right) \Delta\bar{E} = {}_{L}\bar{E}\ .$$

B . Equation (3) above means, in effect, that the number of repetitions just necessary to reach the reaction threshold is always one less than the arithmetical average of the mean presentation at the first success and the mean presentation at the last failure. The simplest way to illustrate this is by reference to Figure 22, which is not complicated by variability in the value of the reaction threshold; i.e., $\sigma = 0$. There it may be observed by simple inspection that $R_a = 3$ and $R_c = 4$. Substituting in (5) we have (recalling that in this figure, $\Delta\bar{E} = 1$),

$${}_{L}\bar{E} = \left(\frac{3 + 4}{2} - 1\right) 1$$

i.e.,

$${}_{L}\bar{E} = 3.5 - 1 = 2.5 = {}_{L}R$$

as may be seen to be the case.

A second example and one involving a variable reaction threshold is found in Figure 21, taken in conjunction with Problems III and IV. Problem III gives $R_a = 10.0$, and Problem IV gives $R_c = 7.0 \pm$. Substituting these values in the equation of (5), we have,

$${}_{L}\bar{E} = \left(\frac{10 + 7}{2} - 1\right) \frac{1}{3}$$

$$= \frac{(8.5 - 1)}{3}$$

$$= \frac{7.5}{3}$$

$$= 2.5 = {}_{L}R$$

exactly as shown in Figure 1. Note that the presentation-of-test at which the line of circles crosses L is at 8.5, as the above computation shows should be the case.

C. The application of the procedure outlined above for determining the number of recall failures which are required to cross the central tendency of the reaction threshold (${}_{L}\bar{E}_{g}$) is complicated by the fact that in order to secure enough data to give a stable basis for computation it is always necessary either to pool results from numerous subjects of unknown characteristics with respect to ΔE, ΔK, L,

σ_L, or at the very least to pool results from the same person at different stages of practice. At the present time practically all of the experimental data available involve both the above types of pooling.

The complexity of the above considerations makes it impossible at present to derive with complete rigor the exact theoretical status of such necessarily pooled empirical data because we cannot be sure of the extent to which the distribution of chance variability about the composite $_L\bar{E}$ will be symmetrical. However, such evidence as appears in Table 10 and Figures 26, 27, and 28 shows that pooled data of this kind present a fair approximation to symmetry. We accordingly proceed on this working hypothesis for the purposes of the present preliminary exploration. The last row of values in Table 9 presents a whole series of threshold values computed from empirical data by the formula employed above. These computed threshold values approach very closely the empirical mean number of failures preceding success, though usually they are a little in excess. Theoretically they should be in excess by exactly .5 of a repetition (see Problem VIII B).

PROBLEM VIII

A. *In the learning of rote series, given the mean number of failures to make correct reaction preceding and including the last failure, it is required to find the mean number of repetitions just necessary to bring the effective excitatory potential to the reaction threshold (*L*).*

Proof:

From Problem VII we have,

(1) $$R_i - R_g = 1.$$

If R_b is the mean number of failures before complete success, R_{max} the first repetition for which the probability of failure is zero, and R_d the mean number of successes up to and including R_{max}, then assuming complete symmetry of the distribution, we have,

(2) $$R_b = R_d.$$

However, $R_b + R_d$ = the total number of trials. That is,

(3) $$R_b + R_d = R_{max}.$$

From (2) and (3) we know that,

(4) $$R_b = \frac{R_{max}}{2}.$$

Now the central tendency of the distribution curve is ${}_{L}R_i$. Therefore,

$$(5) \qquad {}_{L}R_i = \frac{R_{max} + 1}{2} .$$

Therefore (6) $${}_{L}R_i = R_b + \frac{1}{2} .$$

But by (1) $${}_{L}R_g = {}_{L}R_i - 1 .$$

Substituting (6) in (1), we have,

$$(7) \qquad {}_{L}R_g = R_b - \frac{1}{2} .$$

Multiplying each side of equation (7) by $\Delta\bar{E}$,

$${}_{L}R_g \,\Delta\bar{E} = (R_b - .5)\,\Delta\bar{E}$$

i.e., (8) $${}_{L}\bar{E}_g = (R_b - .5)\,\Delta\bar{E}$$

B. The substance of equation (7) above is that if .5 be subtracted from the number of failures required for complete learning, the result will be the number of presentations required to raise the effective excitatory potential just to the reaction threshold. This principle may be illustrated most simply by means of Figure 22, where there is no complication by variability in the reaction threshold, i.e., $\sigma_L = 0$. There it may be seen that the number of failures preceding learning (R_b) is 3. Substituting this in place of R_b in equation (8), we have (recalling that here $\Delta\bar{E}$ is 1),

$${}_{L}\bar{E} = (3 - .5)\,1$$
$$= 2.5$$

as Figure 22 shows to be the case.

A second example, this one involving variability in the reaction threshold, is seen in Figure 21, taken in conjunction with Problem V. By Problem V it was shown that the situation presented in Figure 21 should yield 8 ± failures preceding and including the final failure before learning, i.e., in the present situation $R_b = 8$. Substituting appropriately in equation (8), we have,

$${}_{L}\bar{E} = (8 - .5)\frac{1}{3}$$
$$= \frac{7.5}{3}$$
$$= 2.5,$$

exactly as shown in Figure 21.

C. The considerations concerning the application of the above methodology to pooled empirical data for the purposes of preliminary exploration are the same as those outlined in Problem VII C.

THEOREM XX

A. *At the conclusion of the learning of any rote series by massed practice there will be a finite amount of inhibitory potential (I_n) concurrent with every syllable-exposure interval during which a correct reaction may begin.*

Proof:

By D5 and D38, the period during which a correct reaction may begin is that of the syllable-exposure interval whose number is one less than the number of the syllable reaction in question. By Postulate 4, at every repetition of a rote series there is left a finite amount of inhibitory potential active during the stimulus-trace segment whose number is one less than that of the reaction in question. It follows that after any repetition of a rote series, there will be a finite amount of inhibitory potential concurrent with every syllable-exposure interval during which a correct reaction may begin. But all inhibitory potentials associated with a given stimulus trace segment and resulting from successive repetitions are positive, and under conditions of massed practice summate arithmetically (Postulate 8). From these considerations the theorem follows.

B. This theorem is illustrated by Table 1 and Figure 17, where it may be seen in the case of a 15-syllable series learned by massed practice that, theoretically, each syllable has an appreciable amount of inhibition, as stated in the theorem.

THEOREM XX, COROLLARY 1

A. *In rote series learned by massed practice, the mean number of repetitions required on the average to reach the reaction threshold of any syllable is greater than the number of repetitions required if there is no inhibition present.*

Proof:

By Postulate 16, Corollary 1:

$$_{\mathrm{L}}R = \frac{\mathrm{L}}{\Delta \mathrm{E} - \Delta \mathrm{I}} .$$

But, by Theorem XX,

$$\Delta \mathrm{I} > 0 .$$

$$\therefore \quad _{\mathrm{L}}R = \frac{\mathrm{L}}{\Delta \mathrm{E} - \Delta \mathrm{I}} > \frac{\mathrm{L}}{\Delta \mathrm{E}}$$

and $\frac{L}{\Delta E}$ is precisely the number of repetitions necessary on the average to reach the reaction threshold when there is no inhibition.

B. This corollary may be illustrated by values obtained from the concrete analysis of Problem I, Example A, where L equals 2.32. If, now, there should be no inhibition involved in the learning, i.e., if $\Delta I = 0$ and ΔE is the unit of measurement, the equation (Postulate 16, Corollary 1),

$$_{L}R = \frac{L}{\Delta E - \Delta I}$$

becomes

$$_{L}R = \frac{L}{1 - 0}$$

$$= L$$

which means that at all syllable positions in the rote series the number of repetitions required for learning will be the same, i.e., L. In the case of the first 15-syllable series of Problem I, Example A, this would be 2.32 repetitions. Table 2 and Figure 18 show that, theoretically, when learned by massed practice every syllable requires appreciably more than 2.32 repetitions to reach the threshold of recall, exactly as stated in the corollary.

THEOREM XXI

A. *In rote series learned by massed practice the amount of inhibition increases progressively from each end of the series toward a maximum.*

To prove:

As n goes from 1 to N, that J_n increases steadily to a maximum and then decreases steadily. To prove this we show that $\Delta'_{n}J_n$ begins by being positive and then becomes and *remains* negative.

The proof consists of two parts:

(A) $\Delta' J_n$ is positive for $n' \leq \frac{N}{2}$.

(B) $\Delta' J_n$ decreases steadily for $n' < \frac{N}{2}$, reaching a final value $\Delta' J_{N-1}$ which is negative.

Proof A:

We write the formula for $\Delta'_{n}J_n$ (Theorem V, Corollary 1)

$$\Delta' J_n = (n' + 1) \left[\frac{1}{(n' + 1) F^{n'}} + \cdot \cdot \cdot \cdot \cdot \cdot + \frac{1}{NF^{N-1}}\right]$$

$$+ (N - n') \left[\frac{1}{(N - n' + 1) F^{N-n'}} + \cdot \cdot \cdot \cdot \cdot \cdot + \frac{1}{NF^{N-1}}\right]$$

$$- \frac{1}{F^{N-n'}} - \cdots\cdots - \frac{1}{F^{N-1}}$$

We note that if $n' \leq \frac{N}{2}$, then certainly $2n' \leq N$ and $n' \leq N - n$. Hence there are more terms (one for each of $\frac{1}{F^{n'}}, \frac{1}{F^{n'+1}}, \ldots, \frac{1}{F^{N-1}}$) in the first line than in either the second or third lines (one for each of $\frac{1}{F^{N-n'}}, \ldots, \frac{1}{F^{N-1}}$). Consequently, $\Delta' J_n$ is the sum of positive in $\frac{1}{F^{n'}}, \ldots, \frac{1}{F^{N-n'-1}}$ from the first line, and other terms in $\frac{1}{F^{N-n'}}, \ldots, \frac{1}{F^{N-1}}$ one from each line. But the sum of the three terms involving $\frac{1}{F^{C-1}}$ is,

$$\frac{n'+1}{CF^{C-1}} + \frac{N-n'}{CF^{C-1}} - \frac{1}{F^{C-1}} = \frac{N+1-C}{CF^{C-1}},$$

and this is positive since C is at most N. Consequently, for $n' \leq \frac{N}{2}$, $\Delta' J_n$ must be positive.

Proof B:

For $n' > \frac{N}{2}$, from Theorem V, Corollary 2,

$$\Delta_2' J_n = -\frac{1}{F^{n'}} + G(N-n'-1) - G(n'+1).$$

But for $n' > \frac{N}{2}$, we have $n' > N - n'$, and certainly $n' + 1 > N - n' - 1$. Hence, $G(n'+1)$ is greater than $G(N-n'-1)$ and consequently $\Delta_2' J_n$ is negative. Hence, $\Delta' J_n$ decreases steadily for $n' > \frac{N}{2}$. The final value for $\Delta' J_n$ which is $\Delta' J_{N-1}$ is,

$$\Delta' J_{N-1} = N\left(\frac{1}{NF^{n'-1}}\right) + 1\left[\frac{1}{2F} + \frac{1}{3F^2} + \cdots\cdots + \frac{1}{NF^{N-1}}\right]$$

$$- \frac{1}{F} - \frac{1}{F^2} - \cdots\cdots - \frac{1}{F^{N-1}}$$

$$= -\frac{1}{2F} - \frac{2}{3F^2} - \cdots\cdots - \frac{N-2}{(N-1)F^{N-2}} - \frac{N-1}{NF^{N-1}} + \frac{1}{F^{N-1}}.$$

The last three terms may be combined thus:

$$-\frac{N-2}{(N-1)F^{N-2}} - \frac{N-1}{NF^{N-1}} + \frac{1}{F^{N-1}} = \frac{-(N-2)NF - (N-1)^2 + N(N-1)}{(N-1)NF^{N-1}}$$

$$= \frac{-N(N-2)F + N - 1}{(N-1)NF^{N-1}}.$$

Now, by Postulate 5, F must have a value greater than 1, since the terminal inhibitions decrease with remoteness. Hence, as $N > N - 1$, and $(N - 2)F > 1$, whenever N is greater than 2 we have $N(N - 2)F > N - 1$ and consequently the above fraction is negative. Hence, for $N > 2$, $\Delta' J_{N-1}$ is negative. This completes the proof of Part B for $N > 2$.

[*Exceptions:* F or $N = 1$, J_n has only a single value, and we cannot properly speak of a maximum F or $N = 2$; we have $J_1 = 1 + \frac{1}{2F}$ and $J_2 = 1 + \frac{1}{F}$. Here J_2 is the greater.]

B. By Postulate 9, Corollaries 1 and 2 together with D76, the inhibition opposing any given syllable in the learning of a rote series is given by the expression,

$$I_n = R\,J_n \Delta K . \tag{1}$$

Now, since at any given integral number of presentations of the series as a whole, both the R and the ΔK in (1) are constants, it follows that J is the critical value involved in the differences in inhibition among the several syllables of a series, i.e., J_n becomes a true index of the relative amount of inhibition at the several syllable positions in a rote series.

From the preceding considerations it comes about that the present theorem finds adequate illustration in Table 1. Consider the 15-syllable series where F = 1.37 and the corresponding (upper) curve in Figure 17. There it may be seen at a glance that the inhibition increases progressively from each end toward a maximum, which falls somewhere between the 8th and the 9th syllables.

THEOREM XXI, COROLLARY 1

A. *In rote series learned by massed practice, the mean number of failures of correct reaction preceding complete learning increases from each end toward a maximum.*

Proof:

This proposition follows directly from Theorems XXI and VI, and Problem VIII.

C. That the interior of serial lists is more difficult to learn than the end portions was first reported by Ebbinghaus (*8*). His results, based on the number of promptings required after the first exposure, are presented in Table 11.

More recent data on the effect of serial position upon the number of repetitions required for learning will be found in studies by Hull (*29*), Hovland (*22*), and others. Figure 29 gives data based on lists of various lengths from data of Hovland (*26*).

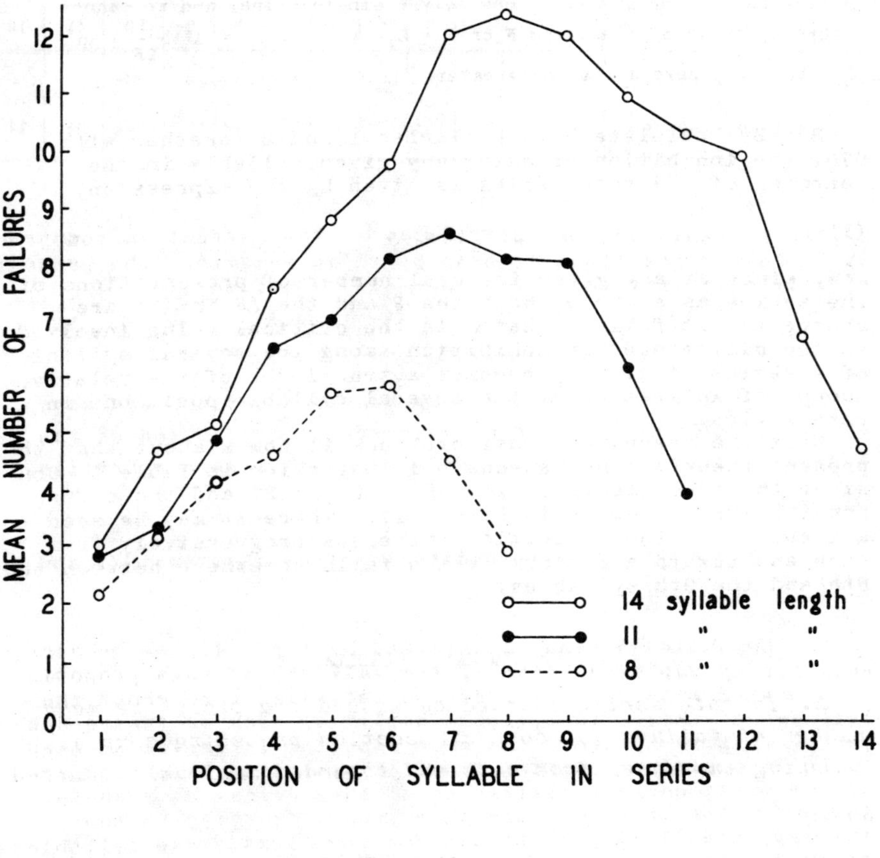

Figure 29. Mean number of failures in the various syllable positions during the learning of 8-unit, 11-unit, and 14-unit lists to a criterion of one perfect recitation. Data from Hovland (26).

TABLE 11

Number of promptings required in various serial positions. Data from Ebbinghaus (7).

	Serial Position											
	1	2	3	4	5	6	7	8	9	10	11	12
10-syllable series	0	3	6	9	23	24	31½	25	23	5½		
12-syllable series	0	11	21	13½	35	36	36	29½	43	37½	34	11

The markedly greater difficulty of the central as compared with the end positions is seen by these results. The present corollary is therefore experimentally confirmed.

THEOREM XXI, COROLLARY 2

A. *In rote series learned by massed practice, the mean reaction latency at the first successful recitation of the series as a whole increases from each end toward a maximum.*

Proof:

This proposition follows directly from Theorems XXI and VIII.

C. Unpublished results obtained by Ward (*64*) serve very well for an empirical test of the validity of this proposition. Ward measured the reaction latencies throughout the learning of twelve 12-syllable series by each of twelve subjects. Figure 30 shows the mean reaction latencies at each syllable position computed from the reactions which occurred at the first correct recitation of each series as a whole. An inspection of this graph shows that, contrary to the theorem, the latency decreases for the first three syllables of the series. After the third syllable, however, the latency rises steadily until a maximum is reached (at the 8th syllable) substantially as the corollary states. The picture rather looks as if some factor not taken into consideration in the theory has entered to over-ride the factors explicitly recognized. The evidence shown by Figure 30 and other unpublished confirming data of a similar nature obtained by Ward (*64*) lead to the conclusion that the corollary finds a distinctly limited or qualified confirmation.

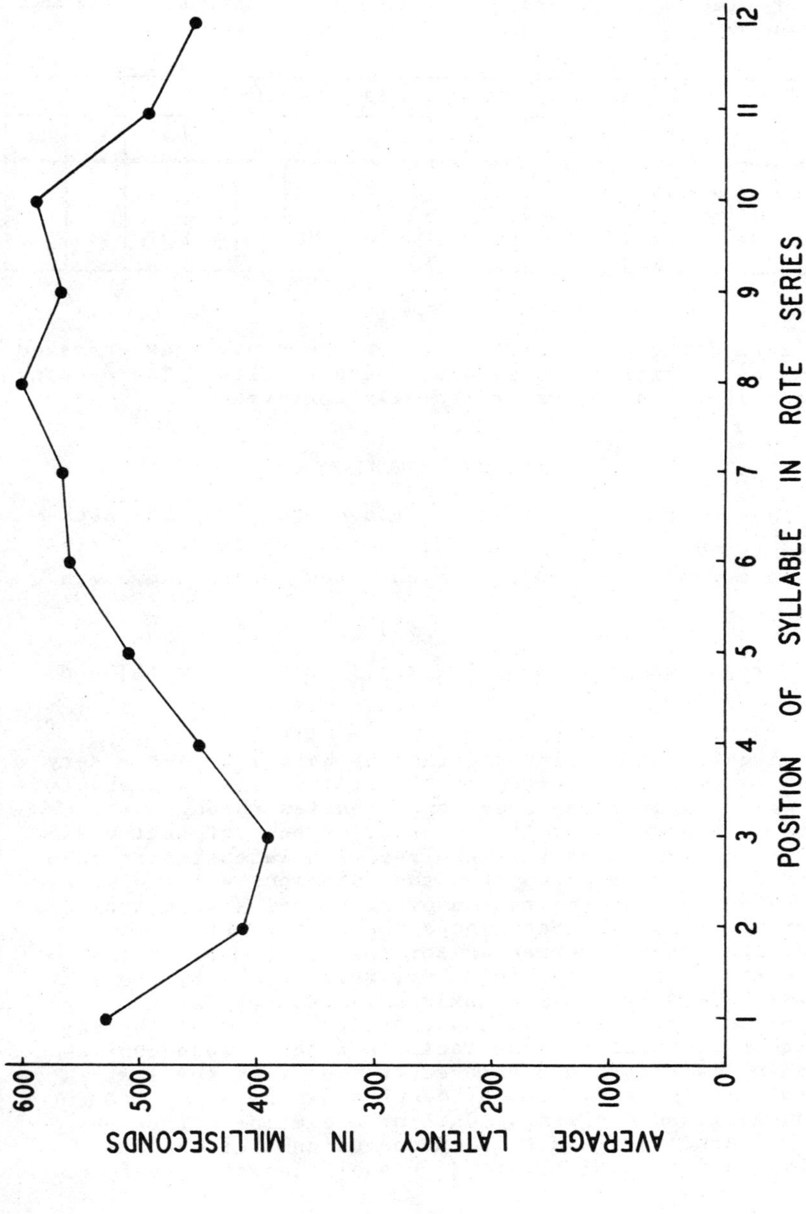

Figure 30. The mean reaction latency at the first perfect recall of 12-syllable rote series as a whole. Adapted from Ward (63).

THEOREM XXI, COROLLARY 3

A. *In the learning of rote series, the advantage of well distributed practice over massed practice, in terms of the mean number of failures of correct reaction preceding complete learning, increases from each end toward a maximum.*

Proof:

This proposition follows directly from Theorems XXI, VI, XIV, and Problem VIII.

C. The experiments of Hovland (*22,26*) and Patten (*46*) have demonstrated that the advantage of distribution of practice in terms of repetitions is greater in the central portions than at the end syllables. Data on this point from Hovland (*22*) are presented in Table 12. If the first presentation of the list is counted as a failure, which is the

TABLE 12

Mean number of failures to respond correctly in different serial positions during learning to mastery by massed and distributed practice. After Hovland (*22*).

	Position of syllable in series											
	1	2	3	4	5	6	7	8	9	10	11	12
Massed practice	1.05	1.77	3.86	6.15	7.36	8.40	10.45	9.41	8.60	6.45	5.41	3.61
Distributed practice	0.96	1.31	2.74	3.90	5.75	6.48	7.23	6.87	5.81	5.32	4.67	3.24
Advantage of distributed over massed practice	0.09	0.46	1.12	2.25	1.61	1.92	3.22	2.54	2.79	1.13	0.74	0.37

convention adopted in the theory, one unit must be added to each value in the first two rows of data in the table.

The increased advantage of distributed practice in the central as compared with the end portions of the list gives experimental verification of this corollary.

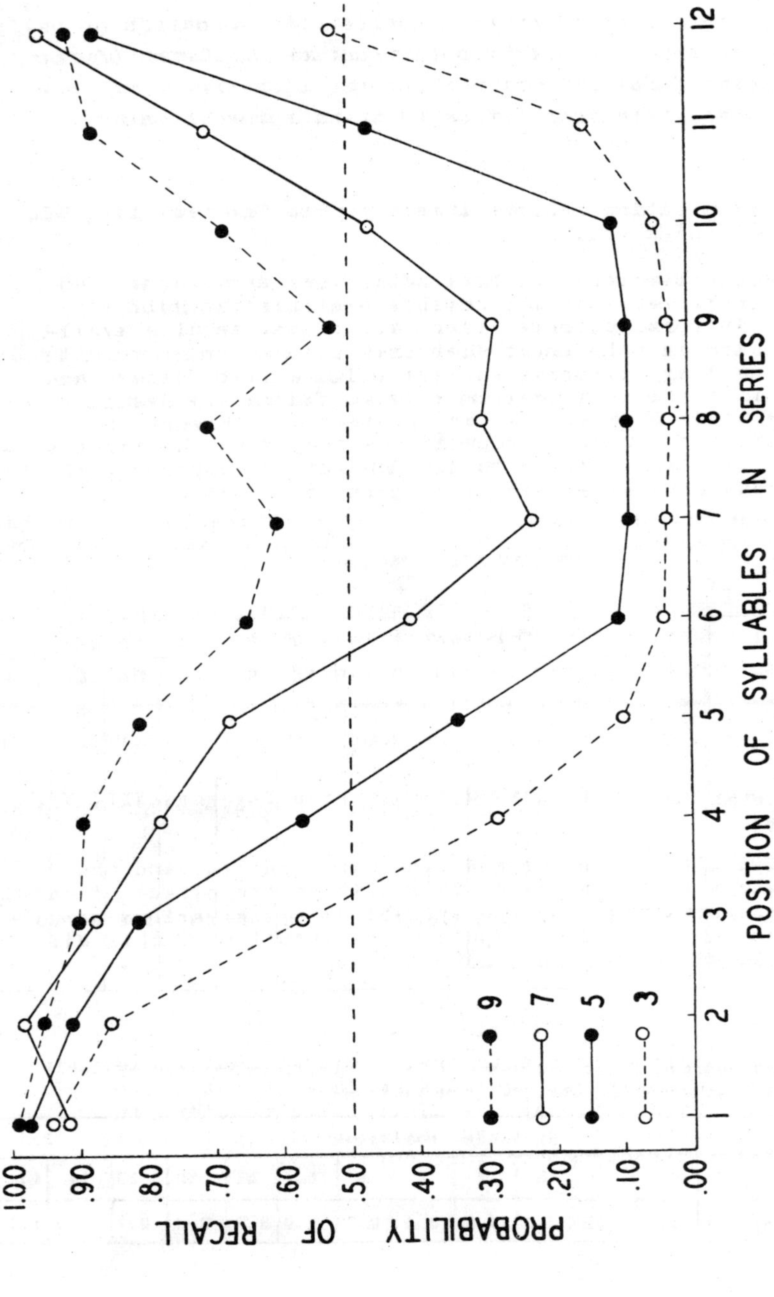

Figure 31. Diagram showing the empirical probability of successful recall at the several syllables of a rote series at four stages in the learning process - 3, 5, 7, and 9 correct reactions. Adapted from Ward (63).

THEOREM XXI, COROLLARY 4

A. *In rote series learned by massed practice, the mean number of repetitions up to and including the first success, and the mean number of repetitions up to and including the last failure, increase from each end toward a maximum.*

Proof:

This proposition follows directly from Theorems XXI, VI, and XVI.

C. Results bearing on this point were first presented by Hull (*27*). More recent results are given by Hull (*29*), Hovland (*22*), and Shipley (*58*). All of the results available to date on this point show that the number of repetitions preceding the first success and the last failure are greater in the central region than at the ends. Results from the study of Hull (*29*) are presented in Figure 23.

The greater number of repetitions preceding the first success and the last failure in the central as compared with the end regions give experimental support to the corollary.

THEOREM XXI, COROLLARY 5

A. *In rote series learned by massed practice, the difference in mean number of presentations up to and including the first success and up to and including the last failure, increases from each end toward a maximum.*

Proof:

This proposition follows directly from Theorems XXI, VI, and XVII.

C. The differences between the first success and the last failure in the study of Hull (*29*) on the effect of caffeine, are presented for the syllables in the various serial positions in Table 13. Graphic representation of this difference is given in Figure 23.

TABLE 13

Difference between first success and last failure in various syllable positions. After Hull, derived from Table 9.

Syllable positions														
1	2	3	4	5	6	7	8	9	10	11	12	13	14	15
.8	2.4	2.4	3.5	3.6	2.9	2.8	3.1	2.7	2.9	3.3	3.2	2.7	1.9	1.3

The increased difference between the first success and last failure in the central as compared with the end positions supports the present corollary.

THEOREM XXI, COROLLARY 6

A. *At any incomplete stage of the learning of a rote series by massed or distributed practice, the probability of recall decreases from each end toward a minimum.*

Proof:

This proposition follows directly from Theorem XXI, Postulate 3, and Theorem XVIII.

C. The principal studies bearing on this point are those of Robinson and Brown (*53*) and of Ward (*63*), shown respectively in Figures 31 and 32. Their results indicate that at all stages of learning the recall scores are lower in the central than in the end regions. Support for these conclusions is found in the experiments of Foucault (*12*), Hovland (*22,26*) and Patten (*46*). Raffel (*50*) has given verification of these results with the anticipation method, but has shown that other ways of testing recall give different results.

The corollary is confirmed by experimental evidence.

THEOREM XXII

A. *In rote series learned by massed practice there will be a greater aggregate amount of inhibition at the posterior syllable than at the anterior syllable.*

To prove:

$$J_{N-1} > J_0$$

Proof:

$$J_0 = G(N) = 1 + \frac{1}{2F} + \frac{1}{3F^2} + \cdot \cdot \cdot \cdot \cdot \cdot + \frac{1}{NF^{N-1}} .$$

$$J_{N-1} = \frac{F-1}{F^{N-1}(F-1)} = 1 + \frac{1}{F} + \frac{1}{F^2} + \cdot \cdot \cdot \cdot \cdot \cdot + \frac{1}{F^{N-1}} .$$

Since each term beyond the first in the sum for J_{N-1} is greater than the corresponding term for J_0, it follows that $J_{N-1} > J_0$.

B. The meaning of this theorem will become clear from a glance at Figure 17 and Table 1. From Table 1 it may be seen that in a 15-syllable series where F = 1.37, $J_{14} = 3.670$ and $J_0 = 1.792$. But since 3.670 > 1.792, it is evident that $I_{14} > I_0$ as the theorem states.

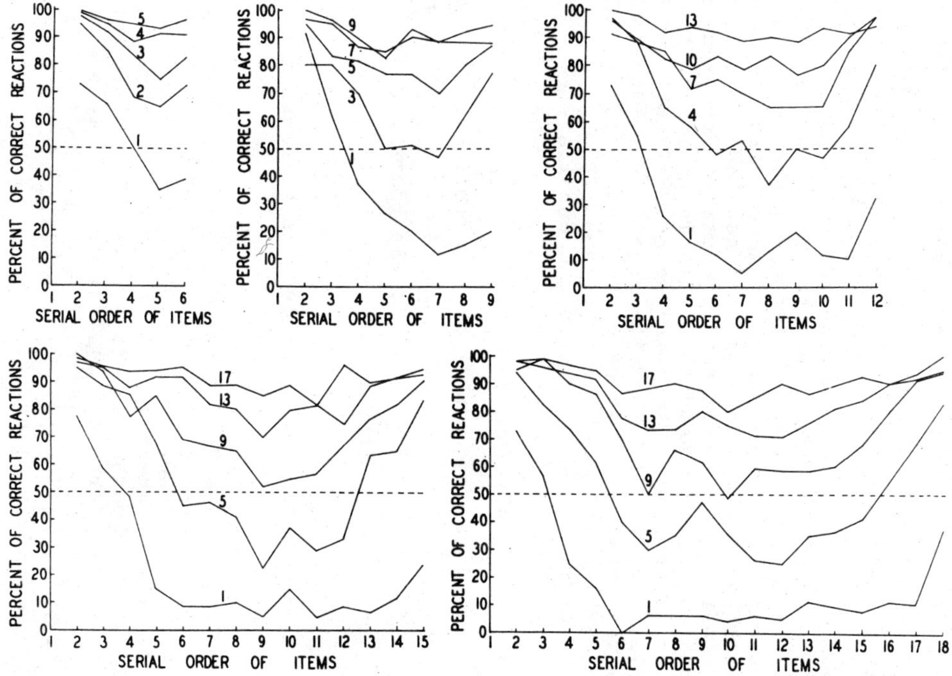

Figure 32. Series of graphs showing the probability of recall at the several syllable positions of rote series of different lengths at various stages of the learning process. The length of series is shown by the last entry in the "Serial Order of Items." The numbers on the curves indicate the number of repetitions preceding the recall score represented by the curves in question. After Robinson and Brown (53).

THEOREM XXII, COROLLARY 1

A. *In rote series learned by massed practice, the mean number of failures of correct reaction preceding complete learning will be greater at the posterior syllable than at the anterior syllable.*

Proof:

This proposition follows directly from Theorems XXII and VI, and Problem VIII.

C. This general problem is one which has frequently been discussed in literature on rote learning as the problem of primacy and recency. In a large number of studies particular methods of testing for recall made the posterior syllable higher in recall, since it was tested with a shorter interval between presentation and recall. Ebbinghaus' results, quoted above under Theorem XXI (Table 11) did not have this difficulty and clearly showed that more promptings were required in the posterior than the anterior syllable position. Some recent results of Hovland (*26*) presented below (Table 18, under Theorem XXVI, Corollary 1) show the same effect. Other data supporting the corollary will be found in Hull (*25*) (cf. Figure 23), Patten (*46*), Lepley (*37*), and Hovland (*22*).
The experimental results confirm the corollary.

THEOREM XXII, COROLLARY 2

A. *In rote series learned by massed practice, the mean reaction latency at the first successful recitation of the series as a whole will be greater at the posterior syllable than at the anterior syllable.*

Proof:

This proposition follows directly from Theorems XXII and VIII.

C. Evidence bearing on the validity of this proposition is found in Ward (*63*). A part of this evidence is presented graphically in Figure 30. Syllable 1 showed a mean latency of 528 milliseconds, whereas syllable 12 showed a mean latency of 455 milliseconds. It has been suggested above (under Theorem XXI, Corollary 2) that other factors are probably present which over-ride those responsible for the primacy-recency relationship. On the basis of the present evidence the corollary is refuted.

THEOREM XXII, COROLLARY 3

A. *In the learning of rote series the advantage of well-distributed practice over massed practice in terms of the mean number of failures of correct reaction preceding complete learning will be greater at the posterior syllable than at the anterior syllable.*

Proof:

This proposition follows directly from Theorems XXII, VI, and XIV, and Problem VIII.

C. In Hovland's experiments (*22,26*) on reminiscence and distribution of practice, data are given on the number of failures in each serial position preceding mastery with massed and distributed practice. These results show that the advantage of distributed practice over massed is greater at the posterior than at the anterior syllable, as predicted. Figure 33, reproduced from Hovland's article (*22*), shows this effect. If the first presentation is to be counted as a failure, one unit must be added to each value given. Substantially similar results are reported by Patten (*46*). In his study, 1.11 errors were made in the first serial position before complete learning with massed practice, and 1.15 with distributed practice, whereas in the posterior syllable position (16th syllable) the corresponding number of errors was 3.97 for massed and 3.11 for distributed. The advantage of distributed practice is thus -.04 syllables in the anterior position and +.86 in the posterior.

Present results furnish confirmation of the corollary.

THEOREM XXII, COROLLARY 4

A. *In rote series learned by massed practice, the mean number of presentations up to and including the first success and up to and including the last failure, is greater at the posterior syllable than at the anterior syllable.*

Proof:

This proposition follows directly from Theorems XXII, VI, and XVI.

C. Figure 23 reproduced from Hull's study of caffeine effects shows that the mean number of repetitions up to and including the first success and the last failure is greater at the posterior than at the anterior syllable. Confirmation of these results was obtained in a study by Hovland (*22*) on distribution of practice ; 1.50 presentations were required before the first success in the first syllable position, and 4.05 in the last portion (syllable 12). Corresponding figures for the number of presentations before the last failure are 2.99 and 6.01 for the anterior and posterior syllables respectively. The first presentation is included as a failure.

The available evidence is in support of the corollary in showing that the number of repetitions to the first success and last failure is greater at the posterior than the anterior position.

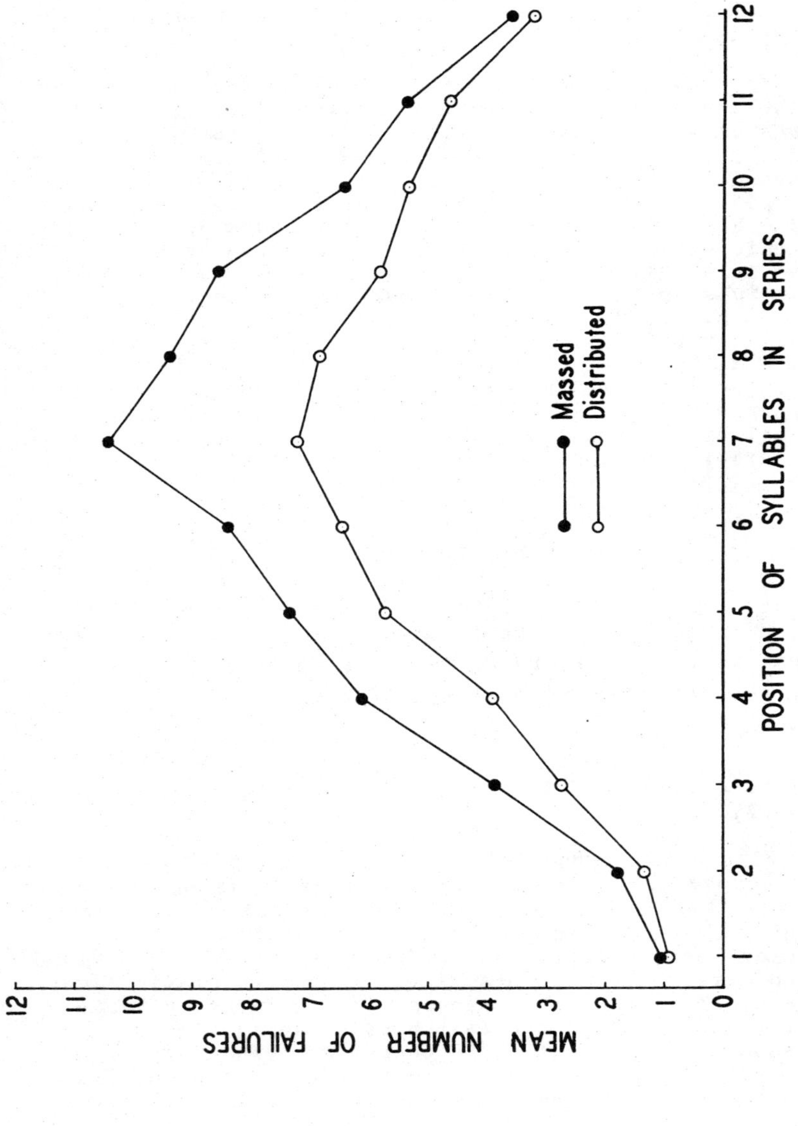

Figure 33. Mean number of failures in the various syllable positions during learning to one perfect recitation by massed and distributed practice. Data from Hovland (22).

THEOREM XXII, COROLLARY 5

A. *In rote series learned by massed practice, the difference between the mean number of presentations at the first success and the mean number of presentations at the last failure is greater at the posterior syllable than at the anterior syllable.*

Proof:

This proposition follows directly from Theorems XXII, VI, and XVII.

C. The data quoted under Corollary 4 show that the difference between the first success and last failure in terms of repetitions is greater at the posterior syllable than at the anterior syllable. In Hovland's study, for example, the difference is 1.49 for the anterior syllable, but 1.96 for the posterior. The corresponding data for Hull's experiment are 1.8 for the anterior syllable and 2.3 for the posterior syllable.

Since the difference at the posterior syllable is larger than at the anterior syllable ($1.96 > 1.49$, and $2.3 > 1.8$), in these studies it appears that the corollary is experimentally verified.

THEOREM XXII, COROLLARY 6

A. *At any incomplete stage of the learning of a rote series by massed practice, the probability of recall is greater at the anterior syllable than at the posterior syllable.*

Proof:

This proposition follows directly from Theorem XXII, Postulate 3, and Theorem XVIII.

C. Robinson and Brown's results based upon massed practice show that after any given number of massed repetitions the anterior syllable has a higher recall than the posterior. Hovland's data confirm this finding and, in addition, show that with distributed practice the same difference is obtained. Data from Ward (*63*) substantiating previous results are given graphically in Figure 31. In all of these studies recall is better at the anterior than at the posterior syllable.

The corollary is confirmed on the basis of available evidence.

THEOREM XXIII

A. *In rote series learned by massed practice the rate of increase of inhibition from the anterior end toward the maximum either (a) proceeds at a progressively lessening rate for sufficiently large values of* F, *or (b) starts with a progressively increasing rate which becomes and stays negative for sufficiently small values of* F.

Proof:

Either (a) $\Delta' J_n$ decreases as n' goes from 1 to the maximal value, provided that F is sufficiently large ($F \geq 1\frac{1}{3}$ is always sufficient; sometimes a smaller value is sufficient); or (b) $\Delta' J_n$ first increases and then decreases before n' reaches the maximal J value.

To prove (a) we show that $\Delta'_2 J$ is always negative for $F \geq 1\frac{1}{3}$ and also for some smaller values of F, depending on the value of N.

To prove (b) we show that whenever (a) does not apply, $\Delta'_2 J_n$ is positive at first and then negative before n' reaches the maximal value; and in particular we show that $\Delta'_2 J$ changes sign *at most once.*

$$\Delta'_2 J_n = -\frac{1}{F^n} + G(N - n' - 1) - G(n' + 1).$$

Here, as $G(n')$ is an increasing function of n', $\Delta'_2 J_n$ is certainly negative whenever $n' + 1 \geq N - n' - 1$; that is, whenever $n' \geq \frac{N}{2} - 1$. But since it was shown in Theorem XXI (A) that $\Delta' J_n$ is positive for $n' \leq \frac{N}{2}$, the maximal value of n' is greater than $\frac{N}{2}$. Hence, in either (a) or (b) $\Delta'_2 J_n$ is negative for some value or values of n' before the maximum. For $n' + 1 < N - n' - 1$, or $n' < \frac{N}{2} - 1$, we have,

$$\Delta'_2 J_n = -\frac{1}{F^{n'}} + \frac{1}{(n'+2) F^{n'+1}} + \frac{1}{(n'+3) F^{n'+2}} + \ldots + \frac{1}{(N-n'-1) F^{N-n'-2}}.$$

Now,

$$\frac{1}{(n'+2) F^{n'+1}} + \frac{1}{(n'+3) F^{n'+2}} + \ldots + \frac{1}{(N-n'-1) F^{N-n'-2}}$$

$$< \frac{1}{(n'+2) F^{n'+1}} + \frac{1}{(n'+2) F^{n'+3}} + \ldots + \frac{1}{(n'+2) F^{N-n'+2}} + \ldots$$

$$= \frac{1}{(n'+2) F^{n'+1}} \cdot \frac{F}{F-1}.$$

Hence,

$$\Delta'_2 J_n < -\frac{1}{F^{n'}} + \frac{1}{(n'+2)F^{n'+1}} \cdot \frac{F}{F-1} = -\frac{1}{F^{n'}} + \frac{1}{(n'+2)F^{n'}(F-1)}$$

$$= \frac{-(n'+2)(F-1)+1}{(n'+2)F^{n'}(F-1)} .$$

But this is negative when

$$-(n'+2)(F-1) + 1 < 0,$$

or

$$1 < (n'+2)(F-1)$$

$$\frac{1}{n'+2} < F - 1$$

or finally when

$$1 + \frac{1}{n'+2} < F .$$

Hence, $\Delta'_2 J_n < 0$ for all values of n provided that $F \geq 1\frac{1}{3}$. This completes the proof of (a).

It remains to show that (b) may actually arise, and that in this case $\Delta'_2 J_n$ remains negative when it becomes negative.

Example:

Set $F = 1$

$$\Delta'_2 J_1 = -\frac{1}{1} + \frac{1}{3} + \frac{1}{4} + \ldots\ldots + \frac{1}{N-2} .$$

For $N = 9$, this has the value,

$$\Delta'_2 J_1 = +.092857 .$$

Hence, for F sufficiently near 1, say 1.0001, $\Delta'_2 J_1$ is positive. This shows that (b) actually arises. Moreover,

$$F^n \Delta'_2 J_n = -1 + \frac{1}{(n'+2)F} + \ldots\ldots + \frac{1}{(N-n'-1)F^{N-2n'-2}}$$

is positive or negative with $\Delta'_2 J_n$, as F^n is always positive. This is a decreasing function of n, since an increase in n makes the positive terms both smaller and fewer. Hence, when this function becomes negative, it remains negative, and so also must $\Delta'_2 J_n$.

A more detailed, but straightforward, investigation shows that the proper critical value for F (see Table 14) is not $1\frac{1}{3}$, but 1.1614, the root of the equation,

$$-\frac{1}{F} - \frac{3}{2F^2} + \log\left(\frac{F}{F-1}\right) = 0 .$$

TABLE 14

Table giving the conditions on F and N under which the index of inhibitory potential (J_n) starts with a positive acceleration.

	$\Delta_2' J_1$ positive	Greatest value of $\Delta_2' J_1$ (F=1)
N = 8		− .050000
N = 9	$1 \leq F \leq 1.0365$	+ .092857
N = 10	$1 \leq F \leq 1.0721$	+ .217857
N = 11	$1 \leq F \leq 1.0958$	+ .328968
. . .	. . . .	. . . . .
$N = \infty$	$1 \leq F \leq 1.1614$	∞
	$\Delta_2' J_2$ positive	Greatest value of $\Delta_2' J_2$ (F=1)
N = 12		− .004365
N = 13	$1 \leq F \leq 1.0277$	+ .095635
. . .	. . . .	. . . . .
$N = \infty$	$1 \leq F \leq 1.1154$	∞

$\Delta_2' J_n$ is positive only when N is sufficiently large (at least $3n' + 4$), and as shown above, only when F is less than $1 + \frac{1}{n'+2}$. Thus for increasing n, $\Delta_2' J_n$ is positive only for larger and larger values of N and for smaller and smaller values of F. On the other hand, the upper bound for $\Delta_2' J_n$ is always infinity.

B. Exemplification of the (a) portion of this theorem is presented by Table 1 and Figure 17. There it may be seen at a glance that the rate of rise of inhibitory potential for two different values of F from the first syllable to the point of maximum inhibition grows progressively less and less steep until between syllables 8 and 9 it ceases to rise altogether, exactly as stated in the theorem.

THEOREM XXIII, COROLLARY 1

A. *In rote series learned by massed practice, in the passage from the anterior syllable to the point requiring the maximum number of presentations to raise the effective excitatory potential to the central tendency of the reaction threshold (L), the rate of increase in the number of such repetitions either (a) proceeds at a progressively lessening rate for sufficiently large values of F and* $\frac{\Delta E}{\Delta K}$*, or (b) for sufficiently small values of F (i.e., r = 1.1614) and sufficiently large values of N, the rate of increase starts with a progressively increasing rate which later becomes and stays negative.*

Proof:

From the formula (Postulate 16, Corollary 1):

$$_{L}R = \frac{L}{\Delta E - \Delta K\, J_n}$$

we find, by direct calculation, that

$$\Delta'_2 R_n = \frac{L \cdot \Delta K}{(\Delta E - \Delta K\, J_n)(\Delta E - \Delta K\, J_{n+1})(\Delta E - \Delta K\, J_{n+2})} [2\, \Delta K\, (\Delta' J_n)^2$$

$$+ \quad \Delta'_2 J_n\, (\Delta E - \Delta K\, J_n + \Delta K\, \Delta' J_n)].$$

Using the results of Theorem XXIII, we easily derive three important facts from this equation:

(i) If $\Delta'_2 J_n > 0$, then $\Delta'_2 R_n > 0$.

(ii) If $\Delta'_2 R_n < 0$ for some n, then $\Delta'_2 R_m < 0$ for all $m < n$.

(iii) If $\Delta'_2 R_n (N) > 0$ for some N, then $\Delta'_2 R_n (M) > 0$ for all $M > N$.

From these observations it is clear that we need only consider the limit of $\Delta'_2 R_n (N)$ as $N \to \infty$. For if this is negative the (a) part of the theorem holds, and if this is positive the (b) part holds.

We shall show that $\text{Lim}\ \Delta'_2 R_0 (N) < 0$ for $F > 1\frac{1}{3}$ and

$$\frac{\Delta E}{\Delta K} > 1 + \frac{3\,(F+1)}{(F-1)(3F-4)}.$$

For, by direct calculation,

$$\frac{L^2 \cdot \text{Lim}\ \Delta'_2 R_0}{\Delta K \cdot R_0 R_1 R_2} = [-\frac{1}{F} + \frac{1}{3F^2} + \ldots][\Delta E - \Delta K\,(1 + \frac{1}{2F} + \ldots)]$$

$$+ \Delta K\, [2\,(\frac{1}{2F} + \frac{1}{3F^2} + \ldots)][2\,(\frac{1}{2F} + \ldots)$$

$$+ 3\,(\frac{1}{3F^2} + \frac{1}{4F^3} + \ldots)]$$

$$< \frac{1}{F}\left[-1 + \frac{1}{3(F-1)}\right][\Delta E - \Delta K] + \Delta K \left[\frac{1}{F-1}\right]\left[\frac{1}{F-1} + \frac{1}{F(F-1)}\right]$$

$$= \frac{\Delta K}{F(F-1)^2}\left[\left(\frac{4}{3}-F\right)\left(\frac{\Delta E}{\Delta K}-1\right)(F-1) + (F+1)\right].$$

But if

$$\frac{\Delta E}{\Delta K} > 1 + \frac{3(F+1)}{(F-1)(3F-4)}$$

then

$$\left(\frac{\Delta E}{\Delta K} - 1\right)(F - 1) > \frac{F+L}{F-\frac{4}{3}}$$

Hence, if $F > \frac{4}{3}$:

$$\left(\frac{4}{3} - F\right)\left(\frac{\Delta E}{\Delta K} - 1\right)(F - 1) < -(F + 1).$$

Hence,

$$\lim_{N\to\infty} \Delta'_2 R_0 (N) < 0.$$

(A more detailed analysis shows that the critical value of F is approximately 1.1614 as in Theorem XXIII, and the critical inequality for $\frac{\Delta E}{\Delta K}$ is:

$$\frac{\Delta E}{\Delta K} > 10 - \frac{12}{F} - 9F \log\frac{F}{F-1} + \frac{18}{F^2\left(1+\frac{3}{2F} - F \log\frac{F}{F-1}\right)}.$$

It may be shown that, whatever the value of F, there are admissible values of $\frac{\Delta E}{\Delta K}$ which do not satisfy this equality. Hence, case (b) of this theorem actually arises for every value of F, with sufficiently large values of $\frac{\Delta E}{\Delta K}$ and N.)

B. Both of the more or less supposititious examples of the theoretical number of repetitions required to reach the mean threshold of recall worked out for 15-syllable series in Table 2 fall clearly under the conditions marked (b) in the above corollary. From Table 2 it may be seen that in the case where F = 1.63 and where ΔK = .163, i.e., $\frac{\Delta E}{\Delta K} = 6.1$, the differences between successive values of R from the beginning of the series are

.98, 1.06, 1.05, .90, .72, .51, .30, and .13,

which shows that the rate of rise in the number of repetitions increases up to the third syllable, after which it

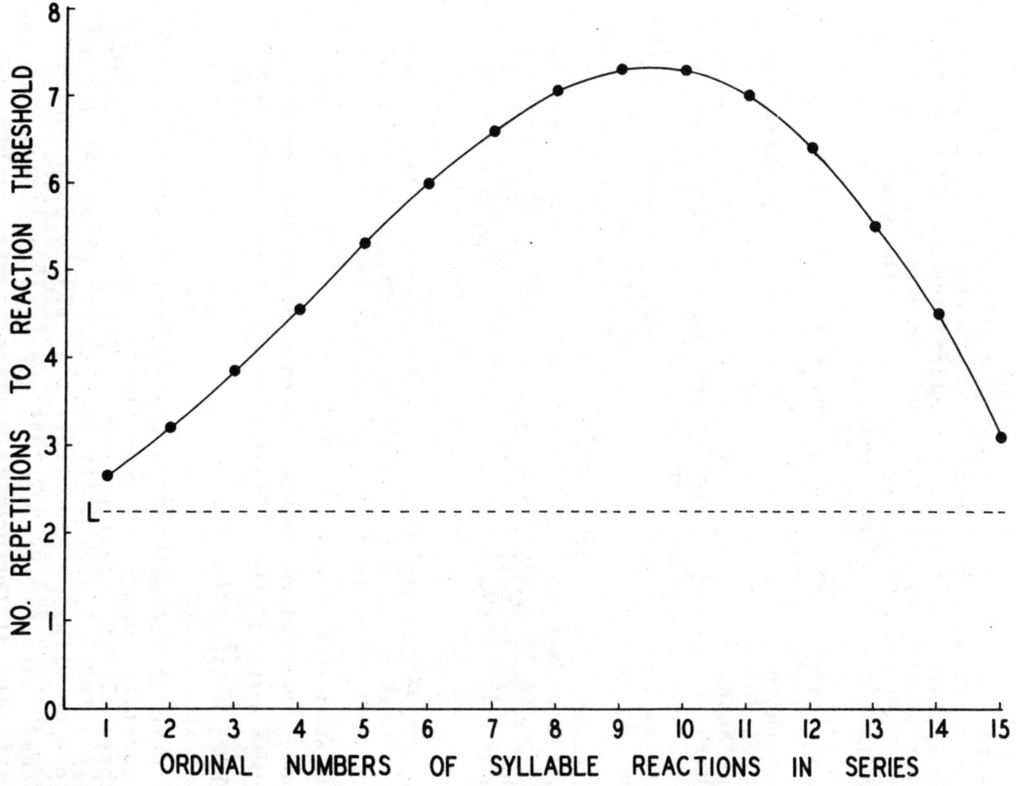

Figure 34. A graphic representation of the theoretical number of presentations required on the average to bring $\bar{E}$ to the reaction threshold, where N = 15, F = 1.37, and ΔK = .09. (Plotted from Table 2.)

begins to fall and continues to fall steadily up to the point of maximum difficulty of learning. In the case where $N = 15$, $F = 1.37$, and $\frac{\Delta E}{\Delta K} = 11.1$ (Table 2), the corresponding differences are:

.55, .64, .70, .73, .68, .61, .47, and .25.

This series of values shows that the rate of increase in learning difficulty increases at an increasing rate up to the 5th syllable, after which it decreases steadily up to the point of maximum difficulty of learning. A close examination of Figures 18 and 34 will reveal the same situation as that just presented numerically.

C. An examination of the results of a considerable number of investigations yielding data from which the number of presentations required to reach the reaction threshold may be determined by the method of either Problem VII or VIII, or both, discloses much inconsistency in this respect even where the length of the series and the nature of the material learned are substantially the same. These results as a whole, however, present the picture represented by the (b) portion of the corollary and shown graphically in Figure 34. Examples of the results of such empirical investigations are presented in Figures 18, 29, and 33.

In this connection it is important to note that for an investigation to present a smooth contour in this respect and thus to yield reliable results, the syllables in the series should be systematically rotated in such a way that each syllable of every series occupies each position an equal number of times and is learned with equal frequency in the various stages of practice of the subjects. It is probable that no investigation has yet fully satisfied all the required conditions, though several have done so to a degree sufficient to yield a fair indication of the situation.

A second consideration in the attempt to determine the validity of the above corollary is the actual value of F and of ΔK of the series in question. Preliminary attempts at such determinations have yielded somewhat discordant results, which suggests that the equation for ΔI_n (Theorem V) is not sufficiently perfect to serve such an exacting purpose.

In general it seems fair to conclude that the corollary meets with complete empirical confirmation as to the rate of increase in the number of repetitions required to pass the reaction threshold as the point of maximum difficulty is approached, but its statement of the rate of increase in the early syllables of the series meets with distinctly uncertain and qualified confirmation.

THEOREM XXIII, COROLLARY 2

A. *In rote series partially learned by massed practice, the rate of decrease in the probability of recall proceeds from the anterior end to the point of maximum difficulty at a progressively lessening rate for sufficiently large values of* F *(*F > 1.1614), *provided that the effective excitatory potential at each syllable involved is less than* $\frac{L}{2}$.

Proof:

Since $\bar{E} = E - I$, Theorem XVIII states that recall is a decreasing function of inhibition. Now, Theorem XIX implies that recall is a concave function of inhibition, provided $\bar{E} < L$. But from the differential identity:

$$\frac{d^2 z}{dx^2} = \frac{d^2 z}{dy^2}\left(\frac{dy}{dx}\right)^2 + \frac{dz}{dy}\cdot\frac{d^2 y}{dx^2}$$

it is clear that a decreasing, concave function of a convex function is concave. Hence, the corollary follows from Theorem XXIII (A).

C. No exactly relevant data for the empirical test of the validity of this corollary is at present available, for the reason that owing to the small amount of ΔI at the first syllable, it is very likely to have passed the mean reaction threshold even after the first presentation, as is clearly the case with the results of Robinson and Brown reproduced as Figure 32. However, it is clear that the probability curves, both of Robinson and Brown and of Ward, show a negative acceleration from the point at which they cross the 50 per cent level to their minimum level. There is here, of course, the same problem of knowing the F and $\frac{\Delta E}{\Delta K}$ values of the data under consideration, as was encountered in connection with Theorem XXIII, Corollary 1.

In view of the various uncertainties mentioned above, we can conclude no more than that this theoretical proposition finds probable confirmation. It is likely that it will be necessary to set up an experiment in which very slow learners will be employed, before a clear test of this proposition can be secured.

THEOREM XXIV

A. *In rote series learned by massed practice, the rate of decrease of inhibition* ($\Delta' J_n$) *from the maximum to the posterior end becomes progressively greater.*

To prove:

$\Delta' J_n$ decreases for values of n' from the maximum to the posterior end.

Proof:

From Theorem XXI (A), the maximum n' is greater than $\frac{N}{2}$. From Theorem XXI (B), $\Delta' J_n$ decreases steadily for $n' > \frac{N}{2}$. Hence, *a fortiori*, $\Delta' J_n$ decreases steadily from the maximum to the posterior end.

B. The present theorem finds its illustration in Table 1 and Figure 17. In the latter, a glance will show that the fall of the curve of J from $n = 9$ to $n = 15$ becomes progressively steeper as the end of the series is approached, exactly as stated in the theorem.

THEOREM XXIV, COROLLARY 1

A. *In rote series partially learned by massed practice, the rate of decrease in probability of recall, at any stage of learning, proceeds from the point of maximum difficulty at a progressively increasing rate, provided that the effective excitation at each syllable is below the mean threshold.*

Proof:

The proof of this proposition is exactly analogous to that presented for Theorem XXIII, Corollary 2.

C. Extensive data on this point are furnished by Robinson and Brown (*53*). As a sample of typical results, computations have been made in Table 15 from the results of Ward (*63*) quoted above. These results and those of other investigators show very clearly that the differences between successive syllables are greater at the end than at the central serial positions, although in Ward's results there is an initial increase in the first syllable positions. This effect is much less in the results of Robinson and Brown.

The experimental evidence supports the corollary.

TABLE 15

Average number of syllables recalled in various syllable positions after 3, 5, and 7 correct anticipations, and the differences in recall for adjacent syllables. Data derived from Ward's figures (63).

Syllable position	Level of Mastery					
	3 Anticipations		5 Anticipations		7 Anticipations	
	Per cent recalled	Difference	Per cent recalled	Difference	Per cent recalled	Difference
1	.94		.96		.92	
		.08		.05		.04
2	.86		.91		.96	
		.29		.09		.10
3	.57		.82		.86	
		.29		.25		.08
4	.28		.57		.78	
		.18		.23		.11
5	.10		.34		.67	
		.05		.22		.25
6	.05		.12		.42	
		.00		.02		.19
7	.05		.10		.23	
		.01		.00		.08
8	.04		.10		.31	
		.01		.00		.03
9	.05		.10		.28	
		.01		.03		.19
10	.06		.13		.47	
		.09		.35		.25
11	.15		.48		.72	
		.37		.39		.21
12	.52		.87		.93	

THEOREM XXV

A. *In rote series learned by massed practice, the rate of increase in inhibitory potential at the anterior end, as n increases, will be greater than the rate of increase at the posterior end as n decreases for any given value of N, provided* F *is greater than a certain critical value depending on N and never exceeding* 1.55036.

To find:

When is $| \Delta' J_0 | > | \Delta' J_{N-2} |$?

Or when is $| \Delta' J_0 | - | \Delta' J_{N-2} > 0$?

Proof:

$$\Delta' J_0 = 2\,[G(N) - G(1)] - \frac{1}{NF^{N-1}} \quad [\text{ and is positive }]$$

$$= \frac{2}{2F} + \frac{2}{3F^2} + \cdots\cdots + \frac{2}{(N-1)F^{N-2}} + \frac{1}{NF^{N-1}}$$

$$\Delta' J_{N-2} = N[G(N) - G(N-1)] + [G(N) - G(1)] - \frac{F^{N-1} - 1}{F^{N-1}(F-1)}$$

$$= \frac{N}{NF^{N-1}} + \frac{1}{2F} + \frac{1}{3F^2} + \cdots\cdots + \frac{1}{NF^{N-1}}$$

$$- [\frac{1}{F} + \frac{1}{F^2} + \cdots\cdots + \frac{1}{F^{N-1}}]$$

$$= \frac{1}{2F} + \frac{1}{3F^2} + \cdots\cdots + \frac{1}{NF^{N-1}}$$

$$- \frac{1}{F} - \frac{1}{F^2} - \cdots\cdots - \frac{1}{F^{N-2}} \quad [\text{- and is negative }].$$

To find when

$$|\,\Delta' J_0\,| - |\,\Delta' J_{N-2}\,| = \Delta' J_0 + \Delta' J_{N-2}$$

$$= \frac{3}{2F} + \frac{3}{3F^2} + \frac{3}{4F^3} + \cdots + \frac{3}{(N-1)F^{N-2}} + \frac{2}{NF^{N-1}}$$

$$- \frac{1}{F} - \frac{1}{F^2} - \frac{1}{F^3} - \cdots\cdots - \frac{1}{F^{N-2}} > 0.$$

or when

$$\frac{1}{2F} - \frac{1}{4F^3} - \frac{2}{5F^4} - \cdots\cdots - \frac{N-4}{(N-1)F^{N-2}} + \frac{2}{NF^{N-1}} > 0.$$

For $N = 1, 2, 3$, and 4, this inequality will always hold, for there will be no negative terms. For $N = 5$, we have:

$$\frac{1}{2F} - \frac{1}{4F^3} + \frac{2}{5F^4} > 0$$

for all values of F, since

$$\frac{1}{2F} > \frac{1}{4F^3}$$

so that,

$$\frac{1}{2F} - \frac{1}{4F^3} > 0$$

and, *a fortiori*,

$$\frac{1}{2F} - \frac{1}{4F^3} + \frac{2}{5F^4} > 0.$$

For $N = 6$, we have:

$$\frac{1}{2F} - \frac{1}{4F^3} - \frac{2}{5F^4} + \frac{2}{6F^5} > 0$$

or,

$$\frac{30\,F^4 - 15\,F^2 - 24\,F + 20}{60\,F^5} > 0\,.$$

The numerator is a polynomial $N(F) = 30\,F^4 - 15\,F^2 - 24\,F + 20$, which is positive (and equal to $+11$) for $F = 1$. Consider its derivative $N'(F) = 120\,F^3 - 30\,F - 24 = F\,(120\,F^2 - 30) - 24$. $120\,F^2 - 30$ is always greater than or equal to 90 for $F \geq 1$. Hence, $F\,(120\,F - 30) \geq 90$, and $F\,(120\,F - 30) - 24 \geq 66$, for $F \geq 1$. Since the derivative is always positive for $F \geq 1$, the polynomial has its least value for $F = 1$, and is consequently always positive.

For $N = 7$, if we set $F = 1$, we have:

$$|\Delta' J_1| - |\Delta' J_6| = \frac{1}{2} - \frac{1}{4} - \frac{2}{5} - \frac{3}{6} + \frac{2}{7} = \frac{-51}{140}$$

and consequently $|\Delta' J_1| - |\Delta' J_6|$ is also negative for values of F sufficiently near 1.

In general we consider,

$$T(N) = F\,\{|\Delta' J_0| - |\Delta' J_{N-2}|\}$$

$$= \frac{1}{2} - \frac{1}{4F^2} - \frac{2}{5F^3} - \cdots\cdots - \frac{N-4}{(N-1)\,F^{N-3}} + \frac{2}{NF^{N-2}}$$

T is positive or negative with $|\Delta' J_0| - |\Delta' J_{N-2}|$. The difference of the value of T for $N + 1$ and N when F is kept fixed is,

$$T(N+1) - T(N) = \frac{2N - (N^2-1)F}{N\,(N+1)\,F^{N-1}}$$

which is certainly negative when N is 7 or greater. The change in T when N is kept fixed but F increases is easily observed by writing,

$$T = \frac{1}{2} - \frac{1}{4F^2} - \frac{2}{5F^3} - \cdots\cdots - \frac{N-5}{(N-2)\,F^{N-4}} + \frac{N-4}{(N-1)\,F^{N-3}} + \frac{2}{NF^{N-2}}$$

$$T'(F) = \frac{2}{4F^3} + \frac{3.2}{5F^4} + \cdots\cdots + \frac{(N-4)\,(N-5)}{(N-2)\,F^{N-3}} + \frac{(N-3)\,(N-4)}{(N-1)\,F^{N-2}} + \frac{2\,(N-2)}{NF^{N-1}}$$

Now,

$$\frac{(N-3)(N-4)}{(N-1)F^{N-2}} + \frac{2(N-2)}{NF^{N-1}} = \frac{F(N)(N-3)(N-4) - 2(N-1)(N-2)}{N(N-1)F^{N-1}}$$

is certainly positive, since F $\geq$ 1, and we have,

$$N^3 - 7N^2 + 12N \geq 2N^2 - 6N + 4 ,$$

or,

$$N^3 - 9N^2 + 18N - 4 > 0 ,$$

or,

$$N^2(N-9) + 18N - 4 > 0 .$$

This is certainly true for $N \geq 9$, and moreover,

$$N^3 - 9N^2 + 18N - 4 = +24 \quad \text{for } N = 7$$

and

$$+76 \quad \text{for } N = 8$$

Hence, T (F) > 0 whenever $N > 7$. This means that if $|\Delta' J_0| - |\Delta' J_{N-2}|$ is zero or positive for any value of F, it is positive for all greater values of F. Moreover, as shown above, if $|\Delta' J_0| - |\Delta' J_{N-2}|$ is zero or negative for one value of N, it is certainly negative for greater N. Hence, for each value of $N \geq 7$, there is a critical value for F, such that if F is less than this value, $|\Delta' J_0| - |\Delta' J_{N-2}$ is negative, but if F is greater then $|\Delta' J_0| - |\Delta' J_{N-2}|$ is positive. These critical values of F increase steadily with N, as shown in Table 16.

TABLE 16

Minimum value of F such that $|\Delta' J_0| > |\Delta' J_{N-2}|$

N	Critical value of F	N	Critical value of F
3	1	10	1.492
4	1	11	1.514
5	1	..	.....
6	1	..	.....
7	1.363	..	.....
8	1.379	..	.....
9	1.439		1.55036

B. This theorem may be illustrated by means of the values in Table 1, where N = 15 and F = 1.63, which is slightly greater than the critical value of 1.55. In Table 1 we find that the change in the index of inhibition between 0 and 1 is 1.097 (i.e., 2.645 - 1.548), whereas the change from 14 to 13 is only 1.032 (i.e., 3.613 - 2.581). The same thing is shown graphically in the lower curve of Figure 17.

THEOREM XXV, COROLLARY 1

A. *In rote series learned by massed practice, at any incomplete stage of learning, the rate of decrease of recall at the anterior syllable is greater than the rate of increase of recall at the posterior syllable when, for any given N,* F *is greater than the critical value of* 1.55036, *provided the effective excitation at each syllable is less than* $\frac{L}{2}$.

Proof:

The proof of this proposition is exactly analogous to that for Theorem XXIII, Corollary 2.

C. Such evidence as exists bearing on this question is contained in the reports of Robinson and Brown (*53*) and of Ward (*63*). Typical results of both studies are shown in Figures 31 and 32. Unfortunately (as pointed out above, Theorem XXIII, Corollary 2 C), even the first presentation raises the probability of recall above the 50 per cent level so that under ordinary learning conditions it is impossible to observe what would occur under the conditions assumed in the postulate.

There is still a different difficulty in determining the validity of this corollary; this lies in the fact that (as pointed out above in Theorem XXIII, Corollary 1 C), the equation for the J-values (Theorem V) does not appear to be sufficiently accurate to permit dependable determinations of the value of F from empirical data by the methods of Problem 1.

THEOREM XXVI

A. *In rote series learned by massed practice, as the number (N) of syllables in the series increases **(a)** the index of inhibition* (J_n) *increases for every syllable counting inward from either end toward the point of maximum inhibition; (b) of two syllables on the anterior side of the point of maximum inhibition, the index of inhibition* (J_n) *of the syllable the more remote from the anterior end of the series will increase the more rapidly; (c) of two* *syllables on the posterior side of the point of maximum inhibition, the index of inhibition (J_n) of the syllable the more remote from the posterior end of the series will increase the more rapidly.*

To prove:

(a) J_n is an increasing function of N, or $J_n(N+1) - J_n(N) > 0$.

(b) J_m increases more rapidly than J_n if $m > n$. (The subscript is the number of the syllable counted from the anterior end.

(c) J_{N-n} increases more rapidly than J_{N-m} if $n > m$, $N - n$, and $N - m$ in the second half.

Proof of (a):

$$J_0(N+1) - J_0(N) = \frac{1}{(N+1)F^N} > 0 .$$

Now,

$$\Delta' J_n(N+1) - \Delta' J\ (N) = \frac{1}{(N-n'+1)F^{N-n'}} + \frac{1}{(N-n'+2)F^{N-n'+1}} + \dots + \frac{1}{(N+1)F^N} > 0 .$$

Hence, as

$$J_{n+1} = J_n + \Delta' J_n$$

we have,

$$J_{n+1}(N+1) - J_{n+1}(N) = J_n(N+1) - J_n(N) + \Delta' J_n(N+L) - \Delta' J_n(N),$$

which is greater than $J_n(N+1) - J_n(N)$,

since $\Delta' J_n(N+1) - \Delta' J_n(N)$ is positive. Hence,

$$J_{n+1}(N+1) - J_{n+1}(N) > J_n(N+1) - J_n(N)$$

for all n. But,

$$J_0(N+1) - J_0(N) > 0 ,$$

and so $J_n(N+1) - J_n(N)$ is positive by induction.

Proof of (b):

Since,

$$J_{n+1}(N+1) - J_{n+1}(N) > J_n(N+1) - J_n(N) ,$$

by (a) it follows by induction that

$$J_m(N+1) - J_m(N) \quad > \quad J_n(N+1) - J_n(N)$$

if $m > n$.

Proof of (c):

$$\Delta' J_{n+1}(N+1) - \Delta' J_n(N) = \frac{1}{(n'+1)F^{n'}} + \frac{1}{(n'+2)F^{n'+1}} + \ldots\ldots + \frac{1}{(N+1)F^N} .$$

We note that S_{n+1} in the series of $N+1$ syllables is the same distance from the posterior end as S_n in the series of N syllables. Since these syllables are in the second half, $n'+1 > N-n$. In the above expression $\frac{1}{(n'+1)F^{n'}}$ is greater than any single one of the positive terms, the sum of the positive terms is less than $\frac{n'+1}{(n'+1)F^{n'}}$, and

$$\Delta' J_{n+1}(N+1) - \Delta' J_n(N) < 0 .$$

Here again, as $J_{n+1} = J_n + \Delta' J_n$, we have,

$$J_{n+2}(N+1) - J_{n+1}(N) \quad < \quad J_{n+1}(N+1) - J_n(N).$$

Consequently the difference $J_{n+1}(N+1) - J_n(N)$ increases as n increases, taking on its least value at the posterior end, i.e., $n' = N$. But,

$$J_N(N) = 1 + \frac{1}{F} + \ldots\ldots + \frac{1}{F^{N-1}}$$

and,

$$J_{N+1}(N+1) = 1 + \frac{1}{F} + \ldots\ldots + \frac{1}{F^{N-1}} + \frac{1}{F^N} .$$

Hence,

$$J_{N+1}(N+1) - J_n(N) = \frac{1}{F^N} > 0 .$$

Hence $J_{n+1}(N+1) - J_n(N)$ is positive for all n' in the second half of the series, and decreases as n increases. Hence, of two syllables in the second half, the more remote from the posterior end increases more rapidly; i.e., J_{N-m} increases more rapidly than N_{N-n} if $m > n$.

(Here the restriction to the second half of the series is not absolute, but depends on the relation of the value of F to N. The critical point of the proof is the inequality,

$$-\frac{1}{F^{n'}} + \frac{1}{(n'+2)F^{n'+1}} + \ldots\ldots + \frac{1}{(N+1)F^N} < 0.$$

We note that this is quite similar to the critical inequality of Theorem XXIII, viz.:

$$-\frac{1}{F^{n'}} + \frac{1}{(n'+2)F^{n'+1}} + \cdot \cdot \cdot \cdot \cdot + \frac{1}{(N-n'-1)F^{N-n'-2}} < 0 .$$

In other words, for a pair of values (N, n') in Theorem XXIII, we have the same critical value of F as we will have for a pair of values $(N + n' + 2, n')$ here. Consequently, the same lower bound will hold for F here, viz., $F > 1 + \frac{1}{n'+2}$ will assure us that the theorem holds for the nth syllable and beyond, and $F > 1.1614$ will assure us that the theorem holds for all n.)

B. This theorem as a whole may be illustrated by the theoretical values shown in Table 1 and by those in Table 17 which nas been derived from Table 1 by a series of subtractions.

TABLE 17

Table showing the differences between the indices of inhibition (J_n) at corresponding positions in a 12-syllable series and in a 15-syllable series, where F = 1.37.

Anterior end		Posterior end	
Syllable positions counting from end	15-syllable J_n less 12-syllable J_n	Syllable positions counting from end	15-syllable J_n less 12-syllable J_n
1	.004	1	.052
2	.004	2	.119
3	.031	3	.207
4	.061	4	.322
5	.109	5	.474

Part (a): A comparison of the parallel values in the last two columns of Table 1, counting in from the ends of the respective series, furnishes clear exemplification of this part of the theorem.

Part (b): Also in Table 1 it may be seen that counting in from the anterior end of the series the differences between parallel values in the 15-syllable and the 12-syllable series grow progressively larger. These differences have been calculated and appear in the second column of Table 17.

Part (c): This portion of the theorem is illustrated by Table 1 and by the fourth column of Table 17, in a manner exactly analogous to that by which Part (b) was exemplified.

THEOREM XXVI, COROLLARY 1

A. *In rote series learned by massed practice, as the length (N) of the series increases, the mean number of presentations required to reach the reaction threshold increases at every syllable position.*

Proof:

This proposition follows directly from Theorems XXVI and VI.

C. Data from Hovland (*26*) for serial position effects in the learning by massed practice of three lengths of series have been presented graphically in Figure 29. These data are given in tabular form in Table 18. The results are given

TABLE 18

(A) mean number of failures and (B) average number of repetitions preceding the first success and up to and including the last failure for learning to one perfect recitation with three lengths of lists, all learning by massed practice. Data from Hovland (26).

Syllable	A			B		
	Length of list			Length of list		
	8-syll.	11-syll.	14-syll.	8-syll.	11-syll.	14-syll.
1	2.17	2.80	2.98	2.65	3.13	3.65
2	3.14	3.36	4.64	3.61	3.80	5.15
3	4.14	4.86	5.14	4.82	5.64	6.53
4	4.60	6.51	7.58	4.86	6.56	8.85
5	5.69	7.01	8.78	5.63	7.25	9.03
6	5.82	8.07	9.77	5.45	8.09	10.55
7	4.50	8.54	12.04	4.45	8.42	11.81
8	2.92	8.07	12.37	2.69	7.96	12.35
9		8.01	11.97		8.20	12.06
10		6.14	10.92		6.75	11.61
11		3.92	10.27		4.46	10.70
12			9.92			10.07
13			6.72			7.68
14			4.71			5.39

under (A) in terms of the number of failures made in each syllable position before complete learning, counting from the first presentation as a failure. Under (B) the corresponding data are given for the average of the number of presentations preceding the first success up to and including the last failure.[10]

In Table 18 it will be observed that the number of repetitions increases for every comparable syllable with an increase in the length of the list. Thus, for syllable 1, the mean numbers of repetitions required are 2.17 with an 8-unit list, 2.80 with an 11-unit, and 2.98 with a 14-unit list. For syllable 2, the corresponding figures are 3.14, 3.36, and 4.64.

The results confirm the corollary.

THEOREM XXVI, COROLLARY 2

A. *In rote series learned by massed practice, as the length (N) of the series increases, of two syllables anterior to the point requiring the maximum number of presentations for learning (counting from the anterior end of each series) the mean number of presentations required to reach the reaction threshold will increase more rapidly for the more remote syllable.*

Proof:

This proposition follows directly from Theorems XXVI and VII.

C. In Table 19 there are given the number of additional repetitions required to learn the 11-unit and the 14-unit lists over the 8-unit list in the first four syllable positions.

TABLE 19

Increases in the number of repetitions required for learning 11-unit and 14-unit lists as compared with 8-unit lists for various syllables. Data from Hovland (26).

Syllable position	Increase of 11-unit list as compared with 8-unit	Increase of 14-unit list as compared with 8-unit
Anterior (1st syllable)	+ .63	+ .81
2nd to anterior (2nd syllable)	+ .22	+ 1.50
3rd to anterior (3rd syllable)	+ .72	+ 1.00
4th to anterior (4th syllable)	+ 1.91	+ 2.98

10. It will be recalled from the theoretical portion of Problem VIII that the number of presentations required to raise the effective excitatory potential just to the mean reaction threshold is .5 of a unit less than the number of failures required for complete learning, and 1.0 unit less than the average of the number of repetitions up to and including the first success and the last failure.

The tendency predicted by the theory for the increment to be larger as one approaches the point of maximum difficulty of learning is experimentally confirmed, although the data are somewhat irregular.

THEOREM XXVI, COROLLARY 3

A. *In rote series learned by massed practice, as the length (N) of the series increases, of two syllables posterior to the point requiring the maximum number of presentations for learning (counting from the posterior end of each series), the mean number of presentations required to reach the reaction threshold will increase more rapidly for the more remote syllable.*

Proof:

This proposition follows directly from Theorems XXVI and VII.

C. Corresponding data from Hovland (*26*) for the posterior four syllables are given in Table 20. On this point the data consistently support the corollary in showing that the

TABLE 20

Increases in the number of repetitions required for learning 11-unit and 14-unit lists as compared with 8-unit lists for various syllables. Data from Hovland (*26*).

Syllable position	Increase of 11-unit list as compared with 8-unit	Increase of 14-unit list as compared with 8-unit
Posterior	1.00	1.79
2nd to posterior	1.64	2.22
3rd to posterior	2.19	4.10
4th to posterior	2.38	4.58

increase in number of repetitions with increased length is least rapid at the posterior syllable and becomes progressively greater as one goes towards the central serial positions.

The corollary is thus experimentally verified.

THEOREM XXVI, COROLLARY 4

A. *In the learning of rote series of different lengths by massed practice, assuming the presentations given each series compared to be equal in number and sufficient in number to raise the effective excitatory potential of each above the reaction threshold throughout: as the number of syllables (N) in a series increases, the mean reaction latency will increase for all syllables of the series.*

Proof:

This proposition follows directly from Theorems XXVI and VIII.

C. No data bearing directly on the validity of this proposition have been found.

THEOREM XXVI, COROLLARY 5

A. *In the learning of rote series of different lengths by massed practice, assuming the presentations given each series compared to be equal in number and sufficient in number to raise the effective excitatory potential of each above the reaction threshold throughout: of two syllables anterior to the point in the series showing the maximum number of presentations for learning, as the number of syllables (N) of the series increases, the mean reaction latency at the more remote syllable (counting from the anterior end of the series) will increase the more rapidly.*

Proof:

This proposition follows directly from Theorems XXVI and IX.

C. No data bearing directly on the validity of this proposition have been found.

THEOREM XXVI, COROLLARY 6

A. *In the learning of rote series of different lengths by massed practice, assuming the presentations given each series compared to be equal in number and sufficient to raise the effective excitatory potential of each above the reaction threshold throughout: of two syllables posterior to the point in the series showing the maximum number of presentations for learning, as the number of syllables (N) of the series increases, the mean reaction latency at the more remote syllable (counting from the posterior end of the series) will increase the more rapidly.*

Proof:

This proposition follows directly from Theorems XXVI and IX.

C. No data bearing directly on the validity of this proposition have been found.

THEOREM XXVI, COROLLARY 7

A. *In rote series learned by massed practice, as the number of syllables (N) in the series increases, the advantage of well-distributed practice over massed practice increases for every syllable.*

Proof:

This proposition follows directly from Theorems XXVI, VI, and XIV.

C. The experiment of Hovland (*26*) on the efficiency of massed and distributed practice with varying lengths of series gives the data shown in Table 21, concerning the mean number of failures in the first four and last three syllable positions for three lengths of series (8, 11, and 14 units) with massed and distributed practice.

TABLE 21

Mean number of failures in various syllable positions during learning by massed and distributed practice with three lengths of list. Data from Hovland (*26*).

Syllable Position	8-unit		11-unit		14-unit	
	Massed	Distributed	Massed	Distributed	Massed	Distributed
First	2.17	2.01	2.80	2.07	2.98	2.24
Second	3.14	2.46	3.36	2.50	4.64	3.35
Third	4.14	3.80	4.86	4.28	5.14	4.07
Fourth	4.60	3.95	6.51	5.14	7.58	5.36
3rd to posterior	5.82	4.06	8.01	5.53	9.92	7.64
2nd to posterior	4.50	3.36	6.14	5.21	6.72	6.07
Posterior	2.92	2.71	3.92	3.10	4.71	3.98

Calculation of the advantage of distributed practice over massed for each syllable position with the three lengths gives the figures shown in Table 22. Reading across, each series snows that, as predicted by the corollary, there is a general tendency for an increased advantage of distributed practice in each syllable position, the longer the series. The regularity is greater for the anterior than for the posterior syllables.

TABLE 22

Advantage of distributed over massed practice in each syllable position for three lengths of list. Data from Hovland (*26*).

Syllable position	Length of list		
	8	11	14
Anterior (1st syllable)	.16	.73	.74
2nd to anterior (2nd syllable)	.68	.86	1.29
3rd to anterior (3rd syllable)	.34	.58	1.07
4th to anterior (4th syllable)	.65	1.37	2.22
3rd to posterior	1.76	2.48	2.28
2nd to posterior	1.14	.93	.65
Posterior	.21	.82	.73

Experimental evidence supports the predictions based upon the corollary.

THEOREM XXVI, COROLLARY 8

A. *In the learning of rote series, as the number of syllables (N) in the series increases, of two syllables anterior to the point requiring the maximum number of presentations for learning, the advantage of well-distributed practice*[11] *will increase more rapidly at the syllable which is the more remote, counting from the anterior end of each series compared.*

Proof:

This proposition follows directly from Theorems XXVI, VII, and XV.

C. The data of Hovland (*26*), part of which was quoted under Corollary 7, can be used to test this corollary. The results show a general tendency for the anterior syllable positions to change least in the advantage of distribution with increased length, and the central ones to change most. Thus the average advantages at the anterior and central syllable positions are as follows:

	8-Unit	11-Unit	14-Unit
Anterior	.42	.79	1.02
Central	1.10	1.11	3.71

11. See footnote page 128.

The first two syllables are averaged to give the anterior value. The central two syllables were used in the case of the even-numbered lists (8- and 14-unit) and the central three in the case of the odd-numbered (11-unit) list.

Although the data are somewhat irregular, the present evidence seems to support the corollary.

THEOREM XXVI, COROLLARY 9

A. *In the learning of rote series, as the number (N) of syllables in the series increases, of two syllables posterior to the point requiring the maximum number of presentations for learning, the advantage of well-distributed practice will increase the more rapidly for the syllable which is the more remote, counting from the posterior end of each series compared.*

Proof:

This proposition follows directly from Theorems XXVI, VII, and XV.

C. The data of Hovland (*26*) also show a greater change in the advantage of distributed practice with increased length in the central than in the posterior syllable positions. Using the procedure employed under Corollary 8, the average advantages of distributed practice for the central and posterior positions are as follows:

	8-Unit	11-Unit	14-Unit
Posterior syllables	.67	.87	.69
Central syllables	1.10	1.11	3.71

The general trend for the advantages to be greater with increased length in the central syllable positions supports the corollary.

THEOREM XXVI, COROLLARY 10

A. *In rote series learned by massed practice, as the number (N) of syllables increases, the mean number of repetitions up to and including the first success increases for each syllable of the series.*

Proof:

This proposition follows directly from Theorems XXVI, VI, and XVI.

C. Shipley (*58*) has reported results on the mean number

of exposures preceding the first success and the last failure with 8-unit, 14-unit, and 20-unit lists. In Table 23 his results for the control groups are quoted, showing the mean number of exposures preceding the first success in the first three and the last three syllable positions for the three lengths of list.

TABLE 23

Mean number of failures preceding first success with varying lengths of list. Data from Shipley (*58*).

	Syllable position					
	1st	2nd	3rd	2nd-from-last	next-to-last	last
8-unit list	1.5	2.2	2.8	4.7	4.7	3.2
14-unit list	2.3	3.2	4.5	6.4	4.9	2.8
20-unit list	2.2	3.6	4.2	8.0	6.0	3.2

It will be observed that almost without exception there is an increase in the number of trials preceding the first success in any given syllable position as the length of the series increases. This result is confirmed in Hovland's (*26*) study on distribution of practice with varying lengths of list.

Experimental verification of the corollary is obtained on the basis of available results.

THEOREM XXVI, COROLLARY 11

A. *In rote series learned by massed practice, as the number (N) of syllables increases, the mean number of repetitions up to and including the last failure increases for each syllable of the series.*

Proof:

This proposition follows directly from Theorems XXVI, VI, and XVI.

C. The data of Shipley (*58*) may also be employed to show the mean number of exposures preceding the last failure with 8-unit, 14-unit, and 20-unit lists. These results are given in Table 24.

It will be observed that there is a clear tendency for the number of trials preceding the last failure in each position to increase with the length of the series, and the corollary is therefore confirmed.

TABLE 24

Mean number of failures up to and including last failure with varying lengths of list. Data from Shipley (*58*).

	Syllable position					
	1st	2nd	3rd	2nd-from-last	next-to-last	last
8-unit list	2.5	3.3	5.0	6.0	6.2	4.1
14-unit list	4.3	9.1	10.0	12.6	8.7	6.9
20-unit list	7.4	8.6	9.9	15.7	13.8	9.6

THEOREM XXVI, COROLLARY 12

A. *In rote series learned by massed practice, as the length of the series (N) increases, the mean number of presentations of the series required for learning the series as a whole increases.*

Proof:

By definition and general practice, the mean number of presentations required for learning the series as a whole is the number of presentations required to learn the syllable at the point of maximum difficulty. Now this is the number of presentations up to and including the last failure of the syllable at the point of maximum difficulty. But (Theorem XXVI, Corollary 11) the number of presentations up to and including the last failure increases for all syllables alike as the length (*N*) of the series is increased. From these considerations the proposition follows.

C. Ebbinghaus' (*8*) original monograph on memory contained data on this point. His data are quoted in Table 25.

TABLE 25

Number of repetitions required for mastery of various lengths of lists. Data from Ebbinghaus (*8*).

Number of syllables in a series	Number of repetitions necessary for first errorless reproduction (exclusive of it)	Probable error
7	1	
12	16.6	± 1.1
16	30.0	± 0.4
24	44.0	± 1.7
36	55.0	± 2.8

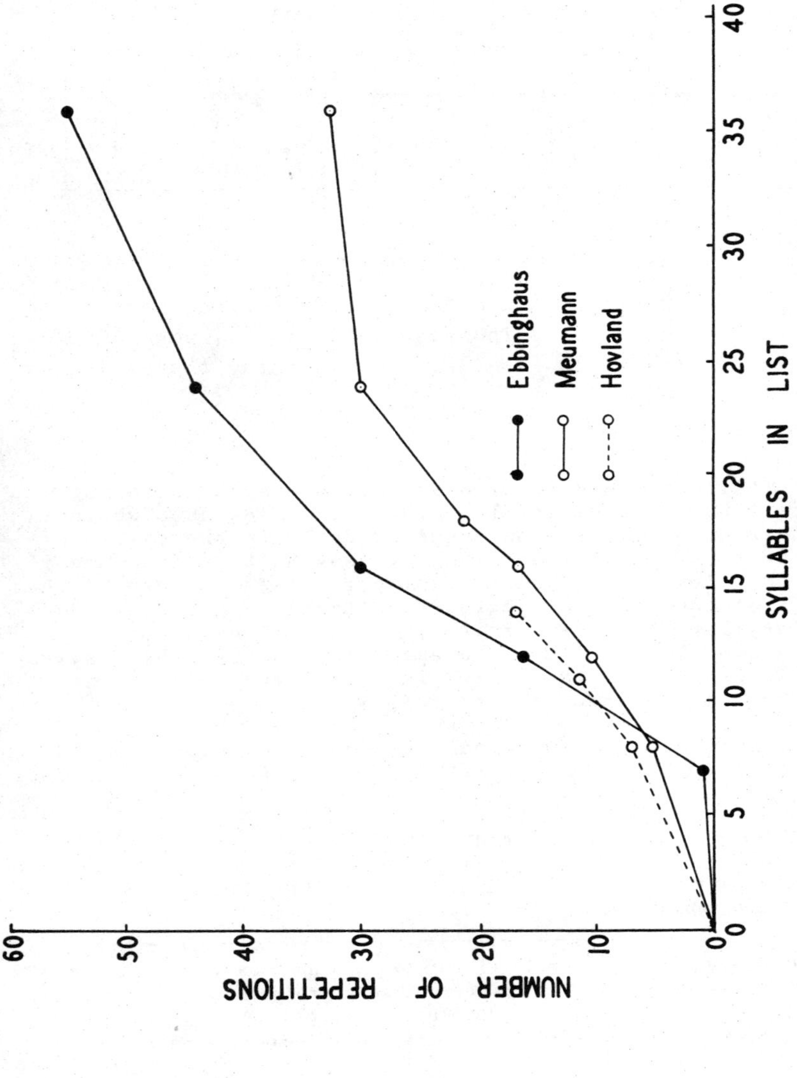

Figure 35. Mean number of repetitions required for learning various lengths of lists. Data from Ebbinghaus (8), Hovland (26), and Meumann (41).

Smaller increases in the number of trials with increased lengths were reported by Meumann (*41*). The data of Lyon

TABLE 26

Number of repetitions required for learning of varying lengths of lists. Data from Meumann (*41*).

Number of syllables	Repetitions
8	5.2
12	10.4
16	17.0
18	21.5
24	30.0
36	32.5

(*39*) and of Hovland (*26*), quoted under Theorem XIV, also show the increase in the number of repetitions with an increase in the length of the series. Extensive data on this problem were obtained by Robinson and Heron (*54*). Data from these various investigators are reproduced in Figure 35.

The experimental evidence supports the present corollary.

THEOREM XXVI, COROLLARY 13

A. *In rote series learned by massed practice, as the number of syllables increases, the mean difference in repetitions between the first success and the last failure increases for each syllable of the series.*

Proof:

This proposition follows directly from Theorems XXVI, VI, and XVII.

C. Calculations based on the results of Shipley quoted above (Corollary 10), demonstrate an increased number of repetitions between the first success and last failure with increased length of list. These data are given in Table 27.

TABLE 27

Difference in number of repetitions between first success and last failure in various syllable positions with three lengths of list. Data from Shipley (*26*).

	1st	2nd	3rd	2nd-from-last	Next-to-last	Last
8-unit list	+ 1.0	+ 1.1	+ 2.2	+ 1.3	+ 1.5	+ 0.9
14-unit list	+ 2.0	+ 5.9	+ 5.5	+ 6.2	+ 3.8	+ 4.1
20-unit list	+ 5.2	+ 5.0	+ 5.7	+ 7.7	+ 7.8	+ 6.4

Hovland's results (*26*) indicate the same trend.

The corollary is experimentally confirmed on the basis of present evidence.

THEOREM XXVI, COROLLARY 14

A. *In rote series, for any given number of massed or distributed repetitions, as the number of syllables (N) increases, the recall decreases at each syllable.*

Proof:

This proposition follows directly from Theorems XXVI and XVIII.

C. The research of Robinson and Brown (*53*) on the effect of serial position upon memorization gives data relevant to this corollary. Unfortunately, the authors do not report the actual figures involved, but present only the graphs. Reading of the graphs to the nearest figure gives the results shown in Table 28.

TABLE 28

Percentages of correct anticipations on the first recall trial for various syllable positions with five lengths of lists. Data derived from figure of Robinson and Brown (*53*).

	6 Syllables	9 Syllables	12 Syllables	15 Syllables	18 Syllables
Position 2	73	92	73	78	73
Position 3	67	62	55	59	57
Position 4	50	37	26	48	25

These data indicate that there is a tendency for the recall to decrease at every syllable with increase in length of list, but the relationship is far from perfect.

The data bearing on this point are somewhat irregular, but tend to support the corollary.

THEOREM XXVII

A. *In the learning of rote series of different length by massed practice, as the number of syllables (N) in the series increases, the inhibition at the anterior end increases at a slower rate than it does at the maximum point, i.e.,*

$$\Delta' J_0\,(N+1) - \Delta' J_0\,(N) \;>\; \Delta' J_{max}\,(N+1) - \Delta' J_{max}\,(N).$$

To prove:

J_0 increases with N less rapidly than J_{max}.

Proof:

Let J_n be the maximum inhibition in the series with N syllables. By Theorem XXVI (b),

$$J_n(N+1) - J_n(N) > J_0(N+1) - J_0(N).$$

But,

$$J_{max}(N+1) \geq J_n(N+1).$$

Hence,

$$J_{max}(N+1) - J_n(N) > J_0(N+1) - J_0(N),$$

or,

$$J_{max}(N+1) - J_{max}(N) > J_0(N+1) - J_0(N),$$

as was to be shown.

B. The concrete meaning of this theorem may be illustrated by substituting in the inequality:

$$J_{max}(N+1) - J_{max}(N) > J_0(N+1) - J_0(N)$$

the theoretical J-values of Table 1. Taking F at 1.37, we have in Table 1 a 12-syllable series and a 15-syllable series the difference in length being 3, so that $N+3$ must be taken in the above equation for $N+1$. Substituting, then, we have,

$$7.688 - 7.066 > 1.792 - 1.788$$

i.e.,

$$.623 > .004$$

which is quite in accordance with the theorem.

THEOREM XXVII, COROLLARY 1

A. *In the learning of rote series of different lengths by massed practice, as the number of syllables (N) increases, the mean number of presentations required to raise the effective excitatory potential to the mean reaction threshold increases at a slower rate at the anterior end than it does at the point requiring the maximum number of presentations for learning.*

Proof:

This proposition follows directly from Theorems XXVII and VII.

C. Computation based on the data of Hovland (*26*), partly quoted above (Table 19), gives the results presented in Table 29 concerning the increased difficulty in learning an 11-unit and a 14-unit list over that involved in learning an 8-unit list for the central and anterior syllable positions.

TABLE 29

Increased number of repetitions required for learning 11-unit and 14-unit lists over 8-unit list for central and anterior positions. Data from Hovland (*26*).

Position	11-unit over 8	14-unit over 8
Central	2.72	6.55
Anterior syllable	.63	.81

It will be observed that the increase in the number of repetitions is more rapid at the maximum point than at the anterior end, thus,

$$6.55 - 2.72 > .81 - .63$$

i.e.,

$$3.83 > .18 .$$

The experimental results are in accord with the theoretical prediction.

THEOREM XXVII, COROLLARY 2

A. *In the learning of rote series of different lengths by massed practice, assuming the presentations given each series to be equal in number and sufficient in number to raise the effective excitatory potential of each above the reaction threshold throughout: as the number of syllables (N) in a series increases, the mean reaction latency increases more slowly at the anterior end than at the point requiring the maximum number of presentations for learning.*

Proof:

This proposition follows directly from Theorems XXVII and IX.

C. No data bearing directly on the validity of this proposition have been found.

THEOREM XXVII, COROLLARY 3

A. *In the learning of rote series of different lengths, as the number of syllables (N) increases, the advantage of well-distributed practice (under the condition of Theorem XV) over massed practice increases more slowly at the anterior syllable than at the maximum point.*

Proof:

This proposition follows directly from Theorems XXVII, XV, and VII.

C. Computations based on the data of the experiment of Hovland (*26*) indicate that the advantage of distributed over massed practice in the anterior syllable position is .16 for an 8-unit list, .73 for an 11-unit list, and .74 for a 14-unit list. The corresponding data for the advantage at the point of maximum difficulty are: 1.76, 1.94, and 3.53. Thus the advantage of well-distributed practice over massed increases more slowly at the anterior syllable than at the point of maximum difficulty,

$$.74 - .16 < 3.53 - 1.76$$

i.e.,

$$.58 < 1.77\,.$$

The corollary appears to be confirmed on the basis of the present evidence.

THEOREM XXVIII

A. *In the learning of rote series of different lengths by massed practice, as the number of syllables (N) in the series increases, the inhibition at the posterior syllable increases at a slower rate than that at the maximum point, i.e.,*

$$J_{max}(N+1) - J_{max}(N) > J_{N-1}(N+1) - J_{N-1}(N)\,.$$

To prove:

J_{max} increases more rapidly with N than J_{N-1}.

Proof:

By Theorem XXI A, the maximum J_n occurs in the second half of the series. Hence we may apply Theorem XXVI C. Let J_{n_0} be J_{max} for a series of N syllables. Then, by Theorem XXVI C,

$$J_{n_0+1}(N+1) - J_{n_0}(N) > J_N(N+1) - J_{N-1}(N).$$

But,

$$J_{max}(N+1) > J_{n_0+1}(N+1).$$

Hence,

$$J_{max}(N+1) - J_{n_0}(N) > J_N(N+1) - J_{N-1}(N).$$

or,

$$J_{max}(N+1) - J_{max}(N) > J_N(N+1) - J_{N-1}(N).$$

as was to be shown.

B. The concrete meaning of this theorem may be illustrated by substituting in the inequality,

$$J_{max}(N+1) - J_{max}(N) > J_N(N+1) - J_N(N).$$

the theoretical J-values of Table 1. Taking F at 1.37, we have in Table 1 a 12-syllable series and a 15-syllable series, the difference in length being 3. Accordingly, $N + 3$ must be taken in the above equation for $N + 1$. Substituting, then, we have,

$$7.688 - 7.065 > 3.670 - 3.618$$

i.e.,

$$.623 > .052$$

which is quite in accordance with the theorem.

THEOREM XXVIII, COROLLARY 1

A. *In the learning of rote series of different lengths by massed practice, as the number of syllables increases, the mean number of repetitions required for learning increases at a slower rate at the posterior end of the series than it does at the point requiring the maximum number of presentations to raise the effective excitatory potential to the mean reaction threshold.*

Proof:

This proposition follows directly from Theorems XXVIII and VII.

C. The results of Hovland (*26*) give values for the increase in number of repetitions at the posterior syllable with increase in length of list. The 11-unit list requires 1.00 more repetition than the 8-unit list, and the 14-unit list requires 1.79 more than the 8-unit at the posterior syllable. The corresponding figures for the maximum position quoted above under Theorem XXVII, Corollary 1, are 2.72 for the 11-unit and 6.55 for the 14-unit list.

Since the differences at the point of maximum difficulty are greater than those at the posterior syllable (2.72 > 1.00 and 6.55 > 1.79), the corollary is experimentally confirmed.

THEOREM XXVIII, COROLLARY 2

A. *In the learning of rote series of different lengths by massed practice, assuming the presentations given each series to be equal in number and sufficient in number to raise the effective excitatory potential of each above the reaction threshold throughout: as the number of syllables (N) in a series increases, the mean reaction latency increases more slowly at the posterior end of the series than at the point requiring the maximum number of presentations for learning.*

Proof:

This proposition follows directly from Theorems XXVIII and IX.

C. No data bearing directly on the validity of this proposition have been found.

THEOREM XXVIII, COROLLARY 3

A. *In the learning of rote series of different lengths by massed practice, as the number of syllables (N) increases, the advantage of distributed practice (under the condition of Theorem XV) over massed practice increases more slowly at the posterior end of the series than at the point requiring the maximum number of presentations for learning.*

Proof:

This proposition follows directly from Theorems XXVIII, XV, and VII.

C. The advantage of distributed over massed practice at the posterior syllable position was found by Hovland (*26*) to be .21 for an 8-unit list, .82 for an 11-unit list, and .73 for a 14-unit list. These values are each smaller than the corresponding values for the advantage at the point of maximum difficulty, i.e., 1.76 for an 8-unit list, 1.94 for an 11-unit list, and 3.53 for a 14-unit list.

The corollary is supported by the available evidence.

THEOREM XXIX

A. *In the learning of rote series of different lengths by massed practice, as the number of syllables (N) in the series increases, the inhibition at the anterior syllable increases at a slower rate than does that at the posterior syllable, i.e.,*

$$J_N\ (N+1) - J_{N-1}\ (N) > J_0\ (N+1) - J_0\ (N).$$

To Prove: J_0 increases less rapidly with N than J_{N-1}

Proof:

By actual calculation, $J_0\ (N+1) - J_0\ (N) = \frac{1}{(N+1)F^N}$

and, $J_N\ (N+1) - J_{N-1}\ (N) = \frac{1}{F^N}$.

Here, $\frac{1}{(N+1)F^N} < \frac{1}{F^N}$

as was to be shown.

B. The concrete meaning of this theorem may be illustrated by substituting in the inequality,

$$J_N\ (N+1) - J_{N-1}\ (N) > J_0\ (N+1) - J_0\ (N)$$

the theoretical J-values of Table 1. Taking F at 1.37, we have in Table 1 a 12-syllable series and a 15-syllable series, the difference in length being 3. Accordingly, $N+3$ must be taken in the above equation for $N+1$. Substituting, then, we have,

$$3.670 - 3.618 > 1.792 - 1.788$$

i.e., $.052 > .004$

which is quite in accordance with the theorem.

THEOREM XXIX, COROLLARY 1

A. *In the learning of rote series of different lengths by massed practice, as the number of syllables (N) in the series increases, the mean number of repetitions required to raise the effective excitatory potential to the mean reaction threshold increases more slowly at the anterior syllable than at the posterior syllable of the series.*

Proof:

This proposition follows directly from Theorems XXIX and VII.

C. Data of Hovland (*26*) show that .63 more repetitions are required to learn the anterior syllable with an 11-unit list than with the 8-unit. For the same syllable position, .81 more repetitions are required to learn with the 14-unit than with the 8-unit list. Corresponding data for the posterior syllable are 1.00 and 1.79.

Since the increase is more rapid at the posterior than at the anterior syllable (1.79 - 1.00 > .81 - .63), the corollary is experimentally confirmed.

THEOREM XXIX, COROLLARY 2

A. *In the learning of rote series of different lengths by massed practice, assuming the presentations given each series to be equal in number and sufficient in number to raise the effective excitatory potential of each above the reaction threshold throughout: as the number of syllables (N) in a series increases, the mean reaction latency increases more slowly at the anterior syllable than at the posterior syllable of the series.*

Proof:

This proposition follows directly from Theorems XXIX and IX.

C. No data bearing directly on the validity of this proposition have been found.

THEOREM XXIX, COROLLARY 3

A. *In the learning of rote series of different lengths, as the number of syllables increases, the advantage of well-distributed practice (under the condition of Theorem XV) over massed practice increases more slowly at the anterior syllable than at the posterior syllable of the series.*

Proof:

This proposition follows directly from Theorems XXIX, XV, and VII.

C. The increase in the advantage of distributed over massed practice at the anterior and posterior ends, with increases in the length of the series, reported by Hovland (*26*), are presented in Table 30.

TABLE 30

Advantage of distributed over massed practice at anterior and posterior syllables with 8-unit, 11-unit, and 14-unit lengths of series. Data from Hovland (*26*).

Syllable position	Length of list		
	8 units	11 units	14 units
Anterior	.16	.73	.74
Posterior	.21	.82	.73

The increases of the repetitions required for learning with the 11-unit and 14-unit lists as compared with the 8-unit list are .57 (.73 -.16) and .58 (.74 - .16) respectively, for the anterior syllable, and .61 (.82 - .21) and .52 (.73 - .21) at the posterior syllable.

The results of the single experiment available appear to be too variable to give an adequate test of the corollary.

THEOREM XXX

A. *In rote series learned by massed practice, the index of inhibition at the beginning, end, and maximum point of the series, as the number of syllables increases, approaches the limits* $F \log_e \left(\frac{F}{F-1}\right)$, $\frac{F}{F-1}$, and $\left(\frac{F}{F-1} + \frac{F}{2(F-1)^2}\right)$ *respectively.*

To prove:

$$\lim_{N\to\infty} J_0(N) = F \log \left(\frac{F}{F-1}\right); \quad \lim_{N\to\infty} J_{N-1}(N) = \frac{F}{F-1};$$

$$\lim_{N\to\infty} J_{max} = \frac{F}{F-1} + \frac{F}{2(F-1)^2}.$$

Proof:

$$J_0(N) = 1 + \frac{1}{2F} + \frac{1}{3F^2} + \dots + \frac{1}{NF^{N-1}}$$

Anterior syllable

$$\lim_{N\to\infty} J_0(N) = 1 + \frac{1}{2F} + \frac{1}{3F^2} + \dots + \frac{1}{rF^{r-1}} + \frac{1}{(r+1)F^r} + \dots$$

$$= F\left(\frac{1}{F} + \frac{1}{2F^2} + \frac{1}{3F^3} + \dots + \frac{1}{rF^r} + \frac{1}{(r+1)F^{r+1}} + \dots\right.$$

The formula $\log(1+x) = x - \frac{x^2}{2} + \frac{x^3}{3} - \frac{x^4}{4} \dots$

is valid for all values of x numerically less than 1. Hence we may set $x = -\frac{1}{F}$ and obtain,

$$\log\left(1 - \frac{1}{F}\right) = -\frac{1}{F} - \frac{1}{2F^2} - \frac{1}{3F^3} \dots$$

Hence,

$$\lim_{N\to\infty} J_0(N) = F\left[-\log\left(1 - \frac{1}{F}\right)\right] = F \log\left(\frac{F}{F-1}\right)$$

Posterior syllable

$$J_{N-1}(N) = 1 + \frac{1}{F} + \frac{1}{F^2} + \dots + \frac{1}{F^{N-1}}$$

$$\lim_{N\to\infty} J_{N-1}(N) = 1 + \frac{1}{F} + \frac{1}{F^2} + \dots + \frac{1}{F^r} + \dots$$

The formula $\frac{1}{1-x} = 1 + x + x^2 + x^3 + \ldots .$
is valid for all values of x numerically less than 1. Hence we may substitute $x = \frac{1}{F}$ and obtain,

$$\lim_{N\to\infty} J_{N-1}(N) = \frac{1}{1-\frac{1}{F}} = \frac{F}{F-1} .$$

Maximum syllable

For the maximum we proceed thus:

Note that,

$$J_0 = 1 + \frac{1}{2F} + \frac{1}{3F^2} + \ldots + \frac{1}{NF^{N-1}} .$$

Moreover,

$$\Delta' J_n = \frac{n'+1}{(n'+1)F^{n'}} + \frac{n'+1}{(n'+2)F^{n'+1}} + \ldots + \frac{n'+1}{NF^{N-1}} .$$

$$- \frac{1}{(N-n'+1)F^{N-n'}} - \frac{2}{(N-n'+2)F^{N-n'+1}} - \ldots - \frac{n'}{NF^{N-1}}$$

Now,

$$J_n = I_0 + \sum_{r=0}^{n-1} \Delta' I_r .$$

from the definition of the difference. Now,

$$\Delta' J_n < \frac{n'+1}{(n'+1)F^{n'}} + \frac{n'+1}{(n'+2)F^{n'+1}} + \ldots + \frac{n'+1}{NF^{N-1}} .$$

Hence, J_n (any n) is such that

$$J_n < J_0 + \sum_{r=1}^{n} \frac{r+1}{(r+1)F^r} + \ldots + \frac{r+1}{NF^{N-1}}$$

$$= 1 + \frac{3}{2F} + \frac{6}{3F^2} + \ldots + \frac{1+2+3+\ldots+(n+1)}{(n+1)F^n}$$

$$+ \frac{1+2+3+\ldots(n+1)}{(n+2)F^{n+1}} + \ldots + \frac{1+2+3+\ldots+(n+1)}{NF^{N-1}} .$$

Now,

$$1 + 2 + 3 + \ldots + (n+1) = \frac{(n+1)(n+2)}{2} .$$

Hence,

$$J_n < 1 + \frac{3}{2F} + \frac{6}{3F^2} + \ldots + \frac{\frac{(n+1)(n+2)}{2}}{(n+1)F^n} .$$

Now,

$$\frac{\frac{(n+1)(n+2)}{2}}{n+1} = \frac{n+2}{2} .$$

Hence,

$$J_n < 1 + \frac{1+2}{2F} + \frac{2+2}{2F^2} + \frac{3+2}{2F^3} + \ldots\ldots + \frac{r+2}{2F^r} + \ldots\ldots$$

Hence,

$$J_n < 1 + \sum_{r=1}^{\infty} \frac{r+2}{2F^r} .$$

On the other hand, consider $\Delta' J_n$ for $n' = 1, 2, \ldots \frac{N}{2}$. The negative terms are less than the corresponding positive terms, and the terms affected have denominators CF^{C-1} with $C > \frac{N}{2}$ since $N - n' \geq \frac{N}{2}$. That is:

$$\Delta' J_n = \frac{n'+1}{(n'+1)F^{n'}} + \ldots\ldots + \frac{n'+1}{\frac{N}{2}F^{\frac{N}{2}-1}} + \ldots\ldots$$

$$+ \frac{n'}{(N-n'-1)F^{N-n'}} + \frac{n'-1}{(N-n'-2)F^{N-n'+1}} + \ldots\ldots + \frac{1}{NF^{N-1}} .$$

Hence,

$$\Delta' J_n = \frac{n'+1}{(n'+1)F^n} + \ldots\ldots + \frac{n'+1}{\frac{N}{2}F^{\frac{N}{2}-1}} \quad \text{for} \quad n' = 1, \ldots \frac{N}{2} .$$

Now,

$$J_{\frac{N}{2}-1} = J_0 + \sum_{r=1}^{\frac{N}{2}-1} \Delta I_r .$$

And so,

$$J_{\frac{N}{2}-1} > J_0 = \sum_{r=1}^{\frac{N}{2}} \left[\frac{r+1}{(r+1)F^r} + \ldots\ldots + \frac{r+1}{\frac{N}{2}F^{\frac{N}{2}-1}} \right]$$

Hence,

$$J_{\frac{N}{2}-1} > 1 + \frac{3}{2F} + \frac{6}{3F^2} + \frac{10}{4F^3} + \ldots\ldots + \frac{\frac{\frac{N}{2}(\frac{N}{2}+1)}{2}}{\frac{N}{2}F^{\frac{N}{2}-1}}$$

or,

$$J_{\frac{N}{2}-1} > 1 + \frac{3}{2F} + \frac{4}{2F^2} + \frac{5}{2F^3} + \ldots\ldots + \frac{\frac{N}{2}-1}{2F^{\frac{N}{2}-1}}$$

This shows that $J_{\frac{N}{2}-1}$ is less than the sum of an infinite series but is greater than the sum of the first $\frac{N}{2}$ terms. Hence, since J_{max} is also less than the sum of this infinite series and is greater than $J_{\frac{N}{2}}$, we have,

$$1 + \sum_{r=1}^{\frac{N}{2}} \frac{r+2}{2F^r} < J_{max} < 1 + \sum_{r=1}^{\infty} \frac{r+2}{2F^r} .$$

Hence,

$$\lim_{N\to\infty} J_{max} = 1 + \sum_{r=1}^{\infty} \frac{r+2}{2F^r} = 1 + \frac{3}{2F} + \frac{4}{2F^2} + \frac{5}{2F^3} + \frac{6}{2F^4} + \cdots$$

$$\frac{1}{F} \lim_{N\to\infty} J_{max} = \frac{1}{F} + \frac{3}{F^2} + \frac{4}{2F^3} + \frac{5}{2F^4} + \cdots$$

$$(1 - \frac{1}{F}) \lim_{N\to\infty} J_{max} = 1 + \frac{1}{2F} + \frac{1}{2F^2} + \frac{1}{2F^3} + \frac{1}{2F^4} + \cdots$$

$$= 1 + \frac{1}{2F} \left(\frac{1}{1 - \frac{1}{F}}\right)$$

or,

$$\frac{F-1}{F} \lim_{N\to\infty} J_{max} = 1 + \frac{1}{2(F-1)}$$

and

$$\lim_{N\to\infty} J_{max} = \frac{F}{F-1} + \frac{F}{2(F-1)^2} .$$

B. These equations showing the limiting values of the amount of inhibition generated by one repetition as the length of the series increases indefinitely may be illustrated by means of the F-values and K-values derived from the concrete analysis of Problem I. Accordingly, remembering (D76) that $\Delta I_n = J_n \Delta K$, and letting $F = 1.37$ and $K = .09$, we have for the limit of the first syllable, by the equation,

$$\Delta I_0 = F \log_e \left(\frac{F}{F-1}\right) \Delta K$$

$$= 1.37 \log_e \frac{1.37}{1.37 - 1} \Delta K$$

$$= 1.37 \log_e 3.7027 \Delta K$$

$$= 1.37 \times 1.309 \Delta K$$

$$= 1.793 \Delta K$$

$$= 1.793 \times .09$$

$$= .1614$$

It is to be noted that the expression,

$$F \log_e (\frac{F}{F - 1})$$

gives the limiting value of J_0, which by this computation comes out theoretically at 1.793, a value very close to that already attained for a 15-syllable series which is 1.792.

In the case of the limiting value for the final syllable by the equation,

$$\Delta I_{N-1} = (\frac{F}{F - 1}) \Delta K$$

we have by substituting,

$$= \frac{1.37}{1.37 - 1} \Delta K$$

$$= 3.703 \Delta K$$

$$= 3.703 \times .09$$

$$= .3333$$

It is to be noted that the expression,

$$\frac{F}{F - 1}$$

gives the limiting value of J which by this computation comes out theoretically at 3.703, a value very close to that already attained by a 15-syllable series which is 3.670.

In the case of the limiting value for the syllable showing the maximum amount of inhibition by the equation,

$$\Delta I_{max} = (\frac{F}{F-1} + \frac{F}{2(F-1)^2}) \Delta K$$

we have by substituting,

$$\Delta I_{max} = \left(\frac{1.37}{1.37 - 1} + \frac{1.37}{2(1.37 - 1)^2}\right) \Delta K$$

$$= \left(\frac{1.37}{.37} + \frac{1.37}{2 \times 1.369}\right) \Delta K$$

$$= (3.703 + 5.004)\ \Delta K$$

$$= 8.707\ \Delta K$$

$$= 8.707 \times .09$$

$$= .7836$$

In the above computation it is to be noted that the portion of the equation,

$$\frac{F}{F-1} + \frac{F}{2(F-1)^2}$$

gives the theoretical maximal possible value for J_{max}.

THEOREM XXX, COROLLARY 1

A. *In rote series learned by massed practice, as the number of syllables in the series increases without limit, the mean number of presentations required to reach the mean reaction threshold at the beginning, end, and maximum points of the series approaches the limits:*

(1) $$\frac{L}{\Delta E - \Delta KF \log \frac{F}{F-1}}$$

(2) $$\frac{L(F-1)}{(\Delta E - \Delta K)F - \Delta E}$$

(3) $$\frac{L(F-1)}{(\Delta E - \Delta K)F - \Delta E} + \frac{2L(F-1)^2}{2\Delta E(F-1)^2 - \Delta K \cdot F}$$

Proof:

From Theorem XXI, the maximum mean number of repetitions occurs at the same syllable as the maximum inhibition. Hence all three of the required limits may be obtained by substituting the limits of Theorem XXX in the formula,

$$_L R = \frac{L}{\Delta E - \Delta K \cdot J_n}$$

C. It is interesting to observe in this connection the theoretical number of repetitions which would be required to learn a series having the maximum amount of inhibition at its point of maximum difficulty. This may be found by the equation (Postulate 16, Corollary 1):

$$_{L}R = \frac{L}{\Delta E - \Delta J_{max}\,\Delta K}$$

Substituting in this equation we have,

$$_{L}R = \frac{2.25}{1 - 8.707 \times .09}$$

$$= \frac{2.25}{1 - .7836}$$

$$= \frac{2.25}{.2164}$$

$$= 10.39$$

This means that with the constants used, no syllable series would ever require more than about 11 syllable-presentation series to reach the mean reaction threshold. Actually, of course, far larger numbers of presentations are required for long series.

THEOREM XXXI

A. *In the learning of different lengths of rote series by massed practice, the maximum inhibition occurs at syllable position* $\bar{n}'$ *where*

$$\bar{n}' = \frac{N}{2} + \frac{\log\left(\frac{(F-1)\,N}{2F}\right)}{2 \log F}$$

provided that N exceeds a certain limit, depending on F.

Proof:

Since $J_{n+1} = J_n + \Delta' J_n$, we have $J_{n+1} > J_n$ if $\Delta' J_n > 0$, and $J_{n+1} < J_n$ if $\Delta' J_n < 0$. Hence, $J_{\bar{n}+1}$ is a maximum if $\Delta' J_{\bar{n}} > 0$ and $\Delta' J_{\bar{n}+1} < 0$. Hence $\bar{n}$ is the largest value of $\bar{n}$ for which $\Delta' J_n$ is positive. By Theorem XXI, there is a single maximum. We prove first:

(A) $$\bar{n} \geq \left[\frac{N}{2}\right], \quad \bar{n}' < \frac{2N}{3}.$$

(where $\bar{n}' = \bar{n} + 1$). Here $\left[\frac{N}{2}\right]$ means the largest integer less than $\frac{N}{2}$. Thus for $N = 7$, $\left[\frac{N}{2}\right] = 3$. Now,

$$\Delta' J_n = (n'+1)\,[G(N) - G(n')] + (N-n') - [G(N-n')] - \frac{F^{n'} - 1}{F^{N-1}\,(F-1)}.$$

Writing this at length,

$$\Delta' J_n = \frac{n'+1}{(n'+1)F^{n'}} + \frac{n'+1}{(n'+2)F^{n'+1}} + \cdots + \frac{n'+1}{NF^{N-1}}$$

$$+ \frac{N-n'}{(N-n'+1)F^{N-n'}} + \frac{N-n'}{(N-n'+2)F^{N-n'+1}} + \cdots + \frac{N-n'}{NF^{N-1}}$$

$$- \frac{1}{F^{N-n'}} - \frac{1}{F^{N-n'+1}} - \cdots - \frac{1}{F^{N-1}}$$

$$= \frac{n'+1}{(n'+1)F^{n'}} + \frac{n'+1}{(n'+2)F^{n'+1}} + \cdots + \frac{n'+1}{NF^{N-1}}$$

$$- \frac{1}{(N-n'+1)F^{N-n'}} - \frac{2}{(N-n-2)F^{N-n'+1}} - \cdots - \frac{n'}{NF^{N-1}} .$$

Here every positive term is greater than the corresponding negative term of the same denominator. Hence, if there are as many or more positive than negative terms, $\Delta' J_n > 0$. But there are $N - n'$ positive and n' negative terms. Hence, when $N - n' \geq n'$ or $n' \leq \frac{N}{2}$, then $\Delta' J_n > 0$. Hence, $\bar{n}' \geq [\frac{N}{2}]$. Henceforth it is assumed that $n' > \frac{N}{2}$. Now $\Delta' J_n$ may be re-written, combining terms, as

$$\Delta' J_n = - \frac{1}{(N-n'+1)F^{N-n'}} - \frac{2}{(N-n'+2)F^{N-n'+1}} - \cdots - \frac{2n'-N}{nF^{n'-1}}$$

$$+ \frac{N-n'}{(n'+1)F^{n'}} + \cdots + \frac{2}{(N-1)F^{N-2}} + \frac{1}{NF^{N-1}}$$

Here, as $N - n' < N$, we have,

$$\frac{1}{(N-n'+1)F^{N-n'}} > \frac{1}{NF^{N-1}}$$

$$\frac{2}{(N-n'+2)F^{N-n'+1}} > \frac{2}{(N-1)F^{N-2}}$$

etc. Hence, if there are as many or more negative terms than positive, $\Delta' J_n < 0$; i.e., if $2n' - N \geq N - n'$, or $n' \geq \frac{2N}{3}$, then $\Delta' J_n < 0$. In conclusion, then, $[\frac{N}{2}] \leq \bar{n}' < \frac{2N}{3}$. In all further discussions of the maximum, n' will be restricted to the range $\frac{N}{2} < n' < \frac{2N}{3}$. Thus,

$$N - n' > N, \quad \frac{N}{3} < N - n' < \frac{N}{2}, \quad 0 < 2n' - N < \frac{N}{3} .$$

We note in passing that the lower limit for n, *viz.*, $[\frac{N}{2}]$, is actually attained. For in

$$F^{N-n'}\,\Delta' J_n = -\frac{1}{(N-n'+1)} - \frac{2}{(N-n'+2)F} - \cdots - \frac{n'-N}{nF^{2n'-N-1}} + \frac{N-n'}{(n'+1)F^{2n'-N}} + \cdots$$

if F is sufficiently large the positive terms become arbitrarily small and $F^{N-n'}\,\Delta' J_n < 0$. Whence, $\Delta' J_n < 0$, provided that there are any negative terms, i.e., $n' > \frac{N}{2}$. For odd N, $F > \frac{N+1}{2}$ is sufficient, while for even N, $F > \frac{\sqrt{N}}{2}$ is large enough, as may be verified by substitution.

The upper limit, $\frac{2N}{3}$, will be improved now to the exact value .6484 N.

(B) Lemma α: As F decreases towards 1, $\bar{n}'$ increases towards .6484 N. For consider $F^{N-n'}\,\Delta' J_n$, as given above. A decrease in F is an increase of $\frac{1}{F}$. Hence, a decrease in F increases $\frac{1}{F}, \frac{1}{F^2}, \ldots,$ each more than the preceding. Hence, the absolute value of the positive terms of $F^{N-n'}\,\Delta' J_n$ is increased by a greater proportion than the absolute value of the negative terms. Consequently, if $\Delta' J_n$ is positive for some value of F, it is positive for all smaller values of F, down to 1. Hence, $\bar{n}'$ cannot decrease as F decreases. That $\bar{n}'$ actually increases is shown by its attainment of the value .6484 N. This is attained for F = 1 when

$$\Delta' J_n = \frac{1}{N-n'+1} - \cdots - \frac{2n'-N}{n'} + \frac{N-n'}{n'+1} + \frac{N-n'-1}{n'+2} + \cdots + \frac{2}{N-1} + \frac{1}{N}.$$

To evaluate this we use the comparison integrals,

$$\Delta' J_n \approx -\int_0^{2n'-N} \frac{x}{N-n'+x}\,dx + \int_0^{N-n'} \frac{x}{N+1-x}\,dx$$

The error of the first integral is less than $\frac{2n'-N}{n}$ and the second, less than $\frac{N-n'}{n'+1}$, each of these certainly being less than $\frac{1}{2}$ in the range $(\frac{N}{2}, \frac{2N}{3})$. Evaluating the integrals we have,

$$\Delta' J_n \approx - (2n'-N) + (N-n') \log \left(\frac{n'}{N-n'}\right) - (N-n') + (N+1) \log \left(\frac{N+1}{n'+1}\right)$$

$$= - n' + (N-n') \log \left(\frac{n'}{N-n'}\right) + (N+1) \log \left(\frac{N+1}{n'+1}\right) .$$

If we now replace $(N+1) \log \left(\frac{N+1}{n'+1}\right) \log N \log\left(\left(\frac{N}{n'}\right)\right.$ the error is less than $\log \left(\frac{N+1}{n'+1}\right) + \log \left(\frac{(N+1)\cdot(n'+1)}{N\cdot n'}\right)$ which cannot affect the order of $\frac{n'}{N}$. If $\overline{\Delta' J_n} = - n' + (N-n') \log \left(\frac{n'}{N-n'}\right) + N \log \left(\frac{N}{n'}\right)$ then, restricting ourselves to $n' \geq .63\, N$, we may calculate that,

$$\overline{\Delta' J_n} - 1.0433 < \Delta' J_n < \overline{\Delta' J_n} + .9057$$

Now if we set $\lambda = \frac{n'}{N}$, we have $\overline{\Delta' J_n} = N\, F(\lambda)$ where

$$F(\lambda) = - \lambda + (1-\lambda) \log \left(\frac{\lambda}{1-\lambda}\right) \log \left(\frac{1}{\lambda}\right)$$

For N sufficiently large, the sign of $\Delta' J_n$ will be that of $F(\lambda)$. Consequently, $\overline{\lim_{N\to\infty}} \frac{\bar{n}'}{N} = \bar{\lambda}$ where $F(\bar{\lambda}) = 0$. The value obtained in this way is $\bar{\lambda} = .6484$. Hence, $\bar{n}' < .6484\, N$ is the upper limit for $\bar{n}'$, and is attained for $F = 1$

Lemma β: As N increases, by multiples of 2, the maximum never moves toward the middle of the series.

To prove:

$n' - \frac{N}{2}$ never decreases, as N increases by multiples of 2.

By Theorem XXVI,

$$\Delta' J_n (N+1) - \Delta' J_n (N) > 0$$

or,

$$\Delta' J_n (N+1) > \Delta' J_n (N).$$

Hence, if $\Delta' J_n (N) > 0$, then $\Delta' J_n (N+1) > 0$, whence $\bar{n}' (N+1) \geq \bar{n} (N)$. Now it may happen that $\bar{n} (N+1) = \bar{n} (N)$ (see examples), and $\bar{n}$ is $\frac{1}{2}$ nearer the middle of a series of $N + 1$ syllables than that of N. But this is only an apparent decrease towards the middle, depending on whether the number of syllables is odd or even. The real situation is expressed by the relation $\bar{n} (N+2) \geq \bar{n} (N) + 1$. In other words, if we increase the number of syllables by twos, the maximum never gets nearer the middle, for $n'+1 - \frac{N+2}{2} = \bar{n}' - \frac{N}{2}$. To prove this critical inequality $n'(N+2) \geq \bar{n}'(N) + 1$ we use,

$$\Delta' J_n(N) = \frac{n'+1}{(n'+1)F^{n'}} + \cdots\cdots + \frac{n'+1}{NF^{N-1}}$$

$$- \frac{1}{(N-n'+1)F^{N-n'}} - \frac{2}{(N-n'+2)F^{N-n'+1}} - \cdots\cdots - \frac{n'}{NF^{N-1}}$$

and,

$$\Delta' J_{n+1}(N+2) = \frac{n'+2}{(n'+2)F^{n'+1}} + \cdots\cdots + \frac{n'+2}{(N+2)F^{N+1}}$$

$$- \frac{1}{(N-n'+2)F^{N-n'+1}} - \frac{2}{(N-n'+3)F^{N-n'+2}} - \cdots\cdots - \frac{n'+1}{(N+2)F^{N+1}}$$

Hence,

$$F\,\Delta' J_{n+1}(N+2) - \Delta' J_n(N) = \frac{n'+2}{(n'+2)F^{n}}$$

$$+ \frac{n'+2}{(n'+3)F^{n'+1}} + \cdots\cdots + \frac{n'+2}{(N+1)F^{N-1}} + \frac{n'+2}{(N+2)F^{N}}$$

$$- \frac{n'+1}{(n'+1)F^{n'}} - \frac{n'+1}{(n'+2)F^{n'+1}} - \cdots\cdots - \frac{n'+1}{NF^{N-1}}$$

$$+ \frac{1}{(N-n'+1)F^{N-n'}} + \frac{2}{(N-n'+2)F^{N-n'+1}} + \cdots\cdots + \frac{n'}{NF^{N-1}}$$

$$- \frac{1}{(N-n'+2)F^{N-n'}} - \frac{2}{(N-n'+3)F^{N-n'+1}} - \cdots\cdots - \frac{n'}{(N+1)F^{N-1}} - \frac{n'+1}{(N+2)F^{N}}$$

Here every positive term is numerically greater than the negative term below it, thus,

$$\frac{n'+2}{(n'+i+1)F^{n'+i-1}} - \frac{n'+1}{(n'+i)F^{n'+i-1}} = \frac{i-1}{(n'+i)(n'+i+1)F^{n'+i-1}} > 0$$

for $i > 1$, and,

$$\frac{i}{(N-n'+i)F^{N-n'+i-1}} - \frac{i}{(N-n'+i+1)F^{N-n'+i-1}} = \frac{i}{(N-n'+i)(N-n'+i+1)F^{N-n'+i-1}} > 0$$

and,

$$\frac{n'+2}{(N+2)F^{N}} - \frac{n'+1}{(N+2)F^{N}} = \frac{1}{(N+2)F^{N}} > 0.$$

Hence, $F\,\Delta' J_{n+1}(N+2) - \Delta' J_n(N) > 0$, or $F\,\Delta' J_{n+1}(N+2) > \Delta' J_n(N)$. Thus, whenever $\Delta' J_n(N)$ is positive, then $F\,\Delta' J_{n+1}(N+2)$, and consequently $\Delta' J_{n+1}(N+2)$ is also positive. This yields

$\bar{n}'(N+2) \geq \bar{n}'(N) + 1$.

Lemma γ: As N increases the maximum never gets nearer to a point in the series one-third from the posterior end. [Proved only for $\bar{n}' > .5458\,N$.]

Proof:

From Theorem XXVI, $\Delta' J_{n+1}(N+1) - \Delta' J_n(N) < 0$ for $n' > \frac{N}{2}$, whence $\bar{n}'(N+1) \leq \bar{n}'(N) + 1$, since the first n for which $\Delta' J_n(N+1)$ is negative cannot be more than one greater than the first n for which $\Delta' J_n(N)$ is negative. As may be seen from examples, it frequently happens that $\bar{n}'(N+1) = \bar{n}'(N) + 1$ and thus momentarily moves a little closer to the point one-third from the posterior end. But the real situation is that $\bar{n}'$ may not increase more than two when N increases by three. It is in this sense that the maximum never gets near to the point one-third from the posterior end. The inequality $\bar{n}'(N+1) \leq \bar{n}'(N) + 1$ yields $\bar{n}'(N+3) \leq \bar{n}'(N) + 3$, but we shall sharpen this to the (best possible) result $\bar{n}'(N+3) \leq \bar{n}'(N) + 2$.

To prove this critical inequality we show that in the interval, $.5458\,N < n' < \frac{2}{3}N$, $F^2 \Delta' J_{n+2}(N+3) < \Delta' J_n(N)$. Whence, if $\Delta' J_n(N) < 0$, it follows that $\Delta' J_{n+2}(N+3) < 0$, and so $\bar{n}'(N+3) \leq \bar{n}'(N) + 2$. This may be proved by the methods used in Lemma α.

To find the position of the maximum we shows that ΔI_n is very close to

$$\frac{1}{F} \cdot \frac{1}{1-\frac{1}{F}} - \frac{1}{(N-n'+1)F^{N-n'}} \cdot \frac{1}{(1-\frac{1}{F})^2}$$

and so we may expect the position of the maximum to be given very nearly by the equation,

$$F^{2n'-N} = (N-n'+1)(1-\tfrac{1}{F}).$$

In fact we may show if $F > 1 + \frac{4 \log N}{N}$ that at the maximum

$$F^{2n'-N} = (N-n'+1)(1-\tfrac{1}{F}) + \frac{E_n}{F}$$

with $-1 < E_n < \frac{5}{2}$, and moreover that as a function of N, $E_n \to 0$ as $N \to \infty$. The solution of the critical equation is very close to

$$\bar{n}' = \frac{N}{2} + \frac{\log\left(\frac{(F-1)\,N}{2}\right)}{\log F} .$$

Hence, $J_{\bar{n}'}$ is the maximum J value.

B. This theorem purports to give the syllable position of any rote series at which the maximum inhibition will occur. The meaning of the equation may be illustrated by utilizing it to determine the syllable position of maximum inhibition where F = 1.37 and N = 15.

$$n'_{max} = 1 + \frac{N}{2} + \frac{\log\left(\frac{(F-1)\,N}{2\,F}\right)}{2 \log F}$$

$$= 1 + \frac{15}{2} + \frac{\log\left(\frac{(1.37-1)\,15}{2 \times 1.37}\right)}{2 \log 1.37}$$

$$= 1 + 7.5 + \frac{\log\left(\frac{5.55}{2.74}\right)}{2 \times .1367}$$

$$= 1 + 7.5 + \frac{\log 2.0255}{.27344}$$

$$= 1 + 7.5 + \frac{.30643}{.27344}$$

$$= 1 + 7.5 + 1.12$$

$$= 9.62$$

Referring to Table 2 , it is seen that the maximum difficulty of learning actually falls on syllable 9, though syllable 10 has very nearly the same J-value.

THEOREM XXXI, COROLLARY 1

A. *In the learning of rote series of different lengths by massed practice, the maximum mean reaction latency at the conclusion of learning the series as a whole occurs at the ($\bar{n}'$th) syllable, provided N exceeds the limit of Theorem XXXI.*

Proof:

This proposition follows directly from Theorems XXXI and VIII.

C. In Table 31 data are given from a number of investigations, in order to check how closely the obtained values for the point of maximum difficulty in terms of repetitions correspond with the theoretical predictions.

TABLE 31

Point of maximum mean number of repetitions required for learning as obtained by various investigators with varying lengths of lists.

	Length of list	Predicted maximum	Obtained maximum
Ebbinghaus (7)	10	6.48	7
	12	7.77	9
Hovland (26)	8	5.12	5-6
	11	7.12	7
	14	9.01	8
Hovland (22)	12	7.77	7
Hull (29)	15	9.62	10
Patten (46)	16	10.22	11-12
Foucault (12)	3		3
	4		4
	5		2
	6		4
	7		4
Lepley (37)	11	7.12	7-8

It will be observed that the obtained values are quite close to the theoretical ones, but the present investigations are entirely too variable to determine the point of maximum difficulty with sufficient accuracy to provide a crucial test for the corollary. Pending further investigation no final decision can be made concerning the corollary, but there are indications of its fair adequacy in the present results.

THEOREM XXXI, COROLLARY 2

A. *In the learning of rote series of different lengths by massed practice, the maximum mean reaction latency at the conclusion of learning the series as a whole occurs at the ($\bar{n}'$th) syllable, provided N exceeds the limit of Theorem XXXI.*

Proof:

This proposition follows directly from Theorems XXXI and VIII.

C. The only investigation found bearing on this problem is one by Ward (*63*). His unpublished results shown in Figure 30 indicate that the maximum reaction time occurs at the 8th syllable. The predicted value for the point of maximum reaction time is 7.77. Although the obtained value itself was not determined with great accuracy, it appears that the experimental results and theoretical prediction are in fair agreement.

THEOREM XXXI, COROLLARY 3

A. *In the learning of rote series of different lengths, the advantage of well-distributed practice over massed practice, in terms of presentations required for learning, is maximal at the* ($\bar{n}'$ *th) syllable, provided N exceeds the limit of Theorem XXXI.*

Proof:

This proposition follows directly from Theorems XXXI and XIII.

C. In Table 32 the results of three studies of distributed practice are quoted. The length of the series used, the point where the maximum advantage of distributed practice is obtained, and the point theoretically predicted are given.

TABLE 32

Point of maximum advantage of distributed practice and the theoretical prediction for varying lengths of lists used by various investigators.

	Length of list	Predicted maximum	Obtained maximum
Patten (46)	16	10.22	11
Hovland (22)	12	7.77	7
Hovland (26)	8	5.12	5-6
	11	7.12	9
	14	9.01	7

It will be seen that the obtained values and the theoretical predictions agree within the limits of experimental error, but judgment on the adequacy of the corollary must be withheld until more precise data are available.

THEOREM XXXI, COROLLARY 4

A. *In the learning of rote series of different lengths*

by massed practice, the mean number of presentations up to and including the first success is maximal at the ($\bar{n}'$ th) syllable, provided N exceeds the limit of Theorem XXXI.

Proof:

This proposition follows directly from Theorems XXXI and XVI.

C. In Hull's (*29*) study of caffeine the point where the maximum number of presentations up to and including the first success occurs is between the 9th and 11th syllables. The predicted point of maximum repetitions for this 15-unit list (15 syllables + 1 cue syllable) would be 9.6. With an 8-unit list Shipley (*58*) found the maximum number of presentations preceding the first success to occur between the 5th and 6th syllables. The predicted value would be 5.12. With a 14-unit list the obtained maximum was between the 9th and 10th syllables, the theoretically predicted one at the 9.01th syllable. The corresponding data from the 20-unit list showed the obtained maximum between the 15th and 16th syllables, and the theoretically predicted one at the 12.58th syllable position.

As in the other corollaries under this theorem, there is a fairly close agreement between the theoretical predictions and the obtained values, but the experiments are not sufficiently critical to permit decision as to the exact validity of the corollary.

THEOREM XXXI, COROLLARY 5

A. *In the learning of rote series of different lengths by massed practice, the mean number of presentations up to and including the last failure is maximal at the ($\bar{n}'$ th) syllable, provided N exceeds the limit of Theorem XXXI.*

Proof:

This proposition follows directly from Theorems XXXI and XVI.

C. The data of Hull (*29*) bearing on the maximum number of presentations up to and including the last failure indicate the maximum between the 9th and the 11th syllable positions. It would be predicted that the maximum would occur at the 9.62nd syllable for this length.

In Table 33 results of Shipley (*58*) are given for varying lengths of lists.

The data available at present indicate that the point of maximum number of presentations up to and including the last failure as predicted by the theory is not far from the actually obtained values, but they lack sufficient precision to determine the true values with satisfactory accuracy.

TABLE 33

Syllable positions at which maximum mean number of presentations up to and including last failure occur. Data from Shipley (*58*).

Length of list	Predicted maximum	Obtained maximum
8	5.12	5-6
14	9.01	9-10
20	12.58	13-16

THEOREM XXXI, COROLLARY 6

A. *In the learning of rote series of different lengths by massed practice, the mean difference (in presentations) between the first success and last failure is maximal at the ($\bar{n}'$ th) syllable, provided N exceeds the limit of Theorem XXXI.*

Proof:

This proposition follows directly from Theorems XXXI and XVII.

C. Hull's (*29*) study indicates that the maximum difference between the first success and last failure is between the 5th and 6th syllables, while the predicted maximum should occur at the 9.62th syllable. Similarly, in Hovland's (*22*) study the maximum difference occurred at the 5th syllable with a 12-unit length, whereas the maximum difference should occur at the 7.77th syllable. In Shipley's (*58*) experiment the maximum difference was at the 3rd syllable with an 8-unit list, at the 2nd syllable with a 14-unit list, and at the 6th and 7th syllables with a 20-unit list. The theoretically predicted values would be 5.12, 9.01, and 12.58 for these lengths.

On the basis of the present evidence, which is incidental and not directly intended to test this point, it appears that the point of maximum difference between the first success and last failure is considerably anterior to the point which was theoretically predicted, and the corollary is thus in error.

THEOREM XXXI, COROLLARY 7

A. *At any stage of the learning of a rote series of different lengths by massed or distributed practice, the recall is minimal at the ($\bar{n}'$ th) syllable, provided N exceeds the limit of Theorem XXXI.*

Proof:

This proposition follows directly from Theorems XXXI and XVIII.

C. Ward's (*63*) reminiscence study gives evidence that with a 12-unit list the point of minimum recall is at about the 8th syllable position, although for varying numbers of presentations it is between 7 and 9. It would be predicted from the theory that the maximum should occur at the 7.77th syllable. Results of Robinson and Brown (*53*) are reproduced in Table 34 to show the extent of the agreement between the predicted and obtained points of minimum recall.

TABLE 34

Experimentally obtained and theoretically predicted values for point of minimum recall with varying lengths of lists. Data from Robinson and Brown (*53*).

	Length of list	Predicted minimum	Obtained minimum
Following one repetition	6*	3.53	4
	9	5.12	6
	12	7.12	6
	15	9.01	8-10
	18	10.82	5
Following five repetitions	6	3.53	4
	9	5.12	6
	15	9.01	8
	18	10.82	11

*All these lengths include the cue syllable.

It appears from the data of Table 34 and from the results of Ward (*63*) that the agreement between the theoretical values and the actually obtained values is consistently good with respect to the point of minimum recall, and the corollary appears the best confirmed of the ones contained under this theorem.

THEOREM XXXII

A. *In the learning of rote series of different lengths by massed practice, the inhibition at its maximum point ($\overline{n}'$) increases as the number of syllables in the series increases, the rate of increase possessing a negative acceleration.*

To prove:

$J_{max}(N)$ as a function of N is increasing and has a negative acceleration.

Proof:

That $J_{max}(N)$ is an increasing function of N we already know (Theorems XXVI and XXVII). Considering $J_n(N)$ as a continuous function of n and N, we may show that $J_{max}(N)$ has a negative acceleration with respect to N. But because of a certain irregularity of the distribution of the discrete values on the continuous curve for $J_n(N)$, the discrete $J_{max}(N)$ will not invariably have a negative acceleration, though it usually will. Let us write,

$$A(s) = \frac{1}{2}\sum_{i=1}^{\infty}\frac{i^2+(2s+1)i}{(s+i)F^{s+i-1}}$$

$$B(s) = \frac{1}{2}\sum_{i=1}^{\infty}\frac{i^2-i}{(s+i)F^{s+i-1}}$$

$$C(s) = \sum_{i=1}^{\infty}\frac{i-1}{(s+i)F^{s+i-1}}$$

These are analytic and *a fortiori* continuous functions of s, provided that $|F| > 1$ as we always assume. As functions of F, $F = 1$ is a singularity of all three functions.

It may be shown that for integral values of n and N,

$$J_n(N) = \frac{2F^2-F}{2(F-1)^2} - A(n') - B(N-n') + n'C(N) + B(N).$$

Hence this formula may be considered an extension of the values of $J_n(N)$ to continuous values for n and N. But it is desirable to show that this extension preserves the principal characteristics of the discrete valued function and particularly that it does not oscillate in passing from one set of discrete values to another. We may show,

(1) N constant $\frac{d^2 J_n(N)}{dn^2} < 0 \quad \frac{N}{2} \leq n' = N$

for all values of F (Cf. Theorems XXIV and XXI A);

(2) N constant $\frac{d^2 J_n(N)}{dn^2} < 0 \quad 1 \leq n' \leq N$

provided $F > 1 + \frac{1}{n'+2}$ (Cf. Theorem XXIII);

(3) N constant $\dfrac{d\,J_n(N)}{dN} > 0 \quad 1 \leq n' \leq N$

(Cf. Theorem XXVI);

(4) N constant $\dfrac{d\,J_n(N)}{dn} > 0$

for $n' \leq \frac{N}{2}$ (Cf. Theorem XXI A);

(5) N constant $\dfrac{d\,J_n(N)}{dn} < 0$

for $n' \geq \frac{2N}{3}$ (Cf. Theorem XXXI).

Hence there is a single maximum of $J_n(N)$ with $\frac{N}{2} < n' < \frac{2N}{3}$.

For this value $\dfrac{d\,J_n(N)}{dn} = 0$, or

$$-A'(n') + B'(N-n') + C(N) = 0$$

and this equation defines n' as a function of N determining the position of the maximum.

Differentiating with respect to N,

$$-A''(n')\frac{dn'}{dN} + B''(N-n')\left(1 - \frac{dn'}{dN}\right) + C'(N) = 0.$$

Whence,

$$\frac{dn'}{dN} = \frac{C'(N) + B''(N-n')}{A''(n') + B(N-n')}$$

which gives the way in which the position of the maximum changes with N. It may be shown that

$$\frac{1}{2} < \frac{dn'}{dN} < \frac{2}{3}.$$

Letting n' depend on N so that $J_n(N) = J_{max}(N)$, we have,

$$\frac{d\,J_n(N)}{dN} = -B'(N-n') + n\,C'(N) + B'(N)$$

and

$$\frac{d\,J_n(N)}{dN} > 0$$

or that $J_{max}(N)$ increases with N.

Finally,

$$\frac{d^2 J_n(N)}{dN^2} = -B'(N-n')\left(1-\frac{dn'}{dN}\right) + C'(N)\frac{dn'}{dN} + n\,C''(N) + B''(N)\,.$$

Here the first and second terms are negative while the last two are positive. But the first term dominates the others and $\frac{d^2 J_n(N)}{dN^2} < 0$ or $J_{max}(N)$ has a negative acceleration.

THEOREM XXXIII

A. *In the learning of rote series of different lengths by massed practice, as the number of syllables (N) increases by multiples of 2, the distance from the middle of the series to the point of maximum inhibition (when the distance is measured in units of syllable positions) increases.*

Proof:

By Lemma β of Theorem XXXI, as N increases the maximum never moves toward the middle of the series. When N is sufficiently large for the formula to be applicable, we have,

$$\bar{n}' = \frac{N}{2} + \frac{\log\left(\frac{N(F-1)}{2F}\right)}{2\log F}$$

Here

$$\bar{n}' - \frac{N}{2} = \frac{\log\left(\frac{N(F-1)}{2F}\right)}{2\log F}$$

is an increasing function of N, since $\log N$ is an increasing function of N. Hence, the distance of the maximum point from the middle of the series, $\bar{n}' - \frac{N}{2}$, never decreases and eventually increases, becoming infinite with N as the logarithm of N.

B. This theorem may be illustrated by applying the formula of Theorem XXXII successively to series from 12 to 15 inclusive. The values in question are shown in the second column of Table 35. Subtract from each of these values the position of the midpoint of the series and we have the distance of the point of maximum ΔJ from the middle of the series. These values are shown in the fourth column of Table 35. An inspection of this column will show that as the length of the series increases, the distance from the midpoint of the series to the point of maximum inhibitory increment ($\bar{n}'$), as measured by syllable positions, grows greater and greater exactly as stated in the theorem.

TABLE 35

Table showing theoretical values of n_{max} where F = 1.37 for several rote series of increasing lengths, together with the difference of each from the midpoint of the series and the quotient obtained by dividing this difference by *N*.

Length of series	Position of maximum Δ J	Mid-point of the series $(\frac{N}{2})$	Distance of maximum ΔJ from mid-point of series	Difference divided by *N*
12	7.7662	6.0	1.7662	.1472
13	8.3934	6.5	1.8934	.1456
14	9.0110	7.0	2.0110	.1436
15	9.6206	7.5	2.1206	.1414

THEOREM XXXIII, COROLLARY 1

A. *In the learning of rote series of different lengths by massed practice, as the number of syllables (N) in a series increases by multiples of 2, the distance from the middle of the series to the point requiring the maximum number of presentations to raise the effective excitatory potential to the mean reaction threshold (when the distance is measured in units of syllable position) increases.*

Proof:

This proposition follows directly from Theorems XXXIII and VI.

C. Relatively few studies are available concerning serial position effects with varying lengths of lists. The results of Ebbinghaus (*7*) quoted above are based on only two lengths of list. They are, on the surface, in support of the corollary, since the point of maximum difficulty (in terms of promptings required) is 1.5 syllables from the middle with a 10-unit list, and 2.5 with a 12-unit list. But the irregularity of the values is great, and from inspection it appears that the point of maximum difficulty obtained by smoothing would be closer to the middle with the 12-unit than with the 10-unit list. The results of Hovland (*26*), quoted above, suggest that the point of maximum difficulty is closer to the middle, the longer the list. Thus, in his experiment, the point of maximum difficulty was 2.0 syllables from the middle with an 8-unit list, 1.5 with an 11-unit list, and 1.0 with a 14-unit list.

The present evidence relative to this corollary is not clear-cut, but appears to be in contradiction to the theory. It is possible that the results are complicated by memory span in the case of short lists. An hypothesis based upon a "set" to learn in a forward direction has been advanced by Youtz (*67*) to explain the contradictory results obtained.

THEOREM XXXIII, COROLLARY 2

A. *In the learning of rote series of different lengths by massed practice, assuming the presentations given the series to be sufficient in number to raise the effective excitatory potential above the reaction threshold throughout the series: as the number of syllables (N) in the series increases by multiples of 2, the distance from the middle of the series to the point showing the maximum mean reaction latency (when the distance is measured in units of syllable positions) increases.*

Proof:

This proposition follows directly from Theorems XXXIII and VIII.

C. No data bearing directly on the validity of this proposition have been found.

THEOREM XXXIII, COROLLARY 3

A. *In the learning of rote series of different lengths, as the number of syllables (N) in a series increases by multiples of 2, the distance from the middle of the series to the point showing the maximum advantage from well-distributed practice over massed practice (when the distance is measured in units of syllable positions) increases.*

Proof:

This proposition follows directly from Theorems XXXIII, VI, and XIV.

C. Only one study is available which is relevant to this point. This is the research of Hovland (*26*). His results do not lend themselves to any simple codification. With an 8-unit list the point of maximum advantage of distribution of practice in terms of repetitions is 1.5 syllables beyond the middle of the list. With the 12-unit list, it is 3.0 syllables beyond, but with the 15-unit list it is 0.5 syllables anterior to the middle of the list.

The experimental results, then, do not verify the corollary, although the evidence is still somewhat ambiguous.

THEOREM XXXIII, COROLLARY 4

A. *In the learning of rote series of different lengths by massed practice, as the number of syllables (N) in a series increases by multiples of 2, the distance from the middle of the series to the point showing the maximum number of presentations up to and including the first success (when the distance is measured in units of syllable positions) increases.*

Proof:

This proposition follows directly from Theorems XXXIII, VI, and XVI.

C. In Table 36 results of Shipley (*58*) are analyzed to show the distance from the center of the list to the point of maximum number of exposures preceding the first success with three lengths of list.

TABLE 36

Distance (in syllables) from the middle of the list to the point of maximum repetitions to first success. Data from Shipley (*58*).

	Length of list		
	8-unit	14-unit	20-unit
First success	+ 2.0	+ 2.5	+ 6.5

These results tend to show the maximum number of repetitions preceding the first success shifts toward the posterior end with increased length of list.

The single study available bearing on this problem appears to support the corollary.

THEOREM XXXIII, COROLLARY 5

A. *In the learning of rote series of different lengths by massed practice, as the number of syllables (N) in a series increases by multiples of 2, the distance from the middle of the series to the point requiring the maximum number of presentations up to and including the last failure (when the distance is measured in units of syllable positions) increases.*

Proof:

This proposition follows directly from Theorems XXXIII, VI, and XVI.

C. Data of Shipley (*58*) on the distance from the middle of the list to the point of maximum number of repetitions preceding the last failure are given in Table 37. The results are so irregular that the corollary is not decisively tested. Smoothing of his curves would not give the same values as are obtained by using the actual maximum.

TABLE 37

Distance (in syllables) from the middle of the list to the point of maximum repetitions to last failure. Data from Shipley (*58*).

	Length of list		
	8-unit	14-unit	20-unit
Last failure	+ .5	+ 2.5	- 2.5

Decision as to whether experimental data will support or refute the corollary must be reserved until less ambiguous data are available.

THEOREM XXXIII, COROLLARY 6

A. *In the learning of rote series of different lengths by massed practice, as the number of syllables (N) in the series increases by multiples of 2, the distance from the middle of the series to the point showing the maximum difference between the mean number of presentations up to and including the first success and the mean number up to and including the last failure (when the distance is measured in units of syllable positions) increases.*

Proof:

This proposition follows directly from Theorems XXXIII, VI, and XVII.

C. Shipley's (*58*) results are also relevant to this point. For an 8-unit list, the distance from the center to the point of maximum difference between the first success and last failure is 0.5 units; for a 14-unit list, 4.5; and for a 20-syllable list, -3.5. The data are very irregular, however, and it appears that smoothing of the curves would give different results.

Because of the irregularity of the data in the single study available, no satisfactory decision as to the truth or falsity of the corollary on the basis of experimental evidence is possible, and judgment is suspended.

THEOREM XXXIII, COROLLARY 7

A. *In the learning of rote series of different lengths by massed practice, assuming that each series has been almost learned as a whole, as the number of syllables (N) in the series increases by multiples of 2, the distance from the middle of the series to the point of minimum probability of recall (when the distance is measured in units of syllable positions) increases.*

Proof:

This proposition follows directly from Theorems XXXIII, VI, and XVIII.

C. Extensive data on serial position effects in recall with varying lengths of lists are presented by Robinson and Brown (*53*). Their graphs are reproduced in Figure 32. No great regularity concerning this point appears in the results presented in Table 38.

TABLE 38

Distance (in syllables) of point of minimum recall from middle of list. Data from Robinson and Brown (*53*).

	Length (in syllables)				
	6*	9	12	15	18
1 repetition	+ 1.0	+ 1.5	0.0	+ 0.5 and + 2.5	- 4.0
5 repetitions	+ 1.0	+ 1.5	-	+ 0.5	+ 2.0

*Since the first syllable was used as a cue, the convention adopted in this monograph would designate this as a 5-unit list. In the calculations it is treated in this way. One syllable should be subtracted from the lengths given for each list if comparisons with the predictions of the theory are made.

The results available on this point are irregular and furnish little support for the corollary.

THEOREM XXXIV

A. *In the learning of rote series of different lengths by massed practice, as the number of syllables (N) in the series is increased the distance of the posterior end of the series from the point of maximum inhibition ($\bar{n}'$) (when the distance is measured in syllable positions) increases; furthermore, the corresponding relative distance eventually increases.*

Proof:

By actual calculation,

$$\Delta' J_{n+1}\ (N+1) \; - \; \Delta' J_n\ (N) \;=\; -\frac{1}{F^{n'}} + G\ (N+1) - G\ (n+1)$$

$$= -\frac{1}{F^{n'}} + \frac{1}{(n'+2)\,F^{n'+1}} + \ldots + \frac{1}{(N+1)\,F^{N}}$$

Here each of the positive terms is numerically less than $\frac{1}{(n'+2)\,F^{n'}}$ and there are $N - n'$ positive terms. Hence their sum is less than $\frac{N-n'}{(n'+2)\,F^{n'}}$ which is less than $\frac{1}{F^{n'}}$ when $N - n' < n' + 2$, or $\frac{N}{2} < n' + 1$. Hence when n' is in the second half of the series, the positive terms are less than the negative, and we have $\Delta' J_{n+1}\ (N+1) - \Delta\ J_n\ (N) < 0$, for $n' > \frac{N}{2} - 1$. From this, $\Delta' J_{n+1}\ (N+1) < \Delta' J_n\ (N)$, when $n' > \frac{N}{2} - 1$. If $\Delta' J_n\ (N)$ is the first negative difference in the series of N syllables, then $J_n\ (N)$ is the maximum. The inequality $\Delta' J_{n+1}\ (N+1) < \Delta' J_n\ (N)$ shows us that $\Delta' J_{n+1}\ (N+1)$ must also be negative and so the maximum in the series of $N + 1$ syllables must be at the $n + 1$st syllable or earlier. Hence the point of maximum never gets any nearer the posterior end of the series, and the formula $\bar{n} = \frac{N}{2} + \frac{\log\left(\frac{N\,(F-1)}{2\,F}\right)}{2 \log F}$ shows that it eventually recedes from the end, since $\log N$ increases less rapidly than $\frac{N}{2}$; and, for the same reason, this is a relative recession.

B. This theorem may be illustrated by the same values as those used to illustrated Theorem XXXIII. It is evident that if the relative distance from the end of the series is *increased*, it will be *decreased* when measured from the center of the series ($\frac{N}{2}$). In the fourth column of Table 35 are given the distances of the points of maximal inhibitory increment from the midpoint of the series. Now, if each of these distances is divided by the number of syllables in each series, we shall have a ratio of the one increase to the other. These ratios are given in the fifth column of the table. There it may be seen that in spite of the fact that each longer series has its maximal incremental inhibitory point farther from its midpoint, the proportionate or *relative* distance grows less and less, as the theorem states.

THEOREM XXXIV, COROLLARY 1

A. *In the learning of rote series of different lengths by massed practice, as the number of syllables (N) in the series is increased, the relative distance of the middle of the series from the point of maximum inhibition* (J_{max}) *eventually decreases.*

Proof:

This proposition follows directly from Theorem XXXIV and the fact that if a point lying between the middle and the end of a series falls relatively farther from the end, it must fall relatively closer to the middle.

C. This corollary is not susceptible of direct experimental test.

THEOREM XXXIV, COROLLARY 2

A. *In the learning of rote series of different lengths by massed practice, as the number of syllables (N) in the series is increased the relative distance from the middle of the series to the point requiring the maximum number of presentations to raise the effective excitatory potential to the mean reaction threshold eventually decreases.*

Proof:

This proposition follows directly from Theorem XXXIV, Corollary 1, and Theorem VI.

C. The data already quoted under Theorem XXXIII, Corollary 1 show a decrease in the number of units from the middle of the series to the point of maximum mean number of repetitions, although this is in opposition to the theory. It is obvious, then, that the *relative* distance from the middle of the series to the point of maximum mean number of repetitions likewise decreases as the length of the series increases.

The experimental results are *formally* in support of the corollary, but are meaningless because the distinction made in the corollaries between the results concerning the absolute distance from the middle to the point of maximum difficulty and those concerning the relative distance is irrelevant when the absolute differences fail to support the corresponding corollary.

THEOREM XXXIV, COROLLARY 3

A. *In the learning of rote series of different lengths by massed practice, assuming the presentations given the series to be sufficient in number to raise the effective excitatory potential above the reaction threshold throughout the series: as the number of syllables (N) in the series increases, the relative distance from the middle of the series to the point showing the maximum mean reaction latency eventually decreases.*

Proof:

This proposition follows directly from Theorem XXXIV, Corollary 1 and Theorem VIII.

C. No data bearing directly on the validity of this proposition have been found.

THEOREM XXXIV, COROLLARY 4

A. *In the learning of rote, series of different lengths, as the number of syllables (N) in the series increases, the relative distance from the middle of the series to the point showing the maximum advantage from distributed practice eventually decreases.*

Proof:

This proposition follows directly from Theorem XXXIV, Corollary 1, and Theorems VI and XIV.

C. The number of syllables from the center of the list to the point of maximum advantage of distributed practice divided by the length of the list, gives the *relative* distance from the middle to the point of maximum advantage of distributed practice over massed. The values obtained by Hovland (*26*) for the distance of maximum advantage from the middle of the list were 1.5 syllables with an 8-unit list, 3.0 with an 11-unit list, and 0.5 with a 14-unit list. These values divided by the appropriate lengths of the series give .19, .27, and .04 for the 8-unit, 11-unit, and 14-unit lists respectively.

Since the absolute values failed to confirm the theoretical expectation (Theorem XXXIII, Corollary 3), the *relative* distances are not closely relevant.

THEOREM XXXIV, COROLLARY 5

A. *In the learning of rote series of different lengths by massed practice, as the number of syllables (N) in the series increases, the relative distance from the middle of the series to the point showing the maximum number of presentations up to and including the first success eventually decreases.*

Proof:

This proposition follows directly from Theorem XXXIV, Corollary 1, and Theorems VI and XVI.

C. When the data of Shipley (*58*) quoted above (Theorem XXXIII, Corollary 4) are divided by the corresponding lengths of lists to give the *relative* distance from the center to the point of maximum repetitions preceding the last success, the values are as given in Table 39. The data show a double inflection in the relationship, and if finally supported by other studies will fail to confirm the corollary.

TABLE 39

Relative distances (total distance divided by length) from the point of maximum repetitions preceding the first success for three lengths of list. Data from Shipley (*58*).

8-unit list	14-unit list	20-unit list
+.25	+.18	+.32

THEOREM XXXIV, COROLLARY 6

A. *In the learning of rote series of different lengths by massed practice, as the number of syllables (N) in a series increases, the relative distance from the middle of the series to the point showing the maximum number of presentations up to and including the last failure eventually decreases.*

Proof:

This proposition follows directly from Theorem XXXIV, Corollary 1, and Theorems VI and XVI.

C. If the results of Shipley (*58*) quoted under Theorem XXXIII, Corollary 5 are divided by the corresponding lengths of list, there are obtained the *relative* distances from the middle of the list to the point of maximum repetition preceding the last failure. The results of this procedure are shown in Table 40.

TABLE 40

Relative distances (total distance divided by length) from the point of maximum repetitions preceding the last failure for three lengths of list. Data from Shipley (*58*).

8-unit list	14-unit list	20-unit list
.06	.18	-.12

As in the case of the data concerning the first success, there is a double inflection in the relationship. If this is substantiated by other experimentation the corollary will be refuted.

THEOREM XXXIV, COROLLARY 7

A. *In the learning of rote series of different lengths by massed practice, as the number of syllables (N) in a series increases, the relative distance from the middle of the series to the point showing the maximum difference between the mean number of presentations up to and including the first success and the mean number up to and including the last failure eventually decreases.*

Proof:

This proposition follows directly from Theorem XXXIV, Corollary 1, and Theorems VI and XVII.

C. As mentioned above, the only data on this point are those of Shipley (quoted under Theorem XXXIII, Corollary 5). These are so variable that they neither support nor refute this corollary.

THEOREM XXXIV, COROLLARY 8

A. *In the learning of rote series of different lengths by massed practice, assuming that each series has been almost learned as a whole: as the number of syllables (N) in the series increases, the distance from the middle of the series to the point of minimum probability of recall eventually decreases.*

Proof:

This proposition follows directly from Theorem XXXIV, Corollary 1, and from Theorems VI and XVIII.

C. Since the data on the *absolute* distances from the middle to the point of minimum recall are irregular and fail to support the theoretical prediction (Theorem XXXIII, Corollary 6), the *relative* distances would not be meaningful.

SOME PRELIMINARY PROPOSITIONS CONCERNING THE CHANGES IN EXCITATORY STRENGTH FOLLOWING THE TERMINATION OF LEARNING

In general, it is assumed:

(1) that mean effective excitatory strength is a function of time (Postulate 10);

(2) that mean effective excitatory strength is always positive (Postulate 11, Corollary 1, and Postulate 13; for $E > I$ and $dI > bE$ imply $d > b$, hence $Ee^{-bt} > Ie^{-dt}$ for all t;

(3) that mean effective excitatory strength is the difference between mean excitatory potential and mean inhibitory potential (Postulate 10);

(4) that mean excitatory potential is always positive and decreasing (Postulate 12);

(5) that mean inhibitory potential is always positive and decreasing (Postulate 13);

(6) that the mean excitatory potential is independent of n, the ordinal number of any given syllable (Postulate 3).

The general conditions stipulated above are made more specific by the equations:

$E_t = ae^{-bt}$ (Postulate 12) (1)

$I_t = ce^{-dt}$ (Postulate 13) (2)

(These formulae state nothing more nor less than that the inhibitory potential and excitatory potential decrease with time at a rate proportional to their magnitude.)

and,

$\bar{E}_t = E_t - I_t = ae^{-bt} - ce^{-dt}$ (3)

where E_t is the mean excitatory potential at time t, I_t is the mean inhibitory potential at time t, and $\bar{E}_t = b(t)$ is the mean effective excitatory strength.

$t = 0$ corresponds to the termination of learning

a, b, c, d are constants (3a)

It follows from assumption (4) that $a > 0$ and $b > 0$ (4)

It follows from assumption (5) that $c > 0$ and $d > 0$ (5)

Since (Postulate 11, Corollary 2) the mean effective excitatory potential is always positive, it follows that $a > c$ and $d > b$ (6)

Whence, $a > c > 0$ (7)

$d > b > 0$ (8)

It follows from (6) that a and b depend upon N and F and not upon n (9)

It seems reasonable to assume that b and d depend solely upon F (10)

b measures what we may call *excitatory decay* (11)

d measures what we may call *inhibitory decay* (12)

The initial value of the mean inhibitory potential is given by I_n, i.e., $c = I_n$ (13)

Also, by Theorem V, $I_n > 0$; by Theorem XXII, $I_{N-1} > I_0$ (14)

THEOREM XXXV

A. *Following the termination of the active learning by the method of massed practice, the mean effective excitatory potential ($\bar{E}$) will show an initial increase.*

To prove:

That the slope of the effective excitatory potential is positive at $t = 0$.

Proof:

$$f'(t) = -ab\,e^{-bt} + cd\,e^{-dt} \quad (1)$$

$$f'(0) = -ab + cd \quad (2)$$

which will be positive since, by Postulate 13, $cd > ab$.

B. The principle stated in this theorem is illustrated by Figure 19, where it may be seen that at $t = 0$, $\bar{E}$ begins to rise rather sharply and continues to do so up to a point near $t = 2$. The actual theoretical values may be checked by reference to column $\bar{E}$ in Table 3.

THEOREM XXXV,.COROLLARY 1

A. *In rote series learned by massed practice to the point where some, but not all, of the syllables are recalled correctly, the mean number of syllables susceptible of correct recall will increase for a certain length of time after the termination of active learning.*

Proof:

This proposition follows directly from Theorems XXXV and XVIII .

C. An increase in the number of syllables recalled for a period of time following learning was first demonstrated by Ward (*63, 64*). The subjects learned 12-unit lists of nonsense syllables by the anticipation method. For one group, the criterion of learning was 7 syllables out of 12 correct; for another group, it was correct anticipation of the entire list. The third group was given only a single presentation of the list. On a given day the subjects learned to the criterion and then continued either immediately or after a chosen rest interval. It was found that the average retention scores were higher after a rest period than on the control trials in which no rest was interpolated. In Table 41 results are quoted from Ward, giving retention in terms of average recall with varying lengths of time interval at the three criteria of learning.

TABLE 41

Mean number of syllables recalled at varying time intervals following learning to three criteria of performance. Data from Ward (*63*).

Interval	Stage of learning		
	1 Trial	7 Correct	Mastery
"C" (6 sec.)	1.75	7.00	9.75
.5-minute	1.92	7.54	10.38
2-minute	1.96	7.50	10.42
5-minute	1.29	7.33	10.04
10-minute	1.88	5.92	8.79
20-minute	1.63	4.75	7.71

Results on relearning demonstrated the same trend, i.e., fewer trials were required for relearning when a period of time was interpolated between learning and recall.

Ward's subjects were permitted to read jokes during the

rest pause. In Hovland's reminiscence experiments (*20; 21; 23; 24*), rehearsal was prevented in a more effective way by the introduction of color-naming during the rest pause. The results, however, support those of Ward at the three degrees

TABLE 42

Recall scores with and without 2-minute rest pause. Data from Hovland (*20*).

	Criterion		
	1 presentation	7 correct	12 correct
Number of syllables recalled immediately after presentation	1.18	6.96	11.12
Number of syllables recalled two minutes after presentation	1.57	7.49	11.49

of learning employed. It will be seen in Table 42 that the recall is higher when a 2-minute rest pause is given between learning and recall.

The corollary is experimentally confirmed.

THEOREM XXXV, COROLLARY 2

A. *In rote series being learned by massed practice, the mean reaction latency of those syllables already above the reaction threshold will show a decrease for a certain length of time following the termination of active learning.*

Proof:

This proposition follows directly from Theorems XXV and VIII.

C. Ward (*63*) obtained measurements on length of anticipation time of all items correctly anticipated. This measure was found to produce a curve of average retention very similar to that based upon recall scores. Data on this point are given in Table 43. The anticipation times quoted are the average times in seconds between the response and the appearance of the item given in response. Thus, the complementary values (determined from the two seconds between syllables) would indicate the reaction times.

The results are in accord with the prediction of the corollary, since the subjects were able to anticipate the syllables with a shorter reaction time after the introduction of short rest pauses than on the control trials with only a sixth second interval.

TABLE 43

Average anticipation times and reaction times (2 seconds - anticipation times) after varying lengths of intervals between learning and recall. Data from Ward (*63*).

Length of interval	Average anticipation time	$P.E._M$	Reaction times *
6 seconds	1.06 seconds	0.03	.94
30 "	1.15 "	0.03	.85
2 minutes	1.15 "	0.02	.85
5 "	1.10 "	0.02	.90
10 "	.97 "	0.03	1.03
20 "	.92 "	0.04	1.08

*Reaction time = interval between exposures (2 seconds) - anticipation time.

THEOREM XXXV, COROLLARY 3

A. *In rote series learned by massed practice to the point where some, but not all, of the syllables are recalled correctly, the probability of recall will show a spontaneous increase for a certain length of time following the termination of active learning.*

Proof:

This proposition follows directly from Theorems XXXV and XVIII.

C. The results of Ward (*63*) and of Hovland (*20*; *21*) quoted under Theorem XXXV, Corollary 1, have demonstrated the spontaneous increase in recall following the termination of learning after learning to such a slight degree as obtained after a single presentation of the list, after partial learning, and after complete mastery. The point in learning where maximum reminiscence is obtained has not yet been determined, but an experiment by Hovland and Hill is in progress to determine this relationship.

The corollary has experimental support.

THEOREM XXXVI

A. *The increase in effective excitatory potential immediately following the termination of active learning by massed practice will manifest a negative acceleration.*

To prove:

That the rate of change of slope (velocity of change) is negative at time $t = 0$; i.e., $f''(0) < 0$.

Proof:

$f'(t) = -ab\,e^{-bt} + cd\,e^{-dt}$ by Theorem XXXV (1)

hence,

$f''(t) = ab^2\,e^{-bt} - cd^2\,e^{-dt}$ (1)

$f''(0) = ab^2 - cd^2$.

But,

$cd > ab$ by Theorem XXXV (3)

and,

$d > b$ Equation (8) of "Preliminary Propositions," p. 256

Then,

$cd^2 > ab^2$ (2)

and

$ab^2 - cd^2 < 0$.

Whence,

$f''(0) < 0$. (3)

B. The principle stated in this theorem may be understood by referring to Figure 19. In this figure it may be seen that the curve rises rather sharply at first, but gradually flattens out in the neighborhood of $t = 2$. In Table 3 the same thing may be seen in numerical values: $\bar{E}$ rises .074 points between $t = 0$ and $t = .2$; it rises only .064 points between $t = .2$ and $t = .4$; and so on until it rises .014 points between $t = 1.6$ and $t = 1.8$, and finally only .002 points between $t = 1.8$ and $t = 2.0$, which brings the negative acceleration to the vanishing point. This is all in accordance with the statement in the theorem.

THEOREM XXXVI, COROLLARY 1

A. *In rote series learned by massed practice to a point where either a part, or all, of the syllables are recalled correctly, the mean reaction latency of those syllables already just above the reaction threshold will decrease on the average with a negative acceleration for a finite length of time after the termination of active learning.*

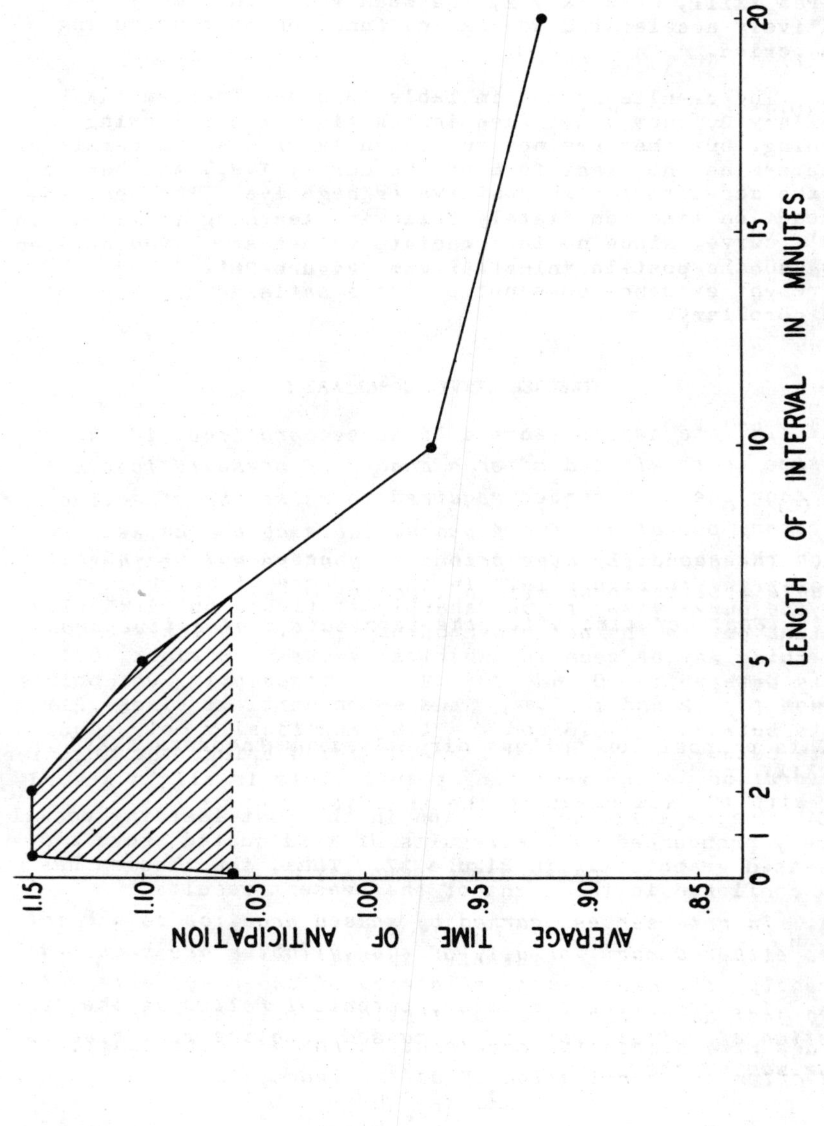

Figure 36. Average time of anticipation for items correctly given in recall. Reaction time = 2 seconds - anticipation time. Data from Ward (63).

Proof:

The mean reaction latency is a decreasing, negatively accelerated function of the mean effective excitatory strength (Postulate 18), and the mean effective excitatory strength is, by Theorem XXXVI, an increasing negatively accelerated function of time for a certain period after the termination of active learning. Hence, by the argument of Theorem XXIII, Corollary 2, the mean reaction time is a negatively accelerated decreasing function of time during this period.

C. The results quoted in Table 43 under Theorem XXXV, Corollary 2, show a decrease in reaction time following learning, but they are not sufficiently precise to permit us to determine the exact form of the curve; i.e., whether or not the acceleration is positive or negative. The decrease in reaction time immediately following learning is linear in Ward's curve, since no intermediate values are given between zero and the post-learning maximum (Figure 36).

Present evidence does not permit a satisfactory test of this corollary.

THEOREM XXXVI, COROLLARY 2

A. *In rote series learned by massed practice, if the practice is terminated after a number of presentations not less than the mean number required to raise the effective excitatory potential of a given syllable above the mean reaction threshold (L), the probability of recall at that syllable will increase with a negative acceleration for a finite length of time after the termination of active practice.*

Proof:

This proposition follows directly from Theorems XXXVI and XIX.

C. The negative acceleration in the post-learning recall is very pronounced in the results of Ward quoted above and presented graphically in Figure 37. Thus, the theorem has been confirmed in the light of the present results.

THEOREM XXXVII

A. *The effective excitatory potential following the termination of active learning by massed practice will rise to a maximum at the point* $t = \bar{t}$, *where,*

$$\bar{t} = \frac{1}{d-b} \log_e \left(\frac{cd}{ab}\right).$$

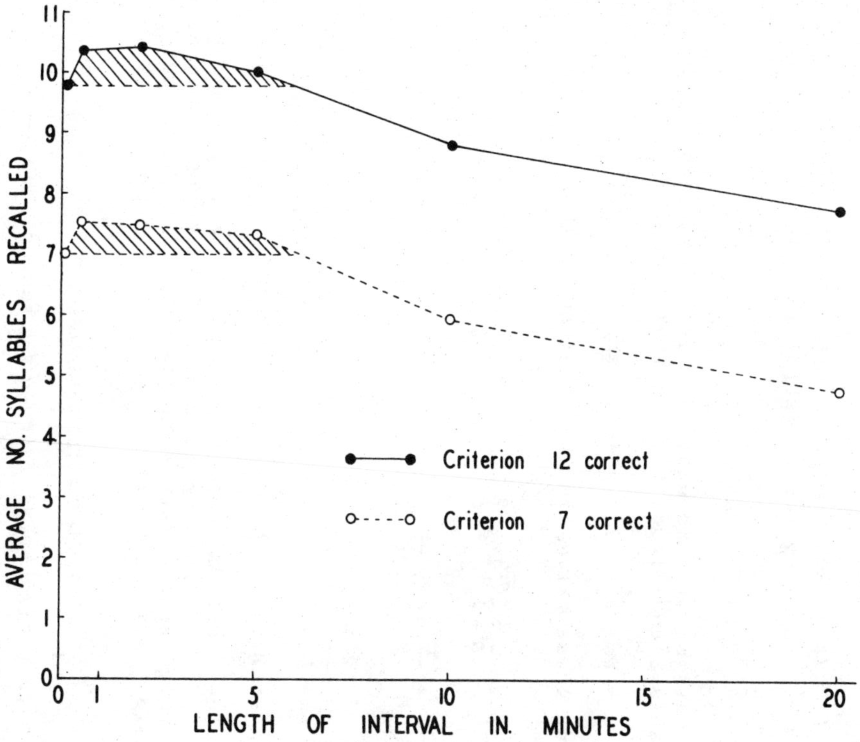

Figure 37. Average number of syllables recalled after various time intervals following learning. In Part I the learning was carried through the first trial on which every item was correctly anticipated, while in Part II the learning was stopped after the first trial on which 7 out of 12 syllables were correctly given in anticipation. Cross-hatching shows post-learning increase in recall. Data from Ward (63).

To prove:

That when $f'(t) = 0$

$$\bar{t} = \frac{1}{d-b} \log \left(\frac{cd}{ab}\right).$$

Proof:

$$f'(t) = -ab\, e^{-b\bar{t}} + cd\, e^{-d\bar{t}}$$

by Theorem XXV (A). Hence $f'(t) = 0$ if and only if $ab\, e^{-b\bar{t}} = cd\, e^{-d\bar{t}}$. Solving for $\bar{t}$, we obtain:

$$\bar{t} = \frac{1}{d-b} \log \left(\frac{cd}{ab}\right).$$

B. The meaning of this equation may be illustrated by means of Table 3 and Figure 19. In that example,

$$a = 8, \quad b = .1, \quad c = 4, \quad \text{and} \quad d = .3$$

Substituting these values in the equation of this theorem, we have,

$$\bar{t} = \frac{1}{.3 - .1} \log_e \frac{4 \times .3}{8 \times .1}$$

$$= \frac{1}{.2} \log_e \frac{1.2}{.8}$$

$$= 5 \log_e 1.5$$

$$= 5 \times .4055$$

$$= 2.03$$

It is to be observed that the value 2.03 is very close to the value of $\bar{t}$ secured for the conditions assumed above by the rather rough approximate method employed in the derivation of Table 3 and Figure 19.

THEOREM XXXVII, COROLLARY 1

A. *In rote series being learned by massed practice, the mean reaction latency at each syllable already just above the threshold will reach its post-learning minimum at the time* $\bar{t}$ *(the value of which will depend on the syllable position).*

Proof:

The mean reaction latency is a decreasing function of $\bar{E}$ (Postulate 18). The theorem follows directly from Theorems XXXVII and VIII.

C. The empirical check on the validity of this proposition must await the empirical determination of the value of the constants *a*, *b*, *c*, and *d*. Presumably these latter determinations will be effected by a method analogous to that of Problem 1.

THEOREM XXXVII, COROLLARY 2

A. *In rote series being learned by massed practice, the recall at each syllable position will reach its post-learning maximum at the time $\bar{t}$ (the value of which will depend on the syllable position).*

Proof:

This proposition follows directly from Theorems XXXVII and XVIII.

C. The empirical check on the validity of this proposition must await the empirical determination of the value of the constants *a*, *b*, *c*, and *d*.

THEOREM XXXVII, COROLLARY 3

A. *The presentations in the learning of rote series may be separated by temporal intervals in which no active learning occurs, of such lengths that all syllables of a series will be learned by a smaller number of presentations than by massed practice.*

Proof:

Let T be the time interval between successive repetitions, and let R_{max} be the mean number of repetitions required to learn the syllable with maximum inhibition. Then, if we choose T so that

$$T \cdot R_{max} < \bar{t}$$

the increments of mean effective excitatory strength will be greater at each syllable of the series, on every repetition, than in the case of massed practice. Hence the corollary follows.

C. The results of Ward (*63*) indicate that the optimal time for the appearance of reminiscence with massed practice of verbal materials is about two minutes. This interval between successive trials with distributed practice was consequently employed in a study by Hovland (*22*). The results show that with a 2-second rate of syllable presentation, distributed practice required fewer trials than massed for the learning of every syllable in the list (Figure 33). The advantage of distribution is, however, greatest in the central portion of the series. Data of Patten (*46*) are very similar. Only two of the syllables (1 and 3) required more trials by distributed than by massed practice.

The corollary is experimentally verified.

THEOREM XXXVII, COROLLARY 4

A. *In rote series learned by evenly distributed practice, the time interval between presentations which, at any stage of the learning, minimizes the mean reaction latency at any given syllable, is a decreasing function of the number of repetitions.*

Proof:

Let T be the time interval between successive repetitions and let r be the number of repetitions. Then, the mean effective excitatory strength at the termination of the rth repetition is:

$$\bar{\mathrm{E}}(T, r) = \sum_{i=0}^{r-1} (ae^{-biT} - ce^{-diT}).$$

$\bar{\mathrm{E}}(T, r)$ will be a maximum for a fixed value of r when:

$$\frac{d\bar{\mathrm{E}}(T, r)}{dT} = \sum_{i=0}^{r-1} \frac{d}{dT} (ae^{-biT} - ce^{-diT}) = 0.$$

But,

$$\frac{d}{dT} (ae^{-biT} - ce^{-diT}) = i \frac{d}{d(iT)} (ae^{-biT} - ce^{-diT}),$$

which, by Theorem XXXVII, is positive for $iT < \bar{t}$, and negative for $iT > \bar{t}$. Hence, if

$$\frac{d\bar{\mathrm{E}}(T, r)}{dT} = 0,$$

$(r-1)T$ must be not less than $\bar{t}$. Then, if r is increased by one, the above derivative becomes negative, and hence the optimum time interval becomes less. Then the corollary follows immediately, since the mean reaction time at any syllable is a decreasing function of the mean effective excitatory strength.

C. No data bearing directly on the validity of this proposition have been found.

THEOREM XXXVII, COROLLARY 5

A. *In rote series learned by evenly distributed practice the time interval between presentations which, at any stage of the learning, maximizes the recall at any given syllable, is a decreasing function of the number of repetitions.*

Proof:

The corollary follows from the proof of Theorem XXXVII, Corollary 4, since recall is an increasing function of the effective excitatory strength (Theorem XVIII).

C. No data bearing directly on the validity of this proposition have been found.

THEOREM XXXVII, COROLLARY 6

A. *In the learning of rote series, after any given number of repetitions, the length of time after the termination of active practice at which the number of syllables susceptible of recall is a maximum is greater when the learning is done by massed practice than when done by distributed practice.*

Proof:

Let r be the number of repetitions, and let T be the time interval between successive repetitions in the case of distributed practice. Then, at a given syllable, the mean effective excitatory strength at the termination of active learning is:

$$\bar{\bar{E}}_m = r(a - c)$$

when done by massed practice;

$$\bar{\bar{E}}_d = \sum_{i=0}^{r-1}(ae^{-biT} - ce^{-diT})$$

when done by distributed practice. At a period of time t, after the termination of active practice, these become, respectively:

$$\bar{\bar{E}}_m(t) = r(ae^{-bt} - ce^{-dt})$$

$$\bar{\bar{E}}_d(t) = \sum_{i=0}^{r-1}(ae^{-b(iT+t)} - ce^{-d(iT+t)})$$

But, by Theorem XXXVII, $\bar{\bar{E}}_m(t)$ is maximum at $t = \bar{t}$, whereas $\bar{\bar{E}}_d(t)$ clearly decreases for all values of t not less than $\bar{t}$.

Hence, $\bar{E}_d(t)$ must attain its maximum after a shorter period of time than $\bar{E}_m(t)$.

This result is true for all syllables of the series, and hence the corollary follows.

C. No data bearing directly on the validity of this proposition have been found.

THEOREM XXXVII, COROLLARY 7

A. *In the learning of rote series, after a given number of repetitions, the length of time after the termination of active practice at which the mean reaction latency at any syllable is a minimum, is greater when the learning is done by massed practice than when done by distributed practice.*

Proof:

The corollary follows from the proof of Theorem XXXVII, Corollary 6, since the mean reaction latency is a decreasing function of mean effective excitatory potential (Theorem VIII).

C. No data bearing directly on the validity of this proposition have been found.

THEOREM XXXVII, COROLLARY 8

A. *In the learning of rote series, after a given number of presentations, the length of time after the termination of active practice at which the recall at any syllable is a maximum is greater when the learning is done by massed practice than when done by distributed practice.*

Proof:

The corollary follows from the proof of Theorem XXXVII, Corollary 6, since recall is an increasing function of mean effective excitatory strength (Theorem XVIII).

C. No data bearing directly on the validity of this proposition have been found.

THEOREM XXXVII, COROLLARY 9

A. *In the learning of rote series, at any stage of learning (in terms of mean effective excitatory strength above the threshold) for a given syllable, the maximum reduction in mean reaction latency at that syllable during the post-learning period will be greater when the practice has been done by massed practice than when done by distributed practice.*

Proof:

Let r_m be the number of repetitions required to reach a given stage of learning by massed practice, and let r_d be the number of repetitions required to reach the same stage by distributed practice. Then, using the expressions of the proof of Theorem XXXVII, Corollary 6,

$$\bar{E}_m(t) = r_m (a-c)$$

$$\bar{E}_d(t) = \sum_{i=0}^{r_d-1} [ae^{-b(iT+t)} - ce^{-d(iT+t)}]$$

$\bar{E}_m(t)$ will be maximum for $t = t$, and $\bar{E}_d(t)$ will be maximum for some smaller value of t, say $t = \bar{t}_d$. Then the maximum increases in $\bar{E}$ will be, respectively:

$$\bar{E}_m(\bar{t}) - \bar{E}_m(0) = r_m [(ae^{-b\bar{t}} - ce^{-d\bar{t}}) - (a-c)]$$

$$\bar{E}_d(\bar{t}_d) - \bar{E}_d(0) = \sum_{i=0}^{r_d-1} [ae^{-b(iT+\bar{t}_d)} - ce^{-d(iT+\bar{t}_d)} - ae^{-biT} + ce^{-diT}] .$$

But, from Theorem XXXVII, the terms of the last sum will be negative for all values of i such that $iT > \bar{t}$. Let r_d' be the first such value of i. Then r_d' must be less than r_m, for otherwise we would have,

$$\bar{E}_d(0) > \bar{E}_m(0)$$

which is contrary to hypothesis. Hence,

$$E_d(t_d) - E_d(0) < \sum_{i=0}^{r_d-1} [ae^{-b(iT+\bar{t}_d)} - ce^{-d(iT+\bar{t}_d)} - ae^{-biT} + ce^{-diT}]$$

$$< \sum_{i=0}^{r_m-1} (ae^{-b(iT+\bar{t}_d)} - ce^{-d(iT+\bar{t}_d)} - ae^{-biT} + ce^{-diT}]$$

$$< r_m [(ae^{-b\bar{t}} - ce^{-d\bar{t}}) - (a-c) = \bar{E}_m(\bar{t}) - \bar{E}_m(0)].$$

Hence, the maximum increase in mean effective excitatory strength during the post-learning period is greater in the case of massed practice than in the case of distributed practice.

Since the mean reaction time at the termination of active practice is the same in both cases, and since mean reaction time is a decreasing function of mean effective excitatory strength (Theorem VIII) the corollary follows directly from the above result.

C. No data bearing directly on the validity of this proposition have been found.

THEOREM XXXVII, COROLLARY 10

A. *In the learning of rote series, at any stage of the learning of a given syllable such that probability of recall is greater than zero and less than perfection, the maximum increase in probability of correct recall at that syllable during the post-learning period will be greater when the learning has been done by massed practice than when done by distributed practice.*

Proof:

This corollary follows from the proof of Theorem XXXVII, Corollary 9, since recall is an increasing function of mean effective excitatory strength (Theorem XVIII).

C. Results on individual syllables with a wide variety of time intervals for recall following massed and distributed practice are not available. Two studies, however, suggest confirmation of this corollary. Hovland's study (*20*) of reminiscence following learning by massed and by distributed practice showed that increases two minutes after learning are much more apt to occur with massed than with distributed practice. With massed practice to a criterion of 7 syllables correct, 6.96 syllables were recalled immediately after the criterion trial. When two minutes were interpolated between learning and recall, the number recalled rose to 7.49. When the learning was to the same criterion by distributed practice, 8.00 syllables were recalled with no rest, and 8.04, with rest. The difference for massed practice has a critical ratio of 3.25, while that for distributed practice is only 0.20. It will be observed from these results that recall is better following distributed practice, both with and without a 2-minute rest pause, but reminiscence is more apt to occur following the massed than following the distributed practice. Since these results were obtained at only a single time interval following learning (2 minutes), they do not necessarily prove that the post-learning recall might be greater with some other time interval for distributed practice. Results of another study by Hovland (*25*) make this appear very doubtful, however. Data were obtained on retention at the following time intervals after learning by massed and distributed practice: 6 seconds, 2 minutes, 10 minutes, 24 hours. No post-learning increase is shown with distributed practice, but a reliable rise at 2 minutes with massed practice is indicated.

The present evidence is not sufficient to confirm the corollary unambiguously, but tends to support the theoretical prediction.

THEOREM XXXVII, COROLLARY 11

A. *In rote series learned by distributed practice, the time of the post-learning minimum in mean reaction latency at any syllable will coincide with the termination of active learning for sufficiently large intervals between successive repetitions.*

Proof:

In the notation of Theorem XXXVII, Corollary 6:

$$\bar{E}_d(t) = \sum_{i=0}^{T-1} [ae^{-b(iT+t)} - ce^{-d(iT+t)}].$$

But if T is larger than $\bar{t}$, then,

$$\bar{E}_d(t) < \sum_{i=0}^{T-1} [ae^{-biT} - ce^{-diT}] = \bar{E}_d(t)$$

for all positive values of t. Hence, for sufficiently large values of T, the mean effective excitatory strength is greater at the termination of learning than it is at any subsequent time. Since mean reaction time is a decreasing function of mean effective excitatory strength (Theorem VIII) the corollary follows.

C. No data bearing directly on the validity of this proposition have been found.

THEOREM XXXVII, COROLLARY 12

A. *In the learning of rote series by evenly distributed practice, at any stage of the learning of a given syllable such that the probability of recall is greater than zero and less than perfection, the time of the post-learning maximum in recall at any such syllable will coincide with the termination of active learning for sufficiently large intervals between successive repetitions.*

Proof:

This corollary follows from the proof of Theorem XXXVII, Corollary 11, since recall is an increasing function of mean effective excitatory strength (Theorem XVIII).

C. No data bearing directly on the validity of this proposition have been found.

THEOREM XXXVII, COROLLARY 13

A. *In rote series learned by a given number of evenly distributed repetitions, the time interval between successive repetitions necessary to make the time of the post-learning minimum in reaction latency at a given syllable coincide with the termination of active learning, increases from each end of the series toward the point of maximum inhibition.*

Proof:

The post-learning minimum in mean reaction time occurs at the same time as the maximum in mean effective excitatory strength. Now (Theorem VIII), as the time interval T between successive repetitions increases, the time of the post-learning maximum moves to the left until it first coincides with the termination of active learning, and for this value of T we will have:

$$\left.\frac{d\bar{E}_d(t)}{dt}\right]_{t=0} = 0 .$$

But,

$$\bar{E}_d\,(t) = \sum_{i=0}^{r-1} (ae^{-b(iT+t)} - ce^{-d(iT+t)})$$

$$\therefore \qquad \left.\frac{d\bar{E}_d(t)}{dt}\right]_{t=0} = \sum_{i=0}^{r-1} (cde^{-biT} - abe^{-diT})$$

which increases with c. Hence, the larger the inhibition, the larger will be the time interval (T) necessary to bring about the required coincidence. The corollary, then, follows from Theorem XXI.

C. No data bearing directly on the validity of this proposition have been found.

THEOREM XXXVII, COROLLARY 14

A. *In rote series learned by a given number of evenly distributed repetitions to a point such that the recall of the series as a whole is greater than zero but less than perfection, the time interval between successive repetitions necessary to make the time of the post-learning maximum in recall at a given syllable coincide with the termination of active practice increases from each end of the series toward the point of maximum inhibition.*

Proof:

This corollary follows from the proof of Theorem XXXVII, Corollary 13, using Theorem XVIII.

C. No data bearing directly on the validity of this proposition have been found.

THEOREM XXXVII, COROLLARY 15

A. *In rote series partially or wholly learned by a given number of evenly distributed repetitions, the time interval between successive repetitions necessary to make the time of the post-learning minimum in mean reaction latency coincide with the termination of active practice is greater at the posterior syllable than at the anterior syllable.*

Proof:

This proposition follows directly from the proof of Theorem XXXVII, Corollary 13, by Theorem XXII.

C. No data bearing directly on the validity of this proposition have been found.

THEOREM XXXVII, COROLLARY 16

A. *In rote series partially learned by a given number of evenly distributed repetitions, the time interval between successive repetitions necessary to make the time of the post-learning maximum in recall coincide with the termination of active practice is greater for the posterior syllable than for the anterior syllable.*

Proof:

This proposition follows directly from the proof of Theorem XXXVII, Corollary 15, using Theorem XVIII.

C. No data bearing directly on the validity of this proposition have been found.

THEOREM XXXVII, COROLLARY 17

A. *The mean difference (in repetitions) between the first success and the last failure at any syllable of a rote series is greater when the practice has been done by massed repetitions than when done by well distributed repetitions.*

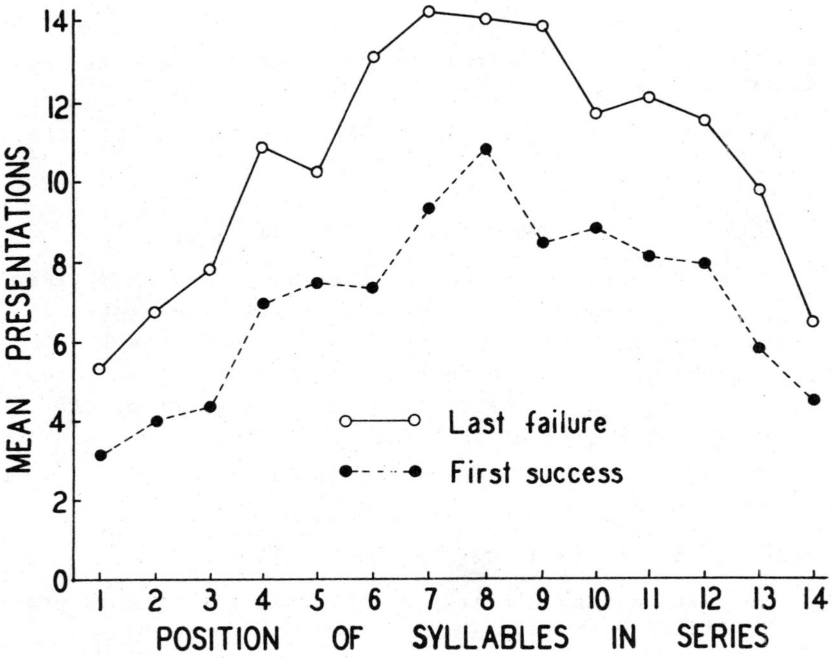

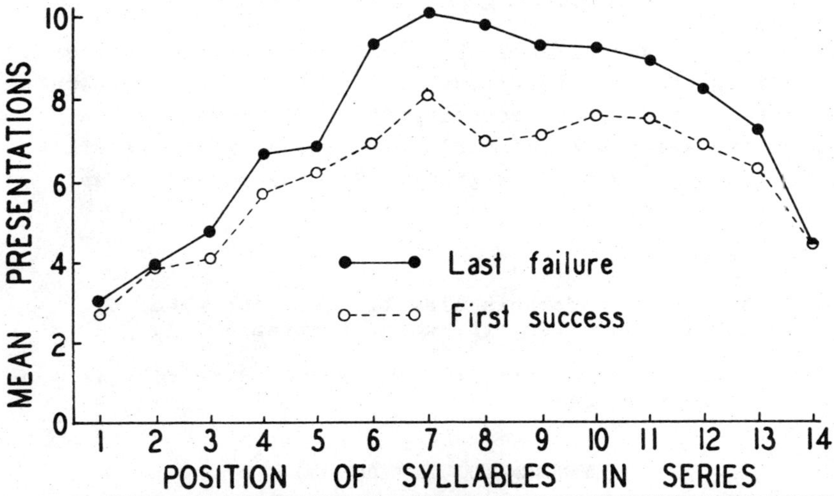

Figure 38. A. Mean number of repetitions up to and including first success and up to and including last failure in various syllable positions during learning by *massed* practice.

B. Mean number of repetitions up to and including first success and up to and including last failure in various positions during learning by *distributed* practice. After Hovland (26).

Proof:

When the distributed practice is well distributed, the increments of mean effective excitatory strength per repetition are greater than when done by massed practice. Hence, the proof of Theorem XVII applies.

C. The accompanying figure (Figure 38) shows the number of trials preceding the first success and the last failure with learning by massed and by distributed practice. The data are from an unpublished study of Hovland. It will be observed that the number of repetitions between the first success and the last failure is considerably greater when the learning has been done by massed practice than when it has been done by distributed practice.

The experimental evidence available at present supports the corollary.

THEOREM XXXVIII

A. *The difference between the effective excitatory potential at the termination of learning by massed practice and that at the post-learning maximum (Ward's reminiscence effect or* $\bar{\bar{E}}$*) is given by the equation:*

$$\bar{\bar{E}} = c\left[1 - \left(\frac{ab}{cd}\right)^{\frac{d}{d-b}}\right] - a\left[1 - \left(\frac{ab}{cd}\right)^{\frac{b}{d-b}}\right]$$

Proof:

$$f(t) = ae^{-bt} - ce^{-dt}$$

$$\bar{t} = \frac{1}{(d-b)} \log_e\left(\frac{cd}{db}\right)$$

by Theorem XXXVII.

$$f(\bar{t}) = ae^{-\frac{b}{d-b}\log_e\left(\frac{cd}{ab}\right)} - ce^{-\frac{d}{d-b}\log_e\left(\frac{cd}{ab}\right)}$$

and

$$f(0) = a - c,$$

whence,

$$\bar{\bar{E}} = f(\bar{t}) - f(0) = ae^{-\frac{b}{d-b}\log_e\left(\frac{cd}{ab}\right)} - ce^{-\frac{d}{d-b}\log_e\left(\frac{cd}{ab}\right)} - a + c,$$

i.e.

$$\bar{\bar{E}} = c\left[1 - \left(\frac{ab}{cd}\right)^{\frac{d}{d-b}}\right] - a\left[1 - \left(\frac{ab}{cd}\right)^{\frac{b}{d-b}}\right]$$

B. The meaning of this equation may be illustrated by means of Table 3 and Figure 19. In that example,

$$a = 8, \quad b = .1, \quad c = 4, \quad \text{and} \quad d = .3 .$$

Substituting these values in the formula, we have,

$$\bar{\bar{E}} = 4\left[1 - \left(\frac{8 \times .1}{4 \times .3}\right)^{\frac{.3}{.3-.1}}\right] - 8\left[1 - \left(\frac{8 \times .1}{4 \times .3}\right)^{\frac{.1}{.3-.1}}\right]$$

$$= 4\left[1 - \left(\frac{.8}{1.2}\right)^{\frac{.3}{.2}}\right] - 8\left[1 - \left(\frac{.8}{1.2}\right)^{\frac{.1}{.2}}\right]$$

$$= 4\left[1 - .6667^{1.5}\right] - 8\left[1 - .6667^{.5}\right]$$

$$= 4\,[1 - .5444] - 8\,[1 - .8165]$$

$$= 4 \times .4556 - 8 \times .835$$

$$= 1.822 - 1.488 = .334$$

It is to be observed that the value .334 for the theoretical amount of spontaneous rise in $\bar{E}$ following the termination of learning is in close agreement with the value secured by the approximate method employed in the derivation of Table 3 and Figure 19.

THEOREM XXXVIII, COROLLARY 1

A. *In the learning of rote series by massed practice, Ward's reminiscence effect ($\bar{\bar{E}}$) in terms of probability of recall will be greater at the syllable of the series requiring the maximum number of presentations to raise the effective excitatory potential to the reaction threshold than at the first syllable of the series, assuming that the probabilities of recall of the syllables compared are equal at the termination of active learning (t_0).*

Proof:

This proposition follows directly from Theorems XXXVIII, XXI, and VI.

C. Ward's results (*63*), reproduced as Figure 39, suggest a confirmation of this corollary but are so complicated by the differences in probability of recall at the termination of active learning that no definite conclusion can be reached.

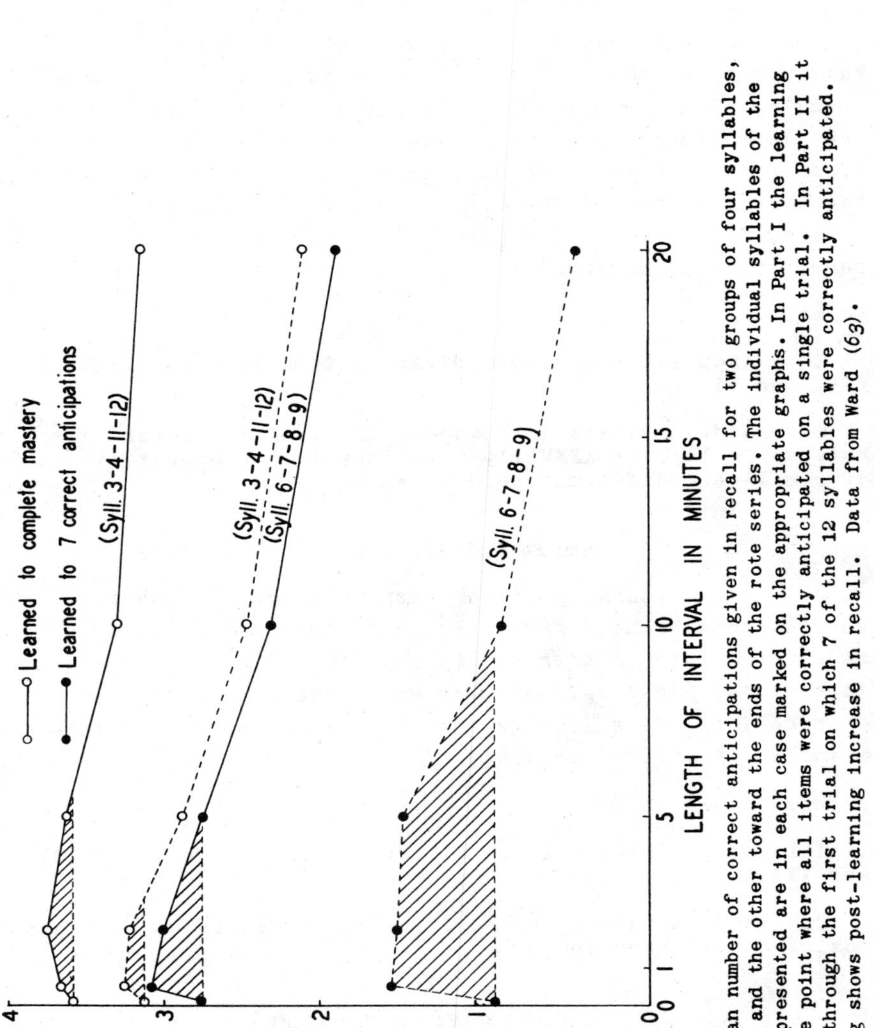

Figure 39. Mean number of correct anticipations given in recall for two groups of four syllables, one at the center and the other toward the ends of the rote series. The individual syllables of the syllable-group represented are in each case marked on the appropriate graphs. In Part I the learning was carried to the point where all items were correctly anticipated on a single trial. In Part II it was carried only through the first trial on which 7 of the 12 syllables were correctly anticipated. The cross-hatching shows post-learning increase in recall. Data from Ward (63).

THEOREM XXXVIII, COROLLARY 2

A. *In the learning of rote series by massed practice, Ward's reminiscence effect ($\bar{\bar{E}}$) in terms of probability of recall will be greater at the syllable of the series requiring the maximum number of presentations to raise the effective excitatory potential to the reaction threshold than at the last syllable of the series, assuming that the probabilities of recall of the syllables compared are equal at the termination of active learning (t_0).*

Proof:

This proposition follows directly from Theorems XXXVIII, XXI, and VI.

C. Ward's results (*63*) appear to support the corollary but, as in Theorem XXXVIII, Corollary 1, the probabilities of recall are difficult to determine.

THEOREM XXXVIII, COROLLARY 3

A. *In the learning of rote series by massed practice, Ward's reminiscence effect ($\bar{\bar{E}}$) in terms of probability of recall, will be greater at the last syllable of the series than at the first syllable, assuming that the probabilities of recall of the syllables compared are equal at the termination of active learning (t_0).*

Proof:

This proposition follows directly from Theorems XXXVIII and XXII.

C. No data bearing directly on the validity of this proposition have been found.

THEOREM XXXVIII, COROLLARY 4

A. *In the learning of rote series by massed practice, Ward's reminiscence effect ($\bar{\bar{E}}$) in terms of probability of recall, will be greater at all syllables of long series than at corresponding syllables of short series, corresponding syllables being determined by counting in from either end of the series compared, to their respective points of maximum learning difficulty, assuming that the probabilities of recall of the syllables compared are equal at the termination* *of active learning* (t_0).

Proof:

This proposition follows directly from Theorems XXXVIII and XXVI.

C. No data bearing directly on the validity of this proposition have been found.

THEOREM XXXIX

A. *In rote series learned by massed practice, as time elapses following the attainment of maximum post-learning effective excitatory potential by a particular syllable, the effective excitatory potential for that syllable will fall, at first with a positive acceleration.*

To prove:

That the proof of the theorem requires the demonstration of three parts:

(a) that there is one and only one point of inflection $(\bar{\bar{t}})$;
(b) that at this point of inflection $\bar{\bar{t}} > \bar{t}$;
(c) that for values of $t < \bar{\bar{t}}$ the acceleration $[f''(t)]$ is negative. (This is equivalent to saying that the curve *falls* with *positive* acceleration.)

Proof:

(a)

$$f''(t) = ab^2 e^{-bt} - ce^2 e^{-dt}$$

by Theorem XXXVI (1), and the point of inflection $(\bar{\bar{t}})$ will be found where $f''(t) = 0$. Whence,

$$ab^2 e^{-bt} - cd^2 e^{-dt} = 0 \qquad (1)$$

$$e^{(d-b)t} = \frac{cd^2}{ab^2}$$

and

$$\bar{\bar{t}} = \frac{1}{(d-b)} \log \left(\frac{cd^2}{ab^2}\right) \qquad (2)$$

and since $f''(t)$ is a continuous function and does not vanish for any other real value of t, this is the only point of inflection.

(b) Since,

$$\bar{\bar{t}} = \frac{1}{(d-b)} \log \left(\frac{cd^2}{ab^2}\right) \qquad \text{by} \quad (2)$$

$$= \frac{1}{(d-b)} \log \left(\frac{cd}{ab}\right) + \frac{1}{(d-b)} \log \left(\frac{d}{b}\right) \qquad (3)$$

but,

$$\bar{t} = \frac{1}{(d-b)} \log \left(\frac{cd}{db}\right)$$

by Theorem XXXVI (3). Hence,

$$\bar{\bar{t}} = \bar{t} + \frac{1}{(d-b)} \log \left(\frac{d}{b}\right) \qquad (4)$$

Now since $d > b$. (Equation (8) following "Preliminary Propositions"),

$$\frac{1}{(d-b)} \log \left(\frac{d}{b}\right) > 0$$

and therefore,

$$\bar{\bar{t}} > \bar{t} \qquad (5)$$

(c) Since $f''(0) < 0$ by Theorem XXXVI (3), and since there is one, and only one, point of inflection at $\bar{\bar{t}}$ by (2), it follows that $f''(t) < 0$ for all values of $t < \bar{\bar{t}}$. (6)

B. The principle of this theorem may be illustrated by the theoretical values shown in Figure 19 and Table 3. In Figure 19 it may be seen that immediately following $t = 2.0$, $\bar{E}$ begins to fall and does so continuously as far as the computation extends.

Secondly, it may be seen by Table 3 that the rate of fall between $t = 2$ and $t = 7.52$ is positively accelerated. Thus, between $t = 2.0$ and $t = 3.0$ the fall is only .054, whereas between $t = 3.0$ and $t = 4.0$ the fall has increased to .142. Between $t = 6.0$ and $t = 7.0$ the amount of fall reaches the value of .246, and between $t = 7.0$ and $t = 8.0$ it has reached its maximum of .251, thus showing a positive acceleration to the latter point, in accordance with the statement of the theorem.

THEOREM XXXIX, COROLLARY 1

A. *In rote series learned by massed practice, at any syllable learned just to the point of consistent correct reaction, there will occur, following the point of post-learning minimum reaction latency, a progressive increase in reaction latency, and this increase will show, at first, a positive acceleration.*

Proof:

This proposition follows from Theorem XXXIX by the argument of Theorem XXIII, Corollary 2, using Postulate 18, Corollary 1.

C. On this point insufficient data are available, but the results of Ward (*63*) quoted above (Table 43) and shown graphically in Figure 35 suggest that the increase in reaction latency following the initial decrease is most likely to be positively accelerated at first, as predicted.

THEOREM XXXIX, COROLLARY 2

A. *In the learning of rote series by massed practice, if practice is terminated at any stage of the learning of a given syllable such that the probability of recall is greater than zero and less than perfection: as time elapses following the point of post-learning maximum of recall at the given syllable, there will occur a progressive decrease in recall, and the decrease will show at first a positive acceleration.*

Proof:

This proposition follows directly from Theorems XXXIX, XVIII, and XIX. (The first part of the corollary is true at any stage of learning.)

C. Data on the form of the post-learning decrease in recall are somewhat irregular, but tend to show the predicted negative acceleration (Ward (*63*), Figure 35). Data on the individual syllables are not reported separately.

Present evidence appears to support the corollary.

THEOREM XL

A. *In rote series learned by massed practice, as time elapses following the attainment of the post-learning maximum effective excitatory potential by any given syllable reaction: at the point of time,*

$$\bar{\bar{t}} = \frac{1}{(d-b)} \log_e \left(\frac{cd^2}{ab^2}\right)$$

the positive acceleration of fall will be replaced by a negative acceleration, thus presenting an inflection in the curve for that syllable.

To prove:

Since a positive acceleration of *fall* is equivalent to a negative acceleration in rise, the theorem will be proved if it can be shown that

$$f''(t) < 0 \quad \text{for} \quad t < \bar{\bar{t}}$$

and

$$f''(t) > 0 \quad \text{for} \quad t > \bar{\bar{t}}$$

Proof:

$$f''(t) = e^{-dt}(ab^2 e^{(d-b)t} - cd^2) \text{ by Theorem XXXVI (1).}$$

Now, e^{-dt} is always positive; $a^2 b > 0$ (7,8, following "Preliminary Propositions"); $d - b > 0$ (8, following "Preliminary Propositions"); and cd^2 is constant. Hence, for a sufficiently large value of t,

$$ab^2 e^{(d-b)t} > cd^2$$

and

$$f''(t) > 0 .$$

Hence, since $f''(t)$ is continuous, $f''(t) > 0$ when $t > \bar{\bar{t}}$.

B. The meaning of this equation may be illustrated by means of Table 3 and Figure 19. In that example,

$$a = 8, \quad b = .1, \quad c = 4, \quad \text{and} \quad d = .3 .$$

Substituting these values in the equation, we have,

$$\bar{\bar{t}} = \frac{1}{.3 - .1} \log_e \left(\frac{4 \times .3^2}{8 \times .1^2}\right)$$

$$= \frac{1}{.2} \log_e \frac{.36}{.08}$$

$$= 5 \log_e 4.5$$

$$= 5 \times 1.5042$$

$$= 7.52$$

It is to be observed that the value 7.52 for the theoretical point of inflection, i.e., the point at which the values of $\bar{E}$ cease to fall at an increasing rate, is in agreement with the value secured by the approximate method necessarily used in its derivation from an array of values such as are presented by Table 3.

THEOREM XL, COROLLARY 1

A. *In rote series learned by massed practice, there will occur at least one inflection point in the mean reaction latency during the post-learning period following the post-learning minimum, but no such inflection point will occur earlier than* $\bar{\bar{t}}$.

Proof:

This proposition follows directly from Theorem XL and the argument of Theorem XXIII, Corollary 1, using Postulate 18, Corollary 1.

C. No data bearing directly on the validity of this proposition have been found.

THEOREM XLI

A. *With the passage of time following the learning of rote series by massed practice, as the magnitude of the inhibitory potential at the termination of learning is diminished, other things equal, the magnitude of the reminiscence effect ($\bar{\bar{E}}$) will decrease.*

Proof:

Let c_1 and c_2 be two different initial inhibitions (see 13, following "Preliminary Propositions") such that,

$$c_1 > c_2 > 0 \tag{1}$$

and let $\bar{t}_1$ and $\bar{t}_2$ be the respective times for the maxima. Then,

$$f_1(t) = ae^{-bt} - c_1 e^{-dt} \tag{2}$$

$$f_2(t) = ae^{-bt} - c_2 e^{-dt} \tag{3}$$

and the theorem states that,

$$f_1(\bar{t}_1) - f_1(0) > f_2(\bar{t}_2) - f_2(0) \,. \tag{4}$$

But,

$$f_1(\bar{t}_1) > f_1(\bar{t}_2)$$

since $\bar{t}_1$ is the maximum for $f_1(t)$. Hence,

$$-f_1(\bar{t}_2) > -f_1(\bar{t}_1)$$

whence, adding $f_2(\bar{t}_2)$ to both sides,

$$f_2(\bar{t}_2) - f_1(\bar{t}_2) > f_2(\bar{t}_2) - f_1(\bar{t}_1) \,. \tag{5}$$

But from (2) and (3),

$$f_1(\bar{t}_2) = ae^{-b\bar{t}_2} - c_1 e^{-d\bar{t}_2}$$

$$f_2(t_2) = ae^{-b\bar{t}_2} - c_2 e^{-d\bar{t}_2}$$

whence,

$$f_2(\bar{t}_2) - f_1(\bar{t}_2) = (c_1 - c_2)\, e^{-d\bar{t}_2} \qquad (6)$$

and by (5)

$$(c_1 - c_2)\, e^{-d\bar{t}_2} > f_2(\bar{t}_2) - f_1(\bar{t}_1) \qquad (7)$$

But,

$$1 > e^{-d\bar{t}_2} \qquad (8)$$

and from (1),

$$c_1 - c_2 > 0.$$

Hence,

$$(c_1 - c_2) > (c_1 - c_2)\, e^{-d\bar{t}_2} > f_2(\bar{t}_2) - f_1(\bar{t}_1) \qquad (9)$$

But,

$$f_1(0) = a - c_1$$

$$f_2(0) = a - c_2$$

$$f_2(0) - f_1(0) = c_1 - c_2 \qquad (10)$$

Therefore from (9),

$$f_2(0) - f_1(0) > f_2(\bar{t}) - f_1(\bar{t}_1) \qquad (11)$$

whence,

$$f_1(\bar{t}_1) - f_1(0) > f_2(\bar{t}_2) - f_2(0). \qquad (12)$$

B. The meaning of this theorem may be seen by comparing the graphs of Figure 40. There it appears that when the amount of inhibition (c) is large, e.g., has a value of 8, say, the amount of the theoretical preliminary rise, or reminiscence effect, is large; where c is 6, the rise is less, and where it is 4 the rise is still less. By computation from the formula in Theorem XXXVIII, the theoretical amounts of rise corresponding to the c values of 8, 6, and 4 are respectively, 3.10, 1.56, and .33, thus indicating theoretically a progressive decrease in the "reminiscence" effect as the amount of inhibition decreases.

It may also be seen in Figure 40 that where I becomes small enough (e.g., I = 2), other factors remaining constant, the "reminiscence" effect does not appear at all.

THEOREM XLI, COROLLARY 1

A. *In rote series learned by massed practice, if active practice is terminated after a number of presentations less*

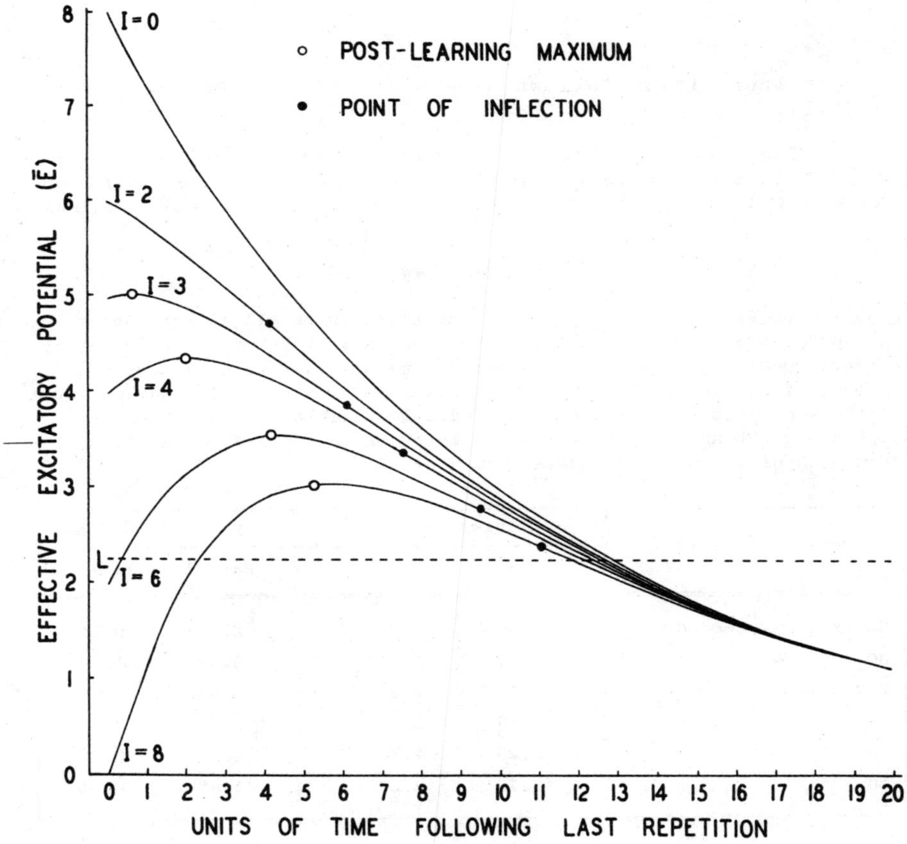

Figure 40. Family of curves showing the spontaneous changes in the strength of the effective excitatory potential following the termination of active learning.

than that required to learn the series as a whole but more than sufficient to yield a fifty per cent probability of recall at that point in the series requiring the maximum number of presentations for complete learning: the difference between the probability of recall at the termination of active learning and that at the post-learning maximum ($\bar{t}$) will increase from each end of the series toward the point of maximum difficulty of learning.

Proof:

This proposition follows directly from Theorems XLI, XIX, and XXI.

C. The study by Ward (*63*) shows the reminiscence effect to be considerably larger for the central syllables than for the end syllables in his list. Table 44 shows these results.

TABLE 44

Average number of correct anticipations given in recall at the center and end sections of rote series. The figures in this table represent the average numbers of anticipations for the groups of four syllables as wholes; i.e., the highest attainable average is 4.00, which would result if every subject correctly anticipated all syllables in one of the two syllable groupings. Each average is based on the data from 24 lists, two from each of 12 subjects. Data from Ward (*63*).

Length of interval	Syllables 3-4-11-12		Syllables 6-7-8-9	
	Part I	Part II	Part I	Part II
Control (6 seconds)	3.58	3.13	2.75	0.88
30 seconds	3.67	3.25	3.08	1.54
2 minutes	3.75	3.21	3.00	1.50
5 "	3.63	2.88	2.75	1.46
10 "	3.29	2.46	2.29	0.83
20 "	3.13	2.21	1.88	0.46

Results in complete accord with those of Ward are reported by Hovland (*20*) for learning by massed practice. The mean number of errors upon recall are decreased more for the central than for the end syllables upon the introduction of a 2-minute rest pause (Figure 38).

The evidence is in support of the corollary.

THEOREM XLI, COROLLARY 2

A. *In rote series learned by massed practice, if active practice is terminated after a number of presentations less than that required to learn the series as a whole but more than sufficient to yield a fifty per cent probability of recall at that point in the series requiring the maximum number of presentations for complete learning: the difference between the reaction latency at the termination of active practice and that at the post-learning maximum ($\bar{t}$) will increase from each end of the series toward the point of maximum difficulty of learning.*

Proof:

By (3) of Postulate 11, Corollary 2,

$$\bar{\bar{E}}_{t=0} = E_{t=0} - I_{t=0}.$$

Also, by Theorem XXI, $I_{t=0}$ increases from each end of the series toward the point of maximum difficulty of learning. From these considerations and Theorem XLI, it follows that $\bar{\bar{E}}$ will increase from each end of the series to the point of maximum difficulty of learning. But by Postulate 18,

$$T = v + (h-v)\,e^{-k\bar{\bar{E}}}.$$

From the characteristics of this equation and the preceding considerations, the corollary follows.

C. No data bearing directly on the validity of this proposition have been found.

THEOREM XLII

A. *With the passage of time following the learning of rote series by massed practice, if the inhibition at termination of learning decreases, the point ($\bar{t}$) of maximum effective excitatory potential ($\bar{\bar{E}}$) will appear closer to the point of time at which learning was terminated.*

To prove:

That since c is the initial inhibition (13 of "Preliminary Propositions") it is to be shown that as c decreases, $\bar{t}$ decreases.

Proof:

$$\bar{t} = \frac{1}{(d-b)} \log_e \left(\frac{cd}{ab}\right) \qquad \text{by Theorem XXXVII (3)}$$

$$= \frac{1}{(d-b)} \log \frac{d}{ab} + \frac{1}{(d-b)} \log c\ .$$

The first term of the equation is constant, $(d-b)$ is positive (8 of "Preliminary Propositions"), and $\log c$ decreases with c. Hence, as c decreases, $\bar{t}$ decreases.

B. The meaning of this theorem will become evident from an inspection of the series of theoretical "retention" curves in Figure 40, in which E is constant at 8, and I has successively the value of 8, 6, 4, 3, 2, and 0. An inspection of these shows that theoretically the crest of the post-learning rise (marked by hollow circles) is farthest from the termination of learning at 8, and appears at progressively shorter distances as I decreases. Computation by means of the equation of Theorem XXXVII shows that at I = 8, $\bar{t}$ = 5.2; at I = 6, $\bar{t}$ = 4.1; at I = 4, $\bar{t}$ = 2.0; at I = 3, $\bar{t}$ = .6; at I = 2 there is theoretically no "reminiscence" effect and, of course, there would be none at I = 1.

THEOREM XLII, COROLLARY 1

A. *In rote series learned by massed practice, the time interval between the termination of active practice and the time of minimum mean reaction latency at the several syllables of the series will increase from each end of the series toward the point of maximum inhibition.*

Proof:

This proposition follows directly from Theorem XLII, and from Theorem XXXVII, Corollary 1.

C. No data bearing directly on the validity of this proposition have been found.

THEOREM XLII, COROLLARY 2

A. *In rote series learned by massed practice, if active practice is terminated after a number of presentations less than that required to learn the series as a whole but more than sufficient to yield a fifty per cent probability of recall at that point in the series requiring the maximum number of presentations for complete learning: the time interval between the termination of active practice and that of maximal recall at the several syllables of the series will increase from each end of the series toward the point of maximum difficulty of learning.*

Proof:

This proposition follows directly from Theorem XLII and from Theorem XXXVII, Corollary 2.

C. The evidence on this corollary is extremely ambiguous. Ward's (*63*) results show the point of maximum post-learning recall for syllables in the central and end portions. In the experiment in which the learning was to complete mastery the maximum recall was at one-half minute for syllables 6, 7, 8, and 9, and at two minutes for syllables 3, 4, 11, and 12. This is in contradiction to the prediction. In the experiment on which the criterion was 7 syllables correct, the maxima were at one-half minute for both the central and end syllables, although the curves make it appear that with further experimentation the maximum for the middle syllables would be more apt to be at a later time.

No decision as to the truth of the corollary can be made on the basis of the present evidence, but the data seem to be in opposition to the corollary.

THEOREM XLIII

A. *With the passage of time following the learning of rote series by massed practice, as the inhibition at the termination of learning decreases, the point of inflection of the fall in excitatory potential following the post-learning maximum ($\overline{\overline{E}}$) will appear closer to the point of time at which learning was terminated.*

To prove:

That $\overline{\overline{t}}$ decreases with decreasing c.

Proof:

$$\overline{\overline{t}} = \frac{1}{(d-b)} \log \left(\frac{cd^2}{ab^2}\right) \quad \text{by Theorem XXXIX (2)}$$

$$= \frac{1}{(d-b)} \log \left(\frac{d^2}{ab^2}\right) + \frac{1}{(d-b)} \log c$$

$(d-b)$ is positive (8 following "Preliminary Propositions"), and $\log c$ decreases with c. Hence $\overline{\overline{t}}$ decreases with decreasing c.

B. The meaning of this theorem will become evident from an inspection of the series of theoretical "retention" curves shown in Figure 40. Since the curvature is so gentle and the exact point of inflection is not evident to ordinary inspection, the several points of inflection have been computed by the equation of Theorem XL, and indicated on each curve by a solid circle. It will be observed that where I = 8, the point of inflection appears at $\bar{\bar{t}} = 11.0$; where I = 6, the point of inflection occurs at $\bar{\bar{t}} = 9.6$; where I = 4, it takes place at $\bar{\bar{t}} = 7.5$; where I = 3, it takes place at $\bar{\bar{t}} = 6.1$; and where I = 2 it takes place at $\bar{\bar{t}} = 4.1$. Thus, as the amount of inhibition decreases the point of inflection appears closer to the time at which active learning terminates, exactly as the theorem states. Where I = 0 there is, of course, no point of inflection.

THEOREM XLIV

A. *With the passage of time following the learning of rote series by massed practice, the interval of time between the point of post-learning maximum effective excitatory potential* ($\bar{\bar{E}}$) *and the point of inflection is constant, i.e., independent of the value of the initial mean inhibitory potential.*

To prove:

That the time interval $\bar{\bar{t}} - \bar{t}$ is independent of the initial mean inhibitory tendency c.

Proof:

$$\bar{\bar{t}} = \frac{1}{(d-b)} \log \frac{cd^2}{ab^2} \quad \text{by Theorem XXXIX (2).}$$

$$\bar{t} = \frac{1}{(d-b)} \log \frac{cd}{ab} \quad \text{by Theorem XXXVII (3).}$$

Hence,

$$\bar{\bar{t}} - \bar{t} = \frac{1}{(d-b)} \log \frac{d}{b} \qquad (1)$$

which is a constant independent of c.

B. This theorem finds a concrete illustration in Figure 40, together with the data presented in Theorems XLII B and XLIII B. If we subtract the theoretical time value for the maximum reminiscence effect ($\bar{t}$) from the theoretical time value for the point of inflection ($\bar{\bar{t}}$), we obtain in all cases a value close to 5.5 time units, thus:

If I = 8, 11.0 - 5.5 = 5.5

If I = 6, 8.6 - 4.1 = 5.5

If I = 4, 7.5 - 2.0 = 5.5

$$\text{If} \quad I = 3, \quad 6.1 - .6 = 5.5$$

which is in accordance with the theorem.

THEOREM XLV

A. *With the passage of time following the learning of rote series by massed practice, the difference between the effective excitatory potential at the termination of learning* $[f(0)]$ *and the point of post-learning maximum effective excitatory potential* $[f(\bar{t})]$ *will be greater for the syllable showing the maximum amount of inhibitory potential* (ΔI_{max}) *than for the terminal syllable (N).*

To prove:

That $f_{\bar{n}}(\bar{t}_n) - f_{\bar{n}}(0) \quad f_N(\bar{t}_N) - f_N(0)$ where $\bar{n}$ is the syllable with the maximum amount of inhibitory potential (ΔI_{max}).

Proof:

It has been shown that if $c_1 > c_2$, then,

$$f_1(\bar{t}_1) - f_1(0) > f_2(\bar{t}_2) - f_2(0)$$

by Theorem XLI, and since $c = I_n$, by (13) of "Preliminary Propositions," and $I_{\bar{n}} > I_N$, it follows that,

$$f_{\bar{n}}(\bar{t}_{\bar{n}}) - f_{\bar{n}}(0) > f_N(\bar{t}_N) - f_N(0).$$

B. This theorem may be illustrated by the use of constants obtained from Problem I. In a 15-syllable series in which $F = 1.37$, $\Delta K = .09\,\Delta E$, and $L = 2.25\,\Delta E$, the syllable (r) of maximum learning difficulty is 9. By Table 2 the theoretical mean number of repetitions necessary for learning at this syllable is 7.31. However, owing to the fluctuations at the recall threshold (Postulate 15), some reaction failures occur after the mean reaction threshold (L) is attained, so that a certain amount of over-learning will occur, on the average before the conventional criterion of "complete" learning is reached. Let us assume that this is enough to make the average repetition per series rise to 9. Taking the amount of increase in excitatory potential at a single repetition as the unit of measurement, i.e., $\Delta E = 1$, it follows that the E at all points in such a series is 9.

The amount of inhibition at a given point in a rote series is given by the expression (Postulate 9, Corollaries 1 and 2 together with D76) $R\,J\,\Delta K$. Table 1 gives J_8 as

7.688. Accordingly,

$$I_8 = 9 \times 7.688 \times .09$$

$$= 6.227$$

With the theoretical values of E and I at the termination of learning thus set at 9 and 6.2 respectively, we proceed to the determination of the amount of maximal spontaneous rise in $\bar{E}$ following the termination of active learning. This is given by Theorem XXXVIII as

$$\bar{\bar{E}} = c\left[1 - \left(\frac{ab}{cd}\right)^{\frac{a}{d-b}}\right] - a\left[1 - \left(\frac{ab}{cd}\right)^{\frac{b}{d-b}}\right]$$

From the preceding, $a = 9$ and $c = 6.2$. By previous assumption, $b = .1$ and $d = .7$. Substituting, we have,

$$\bar{\bar{E}}_{8,9} = 6.2\left[1 - \left(\frac{9 \times .1}{6.2 \times .7}\right)^{\frac{.7}{.7-.1}}\right] - 9\left[1 - \left(\frac{9 \times .1}{6.2 \times .7}\right)^{\frac{.1}{.7-.1}}\right]$$

$$= 6.2\,(1 - .207^{1.167}) - 9\,(1 - .207^{.167})$$

$$= 6.2\,(1 - .159) - 9\,(1 - .769)$$

$$= 5.04 - 2.08$$

$$= 2.96$$

Proceeding in an exactly analogous manner, we find that the inhibition opposing the terminal syllable (I_{14}) is 2.96, and that the post-learning rise in effective excitatory tendency ($\bar{\bar{E}}_{14,15}$) is .68. Since 2.96 > .68 it follows that $\bar{\bar{E}}_{8,9} > \bar{\bar{E}}_{14,15}$ as the theorem states.

THEOREM XLVI

A. *With the passage of time following the learning of rote series by massed practice, the point of post-learning maximum effective excitatory potential ($\bar{E}$) will appear later for the syllable opposed by the maximum amount of inhibitory potential (ΔI_{max}) than for the terminal syllable (N).*

To prove:

That $\bar{t}_{max}\, I_n > \bar{t}_{I_{N-1}}$

Proof:

As c decreases, $\bar{t}$ decreases, by Theorem XLII. But $c = I_n$ (13, of "Preliminary Propositions," p. 256), and if $c_1 = I_{max}$, $c = I$. Then $c_1 > c_2$, since $I_{max} > I_{N-1}$ (14, of "Preliminary Propositions"), from which the theorem follows.

B. This theorem may be illustrated by values drawn from the same situation as that employed in the illustration of Theorem XLV, where the excitatory potential is 9 at all syllable positions, where $I_8 = 6.23$ and $I_{14} = 2.97$; i.e.,

$$a = 9, \quad b = .1, \quad c = 6.23 \text{ or } 2.97, \quad \text{and} \quad d = .7 .$$

Substituting appropriately in the equation of Theorem XXXVII,

$$t = \frac{1}{(d-b)} \log_e \frac{cd}{ab}$$

we have $t_{8,9} = 2.63$ and $t_{14,15} = 1.40$. But since $2.63 > 1.40$, it follows that $t_{8,9} > t_{14,15}$, as the theorem states.

THEOREM XLVII

A. *With the passage of time following the learning of rote series by massed practice, the point of inflection of the effective excitatory potential following the post-learning maximum ($\bar{\bar{E}}$) will appear later for the syllable with the maximum amount of inhibitory potential (ΔI_{max}) than for the terminal syllable (N).*

To prove:

That $\quad \bar{\bar{t}}_{I_{max\,n}} > \bar{\bar{t}}_{I_{N-1}}$

Proof:

$\bar{\bar{t}}$ decreases with c, by Theorem XLIII. And since $c = I_n$ (13, of "Preliminary Propositions," p. 256), and $I_{max\,n} > I_{N-1}$ (14, of "Preliminary Propositions," p. 256), it follows that,

$$\bar{\bar{t}}_{I_{max\,n}} > \bar{\bar{t}}_{I_{N-1}} .$$

B. This theorem may be illustrated by values drawn from the same situation employed in the illustration of Theorems XLV and XLVI, where the excitatory tendency is 9 at all syllable positions, where $I_8 = 6.23$ and $I_{14} = 2.97$, i.e.,

$$a = 9, \quad b = .1, \quad c = 6.23 \text{ or } 2.97, \quad \text{and} \quad d = .7 .$$

Substituting appropriately in the equation of Theorem XL,

$$\bar{\bar{t}} = \frac{1}{(d-b)} \log_e \frac{cd^2}{ab^2}$$

we have, $\bar{\bar{t}}_{8,9} = 5.87$ and $\bar{\bar{t}}_{14,15} = 4.64$. But since $5.87 > 4.64$, it follows that $\bar{\bar{t}}_{8,9} > \bar{\bar{t}}_{14,15}$, as the theorem states.

THEOREM XLVIII

A. *In rote series learned by massed practice, the difference between the effective excitatory potential at the termination of learning [f (0)] and the post-learning maximum effective excitatory potential ($\bar{\bar{E}}$) following the termination of learning [f ($\bar{t}$)] will be less for the first syllable than for the last one in the series (N).*

To prove:

That $f_{N-1}(\bar{t}_{N-1}) - f_{N-1}(0) > [f_0(\bar{t}_0) - f_0(0)]$.

Proof:

Following the argument of Theorem XXXIII, it follows that since $I_{N-1} > I_0$ (14, of "Preliminary Propositions," p. 256),

$$f_{N-1}(\bar{t}_{N-1}) - f_{N-1}(0) > [f_0(\bar{t}_0) - f_0(0).$$

B. This theorem may be illustrated by the use of the set of constants that was employed in Theorem XLV. By Table 1, in a 15-syllable series, $J_0 = 1.79$. Now, $I = R\,J\,\Delta\,K$. Accordingly,

$$I_0 = 9 \times 1.79 \times .09$$
$$= 1.45$$

Taking c as 1.45, using the other values employed in Theorem XLV B, and proceeding in a manner exactly analogous to the method employed there, we obtain from our calculations $\bar{\bar{E}}_{0,1} = .02$, an exceedingly small value.

Now, we have by the computations of Theorem XLV B the value, $\bar{\bar{E}}_{14,15} = .68$. But,

$$.68 > .02.$$

Accordingly,

$$\bar{\bar{E}}_{14,15} > \bar{\bar{E}}_{0,1}.$$

as the theorem states.

THEOREM XLVIII, COROLLARY 1

A. *In rote series learned by massed practice, if practice is terminated after a number of presentations less than enough to raise the terminal syllable to its last failure but more than enough to raise it to the mean reaction threshold (fifty per cent probability of recall, or* L*): the difference between the probability of recall at the termination of active practice and that at the post-learning maximum will be greater for the posterior syllable (N) than for the anterior syllable ($n' = 1$).*

Proof:

This proposition follows directly from Theorems XLVIII and XIX.

C. No results are available with respect to recall at what is known to be definitely the post-learning maximum. If, however, results of Ward (*63*) hold generally and the post-learning maximum is at about two minutes, the results of Hovland (*20*) can be cited. His results show that no errors were made at the anterior syllable with immediate recall, and .016 errors with the recall after two minutes. The corresponding figures for the posterior syllable were 1.219 errors upon immediate recall, and 1.88 with the recall after two minutes. With the great variability of recall scores, it is not surprising to find actual loss in the anterior syllable results, but the difference between the anterior and posterior syllable is definitely in the direction predicted by the corollary.

The evidence appears to support the corollary.

THEOREM XLVIII, COROLLARY 2

A. *In rote series learned by massed practice, if practice is terminated after a number of presentations less than enough to raise the terminal syllable to its last failure but more than enough to raise it to the mean reaction threshold (fifty per cent probability of recall, or* L*): the difference between the mean reaction latency at the termination of learning and that at the post-learning minimum will be greater for the posterior syllable than for the anterior syllable.*

Proof:

As in Theorem XLI, Corollary 2.

C. No data bearing directly on the validity of this proposition have been found.

THEOREM XLIX

A. *With the passage of time following the learning of rote series by massed practice, the point of post-learning maximum effective excitatory potential* ($\bar{\bar{E}}$) *following the termination of learning will appear earlier for the first syllable of a series* ($n' = 1$) *than for the last syllable* (N).

To prove:

That $\bar{t}_{N-1} > \bar{t}_{n=0}$

Proof:

Following the argument of Theorem XLVI, since $I_N > I_1$, it follows that $\bar{t}_{N-1} > \bar{t}_{n=0}$.

B. This theorem may be illustrated by values drawn from the situation employed in the illustration of Theorems XLV, XLVI, etc., where the excitatory potential is assumed to be 9 at all syllable positions, where $I_{14} = 2.97$ and $I_0 = 1.45$, i.e.,

$$a = 9, \quad b = .1, \quad c = 2.97 \text{ or } 1.45, \quad \text{and} \quad d = .7.$$

Substituting appropriately in the equation of Theorem XXXVII,

$$\bar{t} = \frac{1}{(d-b)} \log_e \frac{cd}{ab}$$

we have $\bar{t}_{14,15} = 1.40$ and $\bar{t}_{0,1} = .20$. But since $1.40 > .20$, it follows that $\bar{t}_{14,15} > \bar{t}_{0,1}$, as the theorem states.

THEOREM XLIX, COROLLARY 1

A. *In rote series learned by massed practice, the time interval between the termination of active practice and the time of minimum mean post-learning reaction time is greater at the posterior syllable than at the anterior syllable.*

Proof:

This proposition follows directly from Theorem XLIX and Theorem XXXVII, Corollary 1.

C. No data bearing directly on the validity of this proposition have been found.

THEOREM XLIX, COROLLARY 2

A. *In rote series learned by massed practice, the time interval between the termination of active practice and the time of post-learning maximum recall will be greater at the posterior syllable than at the anterior syllable.*

Proof:

This proposition follows directly from Theorem XLIX and Theorem XXXVII, Corollary 2.

C. No data bearing directly on the validity of this proposition have been found.

THEOREM L

A. *With the passage of time following the learning of rote series by massed practice, the point of inflection following the point of post-learning maximum effective excitatory potential ($\bar{\bar{E}}$) will appear earlier for the first syllable ($n' = 1$) than for the last syllable in a series (N).*

To prove:

That $\bar{\bar{t}}_{N-1} > \bar{\bar{t}}_0$.

Proof:

Following the argument of Theorem XLVII, since $I_{N-1} > I_0$, (14, of "Preliminary Propositions," p. 256), it follows that $\bar{\bar{t}}_{N-1} > \bar{\bar{t}}_0$.

B. This theorem may be illustrated by values drawn from the same situation employed in the illustration of Theorems XLV, XLVI, etc., where the theoretical excitatory potential is 9 at all syllable positions, where $I_{14} = 2.97$ and $I_0 = 1.45$, i.e.,

$$a = 9, \quad b = .1, \quad c = 2.97 \text{ or } 1.45, \text{ and } d = .7.$$

Substituting appropriately in the equation of Theorem XL,

$$\bar{\bar{t}} = \frac{1}{(d-b)} \log_e \frac{cd^2}{ab^2}$$

we have $\bar{\bar{t}}_{14,15} = 4.64$ and $\bar{\bar{t}}_{0,1} = 3.44$. But since $4.64 > 3.44$, it follows that $\bar{\bar{t}}_{14,15} > \bar{\bar{t}}_{0,1}$, as the theorem states.

THEOREM LI

A. *With the passage of time following the learning of rote series by massed practice, the difference between the effective excitatory potential at the termination of learning [f(0)] and the maximum post-learning effective excitatory potential [$f(\bar{t})$] will be greater, other things equal, for long rote series than for short ones.*

Proof:

Following the argument of Theorem XLV, the theorem will be proved if it can be shown that $I_n(N+r) > I_n(N)$; i.e., that the effective excitatory strength at the *n*th syllable is greater in a list of $(N+r)$ syllables than in a list of N syllables. But this follows from Theorem XXVII A.

B. This theorem may be illustrated by comparing the theoretical amount of spontaneous post-learning rise in effective excitatory potential for the syllables of maximum difficulty in the 12-syllable and the 15-syllable series of Table 1 emerging from the analysis of Problem I. By Table 2 the syllable position of maximum difficulty in the 12-syllable series is the 8th, with a J-value (Table 1) of 7.07 and requires 6.18 repetitions for learning. If we allow 1.69 repetitions for over-learning, as was done for the 16-syllable series (Theorem XLV), the 12-syllable series will require on the average 7.87 repetitions to learn. Also, since $I = R\,J\,\Delta\,K$, we have for the 8th syllable position of a 12-syllable series,

$$^{12}I_7 = 7.87 \times 7.07 \times .09$$
$$= 5.0077.$$

Accordingly we have the values,

$$a = 7.87, \quad b = .1, \quad c = 5.008, \text{ and } \quad d = .7\,.$$

Substituting in the equation of Theorem XXXVIII,

$$\bar{\bar{E}} = c\left[1 - \left(\frac{ab}{cd}\right)^{\frac{d}{d-b}}\right] - a\left[1 - \left(\frac{ab}{cd}\right)^{\frac{b}{d-b}}\right]$$

we have $^{12}\bar{\bar{E}}_{7,8} = 2.43$. Now by Theorem XLV B, we have for the point of maximum difficulty of learning in a 15-syllable series, $^{15}\bar{\bar{E}}_{8,9} = 2.96$. But since $2.96 > 2.43$, it follows that the spontaneous post-learning rise in effective excitatory potential is greater at the point of maximum learning difficulty in a 15-syllable series than at the point of maximum difficulty of a 12-syllable series, as the theorem states.

THEOREM LI, COROLLARY 1

A. *In rote series partially learned by massed practice, the difference between the number of syllables susceptible of recall at the termination of learning and the number susceptible of recall at the post-learning maximum will, other things being equal, be greater in long series than in short series.*

Proof:

For the number of syllables susceptible of recall is an increasing function of the mean effective excitation.

C. Shipley (*58*) studied reminiscence employing series 8, 14, and 20 syllables in length. Because of a number of factors discussed in his article, reminiscence did not appear prominently in recall scores. There was, nevertheless, a pronounced tendency for the rest interval to be more advantageous the longer the list. His results show the ratio of the recall after two minutes to the recall on the control trials immediately after learning:

	8 syllables	14 syllables	20 syllables
Ratio 2 minutes to immediate recall	.83	.86	.91

With certain critical values of the equational constants, no reminiscence would be predicted according to the theory, but the high ratio for the longer lengths should still obtain.

The evidence relevant to the corollary is ambiguous, one phase of it supporting and one phase refuting the theoretical prediction.

THEOREM LI, COROLLARY 2

A. *In rote series learned by massed practice, the difference between the mean reaction latency at the termination of active learning and the mean reaction latency at the post-learning minimum will, other things equal, be greater in long than in short series.*

Proof:

For this difference is an increasing function of the difference in mean effective excitatory potential.

C. No data bearing directly on the validity of this proposition have been found.

THEOREM LII

A. *With the passage of time following the learning of rote series by massed practice, the point of maximum effective excitatory potential following the termination of learning* ($\bar{\bar{E}}$) *will appear later in long than in short series.*

To prove:

That $\bar{t}_{(N+r)} > \bar{t}_N$.

Proof:

$\bar{t}$ is an increasing function of c, by Theorem XLII. $c = I_n$. I_n is an increasing function of N, by Theorem XXVII A. Therefore $\bar{t}$ is an increasing function of N, whence,

$$\bar{t}_{(N+r)} > \bar{t}_N .$$

B. This theorem may be illustrated by means of the set of values used to illustrate Theorem LI, which are, for the point of maximum learning difficulty of a 12-syllable series:

$$a = 7.87, \quad b = .1, \quad c = 5.008, \text{ and } \quad d = .7.$$

Substituting these values in the equation of Theorem XXXVII,

$$\bar{t} = \frac{1}{(d-b)} \log_e \frac{cd}{ab}$$

we have for a 12-syllable series ${}^{12}\bar{t}_{7,8} = 2.49$. Now, by Theorem XLVI B, we have found the ${}^{15}\bar{t}_{8,9}$ for the syllable of maximum learning difficulty to be 2.63. But since 2.63 > 2.49, it follows that the point in time at which the spontaneous rise in effective excitatory potential reaches its maximum is greater at the point of maximum difficulty in the 15-syllable series than in the 12-syllable series, as the theorem states.

THEOREM LII, COROLLARY 1

A. *In rote series learned by massed practice, the post-learning maximum of number of syllables susceptible of recall will occur closer to the time of the termination of active learning in the case of short series than in that of long series.*

Proof:

For the time of the post-learning maximum of mean effective excitatory strength is, for each syllable position, an increasing function of N.

C. No data bearing directly on the validity of this proposition have been found.

THEOREM LII, COROLLARY 2

A. *In rote series learned by massed practice, the post-learning minima in reaction latency will occur closer to the time of the termination of active learning in the case of short series than in that of long series.*

Proof:

This proposition follows directly from Theorem LII and from Theorem XXXVII, Corollary 1.

C. No data bearing directly on the validity of this proposition have been found.

THEOREM LIII

A. *With the passage of time following the learning of rote series by massed practice, the point of inflection following the point of maximum effective excitatory potential ($\bar{\bar{E}}$) in the post-learning interval will appear later in long than in short series.*

To prove:

That $\bar{\bar{t}}_{(N+r)} > \bar{\bar{t}}_N$.

Proof:

$\bar{\bar{t}}$ is an increasing function of c, by Theorem XLIII. $c = I_n$ (13, of "Preliminary Propositions," p. 256). I_n is an increasing function of N, by Theorem XXVII A. $\bar{\bar{t}}$ is an increasing function of N, whence,

$$\bar{\bar{t}}_{(N+r)} > \bar{\bar{t}}_N.$$

B. This theorem may be illustrated by means of the set of values used to illustrate Theorems LI and LII, which are, for the point of maximum learning difficulty of a 12-syllable series,

$$a = 7.87, \quad b = .1, \quad c = 5.008, \text{ and } \quad d = .7.$$

Substituting these values in the equation of Theorem XL,

$$\bar{\bar{t}} = \frac{1}{(d-b)} \log_e \frac{cd^2}{ab^2}$$

we have ${}^{12}\bar{\bar{t}}_{7,8} = 5.73$. Now, by Theorem XLVII B, at the syllable of maximum learning difficulty in a 15-syllable series ${}^{15}\bar{\bar{t}}_{8,9} = 5.87$. But since $5.87 > 5.73$, it follows that the inflection point in the spontaneous post-learning shifts of $\bar{E}$ shows a larger t-value for the point of maximum learning difficulty of a 15-syllable series than for the point of maximum learning difficulty of a 12-syllable series, as the theorem states.

THEOREM LIV

A. *In rote series learned by evenly distributed practice, the difference between the mean number of presentations required for learning at the several syllable positions of the series, approaches zero as the length of the temporal interval between the successive repetitions begins to increase.*

Proof:

Let T be the interval between successive repetitions. Then the inhibition, at the several syllables, and the mean number of repetitions required for learning are connected by the relation:

$$\sum_{i=1}^{R} (ae^{-biT} - ce^{-diT}) = L\,.$$

(We assume here that the interval T elapses between the termination of active practice and the testing of the subject, which is the usual case in such an experiment.)

It is necessary here, as in Theorem XIV, to replace this sum by an increasing function $g_1(R,c)$, defined for all positive, real values of R and c, coinciding with the above sum for all integral values of R. We can further assume without inconsistency that $g_1(R,c)$ is a convex for sufficiently large values of R, namely for $RT\,\bar{t}$. Then, the relation between R and c becomes:

$$g_1(R,c) = L\,.$$

Differentiating totally with respect to c, we have,

$$\frac{\partial g_1}{\partial R}\;\frac{dR}{dc}\;\frac{\partial g_1}{c} = 0.$$

$$\therefore \quad \frac{dR}{dc} = -\frac{\partial g_1}{\partial c} \Big/ \frac{\partial g_1}{\partial R}$$

But, again, $\frac{-\partial g_1}{\partial c}$ is to be taken as an increasing, concave function of R, coinciding with the sum:

$$\sum_{i=1}^{R} e^{-diT}$$

for all integral values of R. Hence,

$$\frac{\partial g_1}{\partial c} < \sum_{i=1} e^{-diT}$$

$$\therefore \quad \frac{dR}{dc} < \frac{\sum_{i=1}^{\infty} e^{-diT}}{ae^{-b[R+1]T} - ce^{-d[R+1]T}} \quad \frac{\sum_{i=1}^{\infty} e^{-(d-b)iT}}{a - ce^{-(d-b)[R+1]T}} .$$

But,

$$\sum_{i=1}^{\infty} e^{-(d-b)iT} = \frac{1}{1 - e^{-(d-b)T}} - 1 \rightarrow 0 \quad \text{as} \quad T \rightarrow \infty$$

and

$$a - oe^{-(d-b)[R+1]T} \rightarrow a \quad \text{as} \quad T \rightarrow \infty .$$

$$\therefore \quad \frac{dR}{dc} \rightarrow 0 \quad \text{as} \quad T \rightarrow \infty$$

which is equivalent to the theorem stated.

THEOREM LIV, COROLLARY 1

A. *In rote series learned by evenly distributed practice the difference in the mean number of presentations required for learning long and short series as a whole approaches zero as the temporal interval between successive repetitions increases.*

Proof:

The corollary follows from the proof of Theorem LIV; for, by Theorem XXXIV, the differences in inhibition between long and short series are not larger than a fixed limit, depending on F.

(If we assume, contrary to the usual experimental procedure, that no interval elapses between the termination of active practice and the testing of the subject, then Theorem LIV and Theorem LIV, Corollary 1 are true only in the sense of relative differences.)

C. No data bearing directly on the validity of this proposition have been found.

THEOREM LIV, COROLLARY 2

A. *In rote series learned by distributed practice, the difference between the mean reaction latency at the several syllables of the series approaches zero as the length of the temporal interval between successive presentations increases.*

Proof:

As T increases the mean effective excitatory increment per repetition approaches zero; hence, from Theorem LIV, the difference between the mean effective excitatory strength at the several syllable positions approaches zero; hence, the same is true of mean reaction time.

C. No data bearing directly on the validity of this proposition have been found.

Note: Seven additional theorems, all concerned with anticipatory errors (D43), have been derived but because of the already lengthy nature of the monograph their proofs are omitted. In the main these theorems depend upon the postulated nature of stimulus traces together with the principle of generalization (32, Definition 8), the latter of which has not been included in the present postulate set. The theorems are as follows:

Theorem LV. *In the learning of rote series, anticipatory errors will occur, at some stage of the learning process.*

Theorem LVI. *In the learning of rote series, anticipatory errors (D43) will be of relatively rare occurrence as compared with correct reactions.*

Theorem LVII. *In the learning of rote series, anticipatory errors will appear maximally at the position immediately preceding the correct position and will appear with decreasing frequency the greater the degree of syllable reaction anticipation (D41).*

Theorem LVIII. *In the learning of rote series, if the frequency of anticipatory error is graphed, then the curve is increasingly flat, tend- to zero as the degree of syllable reaction anticipation increases.*

Theorem LIX. *In the learning of rote series, periods of no practice will increase the frequency of anticipatory errors at the point immediately preceding the point of reinforcement.*

Theorem LX. *In the learning of rote series, periods of no practice will increase the maximum distance of syllable reaction displacement.*

Theorem LXI. *In the learning of rote series, anticipatory errors will gradually increase in number during practice, reaching a maximum sometime during the process, after which they will decline to zero.*

CONCLUDING REMARKS

The history of science shows that scientific theory advances by a series of successive approximations, that it is to a certain extent a true trial-and-error process. The indicators of error are primarily failures of the theorems of the system to agree with relevant facts. In general, each successive trial eliminates some of the evidences of error contained in preceding attempts, it extends the system to include a wider range of known fact and, perhaps most important of all, it projects its deductions into new regions where observations have not yet been made. There is no reason to believe that the evolution of theory in the behavioral (social) sciences will be exceptional in this respect. Certainly the development of the present theoretical system has followed this course.[1]

Four years ago the senior author of the present monograph published as a first attempt at a formally systematic theory of rote learning a miniature system consisting of eleven theorems. That preliminary study naturally displayed weaknesses. These weaknesses concerned not only the concepts and postulates employed, but also the deductive procedure utilized. The geometrical method used in deriving the theorems, while not without value in such a preliminary undertaking, proved to be clumsy in practice and limited in the logical rigor attainable. That one or more of the concepts and postulates of the system were defective was shown by two clear failures of the theorems wholly to agree with fact. These latter defects were evident at the time the manuscript was prepared for publication and were duly noted therein (*30*, 501; also footnote, 499). However, even before publication the effort to eliminate these weaknesses had begun. The present monograph gives in some detail the major results of the second attempt.

The most of the fundamental ideas contained in the first draft of the system remain in the present version; many of them, however, have been changed more or less extensively and a number of new principles have been added. Also a serious effort has been made to sharpen the definition of the concepts, chiefly by means of the techniques of symbolic logic. Moreover, the largely qualitative postulates characteristic of the initial miniature system have been recast so as to be capable of metricized mathematical statement; this has resulted in a very great increase in both the rigor and ease of theorem derivation and in the number and range of propositions so derivable. But the fact that improvements over the first attempt have been made does not mean that the present version

1. An authentic and impressive illustration of the trial-and-error aspects of scientific theory is furnished by the history of the evolution of quantum theory. For an account of the many false starts and their successive partial elimination, together with the gradual extension of the theory, see Rieche's work (*51*).

is offered as a perfect system. Partly to emphasize the trial-and-error aspects of systematic scientific theory, but partly also to facilitate the efforts of workers who presumably will carry on the development, two or three major defects of the present system may be pointed out.

1. In the statement of the groundwork of the system, the undefined notions are formally defective in a number of cases in that they do not represent observable objects, processes, or operations, either logical or experimental. More specifically, it is believed that several of the undefined notions (logical signs), including all those representing unobservables, should have appeared among the defined terms. Presumably the next attempt at a formalization of rote-learning theory should make this correction one of its earliest objectives. The techniques of symbolic logic (*66*) should render the task not too difficult of achievement. Here, evidently, is a place where the principles of "operationism," with their very real scientific virtues, should be applied. However, in spite of this formal defect of the system (originating partly in expository difficulties) it is believed that the actual understanding of the meaning of these particular undefined notions has been attained, though in a somewhat roundabout manner, and that real ambiguity as to the application of the present postulates to actual experimental situations will not be serious.

2. A somewhat related defect of the system lies in the fact that circumstances have not permitted the parallel statement of the theorems and corollaries in terms of symbolic logic. It is our belief that only by doing this will it be possible to attain the maximum of clarity as to the relationship of the outcome of the mathematical manipulation of equations to the outcome of experimental procedures. It is true that, so far as we have been able to discover, symbolic logic has never yet been used in this capacity in natural-science theory. Possibly this is because the need for such assistance in the physical sciences is less, since their concepts are less elusive.

3. The indicators of defect in the present system, as in all natural-science systems, lie primarily in the disagreements between its theorems and relevant experimental observations. Just as in the initial miniature system, we have here tried to point out to what extent the known facts agree with, or contradict, each theorem or corollary capable of empirical check. The reader has already seen that evidence concerning many of these propositions is at present non-existent, conflicting, ambiguous, or otherwise inadequate; such propositions offer points of departure for numerous critical experiments of the type already performed on an extensive scale by Hovland. There are, on the other hand, a large number of propositions which have been found to be adequately covered by experiment. The great majority of these are in agreement, but there are still a number of undoubted disagreements which point, of course, to basic defects in the conceptual and postulate system. It must be one of the primary tasks of subsequent theoretical efforts to so recast the concepts and

postulates of the system as to eliminate not only these disagreements but additional ones which further experimentation will doubtless reveal.

The most striking single failure of the system encountered so far is its disagreement with Jost's (*33*) law, one of the most firmly established empirical principles in the field of rote learning. Presumably as a special case of this failure, Hovland (*25*) has shown that the system implies better recall following massed practice than following distributed practice, whereas the experimental facts are exactly opposite. It is suspected that one reason for this failure is that the system makes no distinction in the rate of temporal decay of either excitatory or inhibitory potential as dependent upon the method of learning. It is also especially doubtful whether, as implied by the present system, the magnitude of the increment of excitatory potential at a given point in a rote series is wholly uninfluenced by the amount of inhibitory potential at that point when the repetition occurs. Speaking generally, our impression is quite definite that the weakest and most unsatisfactory portion of the present system is that group of concepts and postulates concerned with inhibitory potential.

Work looking to an improved set of basic assumptions has already begun. This effort naturally has been directed mainly toward this suspected sector of the postulate set. One of the most promising ideas for this revision was put forward independently, and almost simultaneously, by Dr. Eleanor Jack Gibson and Dr. Carl Iver Hovland. Fortunately, Dr. Gibson has already elaborated her version of the hypothesis in a formal system consisting of twenty-four theorems with numerous accompanying corollaries (*13*). Mrs. Gibson's hypothesis, in effect, goes behind the Pavlovian concept of inhibitory potential as put forward in the preceding pages, in that she attempts to derive it from the mutual interference produced by competing incompatible excitatory potentials. She has already carried the analysis far enough to show that it permits the deduction of a wide range of phenomena apparently inaccessible to the present postulate set. Among these may be mentioned the experimentally well-established phenomena of retroactive and proactive inhibition, investigated in recent years especially by Robinson, McGeoch, and their pupils. A second group of the deductions performed by Mrs. Gibson concerns the interesting phenomena depending upon various dishomogeneities in the learned series, which have been reported by certain *Gestalt* psychologists and hitherto interpreted only in terms of concepts characteristic of that school. Whether the promising Gibson-Hovland hypothesis will also be able to mediate the deduction of the phenomena treated in the present monograph more effectively than the set of postulates here employed can be determined only by trial.

In conclusion we shall consider briefly a question relating to the general scientific policy which has produced the present work. The fear has occasionally been expressed that the formalization of such a theoretical approach will "freeze" the concepts, thereby preventing further growth. This seems to us to be a complete inversion of the situation in true scientific theory, and it is our sincere hope that the present investigation will help to dissipate this apprehension. From the logical point of view, formalization is the essential condition which will *prevent* the freezing of a system. Changes in the concepts and postulates of a scientific system are forced by disagreements between its theorems and experimental observations; such disagreements can only force these changes when the theorems are linked to the underlying concepts and postulates by a rigorous logical formalization. Moreover, the empirical facts of actual system construction lead to the same conclusion. It is to make this clear, in part, that we have been at some pains to emphasize in the preceding paragraphs the process of progressive change in the development of the present system.

And so we take leave of this second pioneering attempt at a theory of rote learning. Despite its evident defects, we hope that it will at least serve as a concrete, large-scale demonstration of what we mean when we speak of systematic theory in the social sciences. It expresses our conviction that only by utilizing the logical, as well as the empirical, component of the complete logico-empirical methodology will the social sciences approach the predictive power and practical significance now characteristic of the physical sciences. It is true that behavior phenomena are more complex, and problems involving them are more difficult of solution, than are those of the older disciplines. This but emphasizes the need of the social sciences for the most powerful tools available. It is our hope and expectation that the present monograph will be one of a succession of increasingly rigorous and adequate formalizations, not only of rote-learning behavior, but of behavior throughout the whole range of the social sciences.

APPENDIX A

SUPPLEMENTARY REMARKS CONCERNING SYMBOLIC LOGIC

By Frederic B. Fitch

1. *Introductory*

The symbolic logic in the present monograph may be safely ignored by psychological readers who are not acquainted with this new combination of logic and mathematics. Those who wish to attain sufficient mastery of it to understand what appears on these pages are advised to take at least an elementary course in the subject, since the mere reading of texts may be inadequate without properly directed work in the manipulating of symbols and carrying out of proofs. Among elementary texts in English the following should be useful to the layman: C.I.Lewis and C.H.Langford: *Symbolic Logic* (Century Company); J.C.Cooley: *Outline of Symbolic Logic* (distributed in mimeographed form by the Harvard Co-operative Society); S.C.Langer: *Introduction to Symbolic Logic* (Houghton Mifflin and Company); J.H.Woodger: *The Axiomatic Method in Biology* (Macmillan Company), introductory chapters.

The more advanced student should be familiar with the classic *Principia Mathematica* by A.N.Whitehead and Bertrand Russell, as well as with some of the technical literature indexed in Alonzo Church's *Bibliography of Symbolic Logic* in volume 1 and volume 3 of the *Journal of Symbolic Logic*.

The reason for the use of symbolic logic in the present monograph is to indicate the extent (at least as an ideal) to which rigor and precision may be demanded in connection with the science of psychology. It is supposed that in psychology, in many other sciences and also in philosophy, much energy has been wasted because of the analytical burden placed on such languages as English, French, and German. Verbal confusions and ambiguities often exist unnoticed when an attempt is made to state definitively in ordinary language a scientific hypothesis or postulate. It seems desirable, in formulating a scientific hypothesis, to strive to express it in at least two languages: (1) an ordinary language, such as English, (2) symbolic logic. Even if, as often in the present monograph, the two resulting versions are not completely accurate translations of one another, the attempt will generally serve to elucidate what is really intended to be assumed.

Most important of all, when it once becomes clear what is being assumed, then it also becomes clear whether a given proof of a theorem is really a valid proof or not. A scientific hypothesis is not of much use so long as there is no sure way of demonstrating consequences of the hypothesis; for an empirical hypothesis can be empirically tested only by testing its experimentally verifiable consequences.

The careful analysis of a hypothesis by symbolic logic often reveals, moreover, subtly different (and perhaps better) hypotheses which never would have been imagined had only ordinary language been used. In this way symbolic logic is a tool not only for verification of the old but for discovery and invention of the new.

2. *Defects of the Symbolic Logic in this Monograph*

There are probably several minor errors in the symbolic logic of this monograph, although much effort has been made to eliminate errors of all kinds. Any such errors will not, it is hoped, indicate anything more than a need for a few fairly trivial revisions in the formulations of the definitions and postulates.

Attention is here called to the supplementary postulates in Section 4 of this Appendix. These are required in order that the whole set of postulates should be adequate, but they are not of great psychological significance.

It is not known whether all the postulates are consistent and independent, but it is supposed they are. A consistency proof of the whole set of postulates would presuppose a consistency proof of the logic of *Principia Mathematica*, even with the axioms of reducibility and infinity included. No such proof has yet been offered. (See however, F.B.Fitch: "The Consistency of the Ramified *Principia*," *The Journal of Symbolic Logic*, vol. 3 (1938), pp. 140-149, where a consistency proof is given for *Principia Mathematica* when the axioms of infinity and choice, but not the axiom of reducibility, are included.)

3. *The System of Symbolic Logic*

The symbolism for logical concepts employed in the present monograph is based on that of Whitehead and Russell's *Principia Mathematica*, and the actual system of logic used is that of *Principia Mathematica*, but with the addition of the "axiom of infinity" and the retention of the "axiom of reducibility."

Instead of the *Principia* symbols *B*, *C*, and *H*, the non-italic symbols B, C, and H are respectively used, and some other divergences of notation will be observed, especially as regards variables. (See Index of Logical Symbols.) Symbols for undefined psychological concepts (see U1-U13) are of course here incorporated into the logic of *Principia*. The "type" of each such symbol (or concept) is easily determined from the context in which it is used. It is to be observed that relations of degree > 2 play an important part (triadic relations, tetradic relations, etc.).

The following conventions, V1 - V11, are made concerning variables and differ somewhat from the usage of *Principia Mathematica*.

V1. The italic small letters *a*, *b*, *c*, *d*, *e* are variables having individuals as their values, but *e* is also sometimes used (as in D67', D78', 12A', 13A', 15A', 18A') in expressing the ordinary exponential function e^x. The two uses of *e* clearly do not lead to confusion, as the context determines which use is intended.

V2. The italic small letters *h*, *i*, *j*, *k*, *m*, *n*, *r*, *s*, are variables having the so-called "inductive cardinal numbers" as their values. These numbers may be regarded as classes of classes of individuals, in accordance with the method of Whitehead and Russell; and owing to our present assumption of the axiom of infinity, it will be found that, at least for our purposes, such numbers do not need to be regarded as of ambiguous type. It is possible to regard $(n)\varphi n$ as an abbreviation for $(\varkappa) : \varkappa \varepsilon$ NC induct $. \supset . \varphi\varkappa$. Other uses of these variables may similarly be regarded as abbreviations.

V3. The italic small letters *t*, *u*, *v*, *w*, *x*, *y*, *z*, t_1, t_2, are variables having the so-called "real numbers" as their values. It is known that the system of real numbers is definable with sufficient adequacy in *Principia Mathematica* to make possible a formal treatment of all the more common topics of mathematical analysis. The use of such variables is therefore justified. It is possible to regard $(x)\varphi x$ as an abbreviation for $(\mu) : \mu \ \varepsilon \ C'\Theta_g \ . \supset . \ \varphi\mu$. Other uses of these variables may similarly be regarded as abbreviations.

V4. The italic capital letters *P*, *Q*, *R*, *S* are generally variables having as their values dyadic relations between individuals. The letter *R* is often used for an entirely different purpose outside the symbolic logic; but even within the symbolic logic care should be taken to distinguish *R* from the non-italic R. (See D31'.)

V5. The italic capital letters *P*, *Q*, *R*, *S* are occasionally (as in D55', D56', D58', D59') variables having as their values triadic relations of real numbers to real numbers to individuals or sometimes, as values, dyadic relations of real numbers to individuals or, as values, dyadic relations between real numbers.

V6. The italic capital letters *T*, *U* are generally variables having as their values dyadic relations between dyadic relations between individuals. *T* should not be confused with the non-italic T. (See D61'.)

V7. The italic capital letter *U* is occasionally (as in D66') a variable having as its values triadic relations of real numbers to real numbers to dyadic relations between individuals.

V8. The Greek small letters α, β, γ are variables having as their values classes of individuals.

V9. The Greek small letters λ, μ, ν are generally variables having as their values classes of dyadic relations between individuals.

V10. The Greek small letter λ is occasionally (as in D59′, D60′, D68′) a variable having as its values classes of triadic relations of real numbers to real numbers of individuals.

V11. The Greek small letter ρ is a variable having as its values classes of dyadic relations between dyadic relations between individuals (as in D53′).

V12. The italic small letters f and g are used occasionally as variables which take as values mathematical functions. This usage is justified since such functions can be viewed as relations. See *Principia Mathematica*, Vol.II, Part V, Section C.

4. *Supplementary Postulates*

In addition to Postulates 1A′ - 18A′, the following supplementary Postulates SP1 - SP7 are required in order that all the tacitly assumed properties of the undefined notions shall be made explicit. These supplementary postulates were not included in the main body of the monograph because they are not of any great psychological importance.

SP1. a cn (b,c) . d cn (e,f) . b,e ε slex . c,f ε sb . ⊃ : $a = d$. ≡ . $b = e$. $c = f$

SP2. slcg ε refl ⌒ sym ⌒ trans

SP3. tr ε 1 → 1

SP4. P ε slpncy . $n <$ N′P . ⊃ . E! E′(t,n,P)

SP5. P ε slpncy . $n < r$. s $< r$. $r \leq$ N′P . ⊃ . E! I′(t,n,P,s,r)

SP6. a ε slpn ⊃ E! rnth′a

SP7. a ε slex v a ε slpn v a rn b . ⊃ . E! bg′a . E! nd′a . bg′$a \leq$ nd′a

5. *Supplementary Definitions*

The following definitions do not involve any of the undefined concepts. They are required mainly in order to show how *mathematical* concepts used in various parts of the symbolic logic can be derived from *logical* concepts.

SD1. $< \; = \hat{x}\hat{y}[(\exists z) : x + z = y \;.\; z \neq 0 \;.\; z \,\varepsilon\, \mathrm{C}'\Theta']$

SD2. $> \; = \mathrm{Cnv}'<$

SD3. $\leq \; = \hat{x}\hat{y}[x = y \;.\; \mathrm{v} \;.\; x < y]$

SD4. $\geq \; = \mathrm{Cnv}'\leq$

SD5. $xy = x \times y$

SD6. $n \times x \ .=. \ \vec{\mathrm{H}}{}^{\text{'}}(n/1) \times x$

SD7. $x \times n \ .=. \ n \times x$

SD8. $nx = n \times x$

SD9. $x \div y \ .=. \ (\iota z)\,[z \times y = x]$

SD10. $n \div m \ .=. \ \vec{\mathrm{H}}{}^{\text{'}}(n/m)$

SD11. $x \div n \ .=. \ x \div \vec{\mathrm{H}}{}^{\text{'}}(n/1)$

SD12. $x^2 = x \times x$.

SD13. $(\exists x)\,(P, Q) : \varphi\,(x, P, Q) :. = :. (\exists x) : (P, Q) : \varphi(x, P, Q)$

SD14. $P{}^{\text{'}}(a, \ldots, b) = (\iota c)\,[cP(a, \ldots, b)]$

SD15. Triadic relations of the form $\hat{a}\hat{b}\hat{c}\,[\varphi(a, b, c)]$ are to be dealt with analogously, as far as possible, to dyadic relations. If R is triadic we write $aR(b, c)$. Relations of higher degree are similarly dealt with.

SD16. $\sqrt{x} = (\iota y)\,[y > 0 \,.\, y^2 = x]$

SD17. $\Delta^{♀, Q} = \hat{P}\hat{S}\,[(\exists a, \alpha, \beta, x, y) : P = x \downarrow \alpha \,.\, S = y \downarrow \beta \,.\, a \sim \varepsilon\, \alpha \,.\, \beta = \alpha \cap \iota{}^{\text{'}}a \,.\, y = x♀(Q{}^{\text{'}}a)]$

SD18. $(\sum_{a\,\varepsilon\,\alpha}^{x, ♀})\, Q{}^{\text{'}}a = (\iota y)[(x{\downarrow}\Lambda)\Delta_{*}^{♀, Q}\,(y{\downarrow}\alpha)]$

SD19. $\sum_{\varphi a} f(a) = (\iota x)\,[(\exists\, \alpha, Q) :. (b) : b\varepsilon\alpha \,.\equiv.\, Q{}^{\text{'}}b \neq 0 \,.\, \varphi b : Q{}^{\text{'}}b = f(b) :. x = (\sum_{a\,\varepsilon\,\alpha}^{0, +})\, Q{}^{\text{'}}a\,]$

SD20. $\prod_{\varphi a} f(a) = (\iota x)\,[(\exists\, \alpha, Q) :. (b) : b\,\varepsilon\,\alpha \,.\equiv.\, Q{}^{\text{'}}b \neq 1 \,.\, \varphi b : Q{}^{\text{'}}b = f(b) :. x = (\sum_{a\,\varepsilon\,\alpha}^{1, \times})\, Q{}^{\text{'}}a\,]$

SD21. $\sum_{n=m}^{k} f(n) = \sum_{m \leq n \leq k} f(n)$

SD22. $n! = \prod_{1=m \leq n} m$

SD23. $x^n = \hat{R}\,[(\exists m, k) : m/k \ \varepsilon\ x \,.\, R = m^n/k^n]$

SD24. $\Phi(x) = \hat{y}\hat{n}\,[y = (-1)^{n+1} \times x^{2n-1} \div (2n-1)!]$

SD25. $\Psi(x) = \hat{u}\hat{v}\,[(\exists m) : \vec{\mathrm{H}}{}^{\text{'}}(m/1) \leq v \leq \vec{\mathrm{H}}{}^{\text{'}}((m+1)/1) \,.\, u = \sum_{n=1}^{m} \Phi(x){}^{\text{'}}n\,]$

SD26. $\sin = \Theta_g\, \overline{\Psi(x)}_{sc}\, \Theta_g$ (See also *310.03 and *231.01)

SD27. $\pi = (\iota z)\,[\sin{}^{\text{'}}(z \div 2) = 1 \,.\, 3 < z < 4\,]$

SD28. $\lim_{x \to y} f(x) = (\iota z)\,[(\exists R)\,(w) : f(w) = R{}^{\text{'}}w \,.\, z = R(\Theta_g\, \Theta_g){}^{\text{'}}y = R(\breve{\Theta}_g\, \Theta_g){}^{\text{'}}y = R(\Theta_g\, \breve{\Theta}_g){}^{\text{'}}y = R(\breve{\Theta}_g\, \breve{\Theta}_g){}^{\text{'}}y]$

SD29. $\mathbf{deriv}_x = \hat{z}\hat{P}\{z = \lim_{y \to x} [(P\text{'}y - P\text{'}x) \div (y - x)]\}$

SD30. $f'(x) = \lim_{y \to x} \{[f(y) - f(x)] \div (y - x)\}$

SD31. $\mathrm{e} = (\iota R)[(x) : \mathbf{deriv}_x\text{'}R = R\text{'}x]$

SD32. $e^x = \mathrm{e}\text{'}x$

SD33. $\int_u^v f(x)\,dx = \lim_{t \to 0} \{\sum_{t \leq nt \leq v-u} [t \times f(u + nt)]\}$

SD34. $n = m = k \;.{=}.\; n = m \,.\, m = k$ (The same convention holds also when such symbols as $\leq$ and $<$ are used.)

APPENDIX B

A PROOF IN SYMBOLIC LOGIC OF COROLLARY 1 OF POSTULATE 1

By Frederic B. Fitch

The purpose of this appendix is to indicate briefly how symbolic logic may be used in deducing consequences of A1′-A18′, SP1 - SP7. The proof of Corollary 1 of Postulate 1 (see p. 45) will be given below with a reasonable amount of detail. The symbolic formulation of this corollary may be chosen thus:

$$n \text{ slpnnm } (a, P) \,.\supset.\, \exists ! \parallel_{0\ldots n}\text{tr}_n \parallel {}^{\backprime}P$$

It appears as theorem T51 at the end of the appendix.

In order to avoid too much technicality we shall purposely fail to distinguish below between:

(1) Symbols (of the "object language") which are variables.

(2) Symbols (of the "syntax language") which *denote* variables or constants of some type.

Thus if we say, "$(p \cdot q) \supset p$ is provable," and if p and q are used in sense (2) above, then we would mean also that "$(q \cdot r) \supset q$ is provable," that "$((r \vee s) \cdot q) \supset (r \vee s)$ is provable," and so on, letting p, q, r, s be propositional variables.

We assume, furthermore, that the authors of *Principia* also fail to distinguish between (1) and (2) and that their use of variables may be regarded as being predominantly in sense (2) above.

A more detailed treatment would use different symbols for (2) than for (1), but no fallacy results if the distinction is kept in mind.

The symbols φ and ψ are used in such a way that φa, for example, is to be understood as denoting any propositional expression in which a is a free variable or a constant.

Whatever is said about variables of lowest type is understood to apply also, with appropriate restrictions, to variables of higher types.

The symbol s will often be used below as a propositional variable.

A few symbols occur in this appendix which will not be found in the index of logical signs. In such cases refer to the list of symbols at the end of Vol. I of *Principia*.

Definition. A "proof sequence" will be a finite sequence of propositions $p_1, p_2, \ldots, p_n$ satisfying the following condition: For each member p of the sequence at least one of the following three conditions must hold:

(1) p is by definition an "axiom."

(2) There is some proposition q such that both q and $q \supset p$ precede p in the sequence in question.

(3) p is of the form $(a)\varphi a$ and is preceded in the sequence by the proposition φa. (A variable of different type from a would also be allowable in this condition.)

Definition. Those propositions which are by definition "axioms" are A1′ - A18′, SP1 - SP7, together with all formally established propositions of *Principia Mathematica*, and finally the Axiom of Infinity, defined in *120·02 in *Principia*.

Definition. A proposition p is said to be *provable* if and only if there exists a proof sequence terminating with p.

Using the above three definitions it is well known to logicians that the following useful "rules of procedure" can be derived. They will assist in giving concise indications of proofs. Although not all of them will be used, they are of some interest in their own account.

R1. If p and $p \supset q$ are provable, so is q.

R1′. If $s \supset p$ and $s \supset (p \supset q)$ are provable, so is $s \supset q$.

R2. If $p \supset q$ and $q \supset r$ are provable, so is $p \supset r$.

R2′. If $s \supset (p \supset q)$ is provable and if $s \supset (q \supset r)$ or $q \supset r$ is provable, then $s \supset (p \supset r)$ is provable.

R3. If p and q are provable so is $p \cdot q$.

R3′. If $s \supset p$ and $s \supset q$ are provable, so is $s \supset (p \cdot q)$.

R4. If $p \cdot q$ is provable, so are p and q.

R4′. If $s \supset (p \cdot q)$ is provable, so are $s \supset p$ and $s \supset q$.

R5. If $p \equiv q$ is provable, so are $p \supset q$ and $q \supset p$.

R5′. If $s \supset (p \equiv q)$ is provable, so are $s \supset (p \supset q)$ and $s \supset (q \supset p)$.

R6. If φa is provable, where a is a free ("real") variable, then $(a)\varphi a$ is provable. [We assume here and elsewhere that "collisions" between scopes of variables are suitably avoided. Variables are supposed to be free ("real") unless there is some explicit indication that they are bound ("apparent").]

R6′. If $s \supset \varphi a$ is provable, where a is a free ("real") variable not occurring in s, then $s \supset (a)\varphi a$ is provable.

R7. If $p \equiv q$ and r are provable and if r' is the result of substituting p for q (or vice versa) somewhere in r, then r' is provable. (We might choose r as p. Then r' would be q.)

R7′. If $s \supset (p \equiv q)$ and $s \supset r$ are provable and if r' is the result of substituting p for q (or vice versa) somewhere in r, then $s \supset r'$ is provable.

Propositions L1 - L21 are propositions which are provable without use of A1′ - A18′ or SP1 - SP7 as axioms. In order to save space their proofs will not be indicated. Even for T1 - T35 actual proof sequences will not be given, but the applicability of R1 - R10 will be indicated within braces after each proposition. A second pair of braces often will enclose relevant definitions.

For the sake of simplicity, descriptive expressions such as $B'\breve{P}$ and $bg'a$ are treated as values for variables in the ordinary sense. This is sound, since existence is always provable by use of a relevant hypothesis.

In order to save space the proofs will be only outlined after T35.

L1. $a \varepsilon \hat{b}(\psi b) . \equiv . \psi a$

L2. $6 \leq Nc'C'P . \supset . C'P \varepsilon$ Cls induct $- \iota'\Lambda$

L3. $P \varepsilon ser . C'P \varepsilon$ Cls induct $- \iota'\Lambda : \supset : E!B'P . E!B'\breve{P}$

L4. $P \varepsilon ser . C'P \varepsilon$ Cls induct $- \iota'\Lambda : \supset : P_1``\alpha \subset \alpha . a \varepsilon \alpha . \supset . \overrightarrow{P}'a \subset \alpha$

L5. $(\exists c) : (a, b) : \varphi(a, b, c) :. \supset :. (\exists c) : \varphi(a, b, c)$

L6. $(\exists c) : p . \supset . q . \varphi c :. \supset :. p \supset q$ (c not occurring in p and q.)

L7. $P \varepsilon ser . \supset : a P b . \sim(a P^2 b) . \equiv . a P_1 b$

L8. $p . \supset . q \supset r : \equiv : p . q . \supset . r$

L9. $p . \supset . p \vee q$

L10. $\alpha \subset \beta : \equiv : (a) : a \varepsilon \alpha . \supset . a \varepsilon \beta$

L11. $(a) \varphi a . \supset . \varphi b$

L12. $a P_1 b . \supset . b \varepsilon C'P$

L13. $p \supset q . \supset : r . p . \supset . r . q$

L14. $0 < n - m . n \leq k . \supset . m \leq k$

L15. $p \supset q . \supset . p . r . \supset . q . r$

L16. $n \leq m . m \leq k . \supset . n \leq k$

L17. $(b) . \varphi b \supset p : \equiv : (\exists b) \varphi b . \supset . p$ (b not occurring in p.)

L18. $a \varepsilon (P``\hat{b}(\varphi b)) : \equiv : (\exists b) . a P b . \varphi b$

L19. $n \leq n$

L20. $p . \supset . q \supset p$

L21. $a \varepsilon \overrightarrow{P}'c . \equiv . a P c$

T1. $P \varepsilon slexcy : \equiv :. P \varepsilon ser . C'P \subset slex . P \subseteq \dot{-} slcg . 6 \leq Nc'C'P \leq 100 : (\exists k)(a, b) : a P b . \sim(a P^2 b) . \supset . 0 < bg'b - nd'a = k \leq (du'a \div 10) . 1 \leq du'a = du'b \leq 4$ {L1} {D21'}

T2. $P \varepsilon slexcy . \supset . P \varepsilon ser$ {{T1, R5}, R4' repeatedly}

T3. $P \varepsilon slexcy . \supset . 6 \leq Nc'C'P \leq 100$ {{T1, R5}, R4' repeatedly}

T4. $P \varepsilon slexcy . \supset . C'P \varepsilon$ Cls induct $- \iota'\Lambda$ {T3, L2, R2}

T5. $P \varepsilon slexcy . \supset . P \varepsilon ser . C'P \varepsilon$ Cls induct $- \iota'\Lambda$ {T2, T4, R3'}

T6. $P \varepsilon slexcy . \supset . E! B'P . E! B'\breve{P}$ {T5, L3, R2}

T7. $P \varepsilon slexcy : \supset : P_1``\alpha \subset \alpha . a \varepsilon \alpha . \supset . \overrightarrow{P}'a \subset \alpha$ {T5, L4, R2}

T8. $P \varepsilon \text{slexcy} :\supset: (\exists k) : (a, b) : a\,P\,b\,.\,\sim(a\,P^2 b)\,.\supset.\,0 < \text{bg}'b - \text{nd}'a\,.\,\text{bg}'b - \text{nd}'a = k\,.\,k = (\text{du}'a + 10)\,.\,1 \leq \text{du}'a = \text{du}'b \leq 4$ {{T1, R5}, R4} {SD34}

T9. $P \varepsilon \text{slexcy} :\supset: (\exists k) : a\,P\,b\,.\,\sim(a\,P^2 b)\,.\supset.\,0 < \text{bg}'b - \text{nd}'a\,.\,\text{bg}'b - \text{nd}'a = k\,.\,k = (\text{du}'a \div 10)\,.\,1 \leq \text{du}'a = \text{du}'b \leq 4$ {T8, L5, R2}

T10. $P \varepsilon \text{slexcy} :\supset: a\,P\,b\,.\,\sim(a\,P^2 b)\,.\supset.\,0 < \text{bg}'b - \text{nd}'a$ {T9, L6, R2}

T11. $P \varepsilon \text{slexcy} :\supset: a\,P\,b\,.\,\sim(a\,P^2 b)\,.\equiv.\,a\,P_1\,b$ {T2, L7, R2}

T12. $P \varepsilon \text{slexcy} :\supset: a\,P_1\,b\,.\supset.\,0 < \text{bg}'b - \text{nd}'a$ {T11, T10, R7'}

T13. $P \varepsilon \text{slexcy}\,.\,a\,P_1\,b\,.\supset.\,0 < \text{bg}'b - \text{nd}'a$ {L8, T12, R7}

T14 $b \varepsilon \text{slex}\,.\supset.\,E!\,\text{bg}'b\,.\,E!\,\text{nd}'b\,.\,\text{bg}'b \leq \text{nd}'b$ {L9, {L9, SP7, R2}, R2}

T15. $b \varepsilon \text{slex}\,.\supset.\,\text{bg}'b \leq \text{nd}'b$ {T14, R4'}

T16. $P \varepsilon \text{slexcy}\,.\supset.\,C'P \subset \text{slex}$ {{T1, R5}, R4' repeatedly}

T17. $P \varepsilon \text{slexcy} :\supset: b \varepsilon C'P\,.\supset\;b \varepsilon \text{slex}$ {{L10, T16, R7}, L11, R2}

T18. $P \varepsilon \text{slexcy}\,.\,b \varepsilon C'P\,.\supset.\,b \varepsilon \text{slex}$ {L8, T17, R7}

T19. $P \varepsilon \text{slexcy}\,.\,a\,P_1\,b\,.\supset.\,P \varepsilon \text{slexcy}\,.\,b \varepsilon C'P$ {L12, L13, R1}

T20. $P \varepsilon \text{slexcy}\,.\,a\,P_1\,b\,.\supset.\,\text{bg}'b \leq \text{nd}'b$ {T19, {T18, T15, R2}, R2}

T21. $P \varepsilon \text{slexcy}\,.\,a\,P_1\,b\,.\supset.\,\text{nd}'a \leq \text{nd}'b$ {{T13, T20, R3'}, L14, R2}

T22. $P \varepsilon \text{slexcy} :\supset: a\,P_1\,b\,.\supset.\,\text{nd}'a \leq \text{nd}'b$ {L8, T21, R7}

T23. $P \varepsilon \text{slexcy} :\supset: a\,P_1\,b\,.\,\text{nd}'b \leq \text{nd}'c\,.\supset.\,\text{nd}'a \leq \text{nd}'b\,.\,\text{nd}'b \leq \text{nd}'c$ {T22, L15, R2}

T24. $P \varepsilon \text{slexcy} :\supset: a\,P_1\,b\,.\,\text{nd}'b \leq \text{nd}'c\,.\supset.\,\text{nd}'a \leq \text{nd}'c$ {T23, L16, R2'}

T25. $P \varepsilon \text{slexcy} :\supset: (b) : a\,P_1\,b\,.\,\text{nd}'b \leq \text{nd}'c\,.\supset.\,\text{nd}'a \leq \text{nd}'c$ {T24, R6'}

T26. $P \varepsilon \text{slexcy} :\supset: (\exists b)\,.\,a\,P_1\,b\,.\,\text{nd}'b \leq \text{nd}'c\,.\supset.\,\text{nd}'a \leq \text{nd}'c$ {L17, T25, R7}

T27. $P \varepsilon \text{slexcy} :\supset: (a) : (\exists b)\,.\,a\,P_1\,b\,.\,\text{nd}'b \leq \text{nd}'c\,.\supset.\,\text{nd}'a \leq \text{nd}'c$ {T26, R6'}

T28. $P \varepsilon \text{slexcy} :\supset: (a) : a \varepsilon P_1``\hat{b}\,[\text{nd}'b \leq \text{nd}'c]\,.\supset.\,a \varepsilon \hat{b}\,[\text{nd}'b \leq \text{nd}'c]$ {L1, {L18, T27, R7}, R7}

T29. $P \varepsilon \text{ slexcy} : \supset : P_1 {}^{\prime\prime} \hat{b}[\text{nd}'b \leq \text{nd}'c] \subset \hat{b}[\text{nd}'b \leq \text{nd}'c]$ {L10, L28, R7}

T30. $P \varepsilon \text{ slexcy} : \supset : \text{nd}'c \leq \text{nd}'c$ {L19, L20, R1}

T31. $P \varepsilon \text{ slexcy} : \supset : c \,\varepsilon\, \hat{b}[\text{nd}'b \leq \text{nd}'c]$ {T1, T30, R7}

T32. $P \varepsilon \text{ slexcy} : \supset : P_1 {}^{\prime\prime} \hat{b}[\text{nd}'b \leq \text{nd}'c] \subset \hat{b}[\text{nd}'b \leq \text{nd}'c] \,.\, c \,\varepsilon\, \hat{b}[\text{nd}'b \leq \text{nd}'c]$ {T29, T31, R3'}

T33. $P \varepsilon \text{ slexcy} \,.\supset.\, \vec{P}'c \subset \hat{b}[\text{nd}'b \leq \text{nd}'c]$ {T32, T7, R1'}

T34. $P \varepsilon \text{ slexcy} : \supset : a \,\varepsilon\, \vec{P}'c \,.\supset.\, \text{nd}'a \leq \text{nd}'c$ {L1, {L10, T33, R7}, R7}

T35. $P \varepsilon \text{ slexcy} : \supset : a P c \,.\supset.\, \text{nd}'a \leq \text{nd}'c$ {L21, T34, R7}

By similar methods we can prove T36 and (using D4') T37:

T36. $P \varepsilon \text{ slexcy} \,.\, a P b \,.\supset.\, \text{bg}'\text{B}'P < \text{nd}'a < \text{bg}'b < \text{nd}'\text{B}'\breve{P}$

T37. $P \varepsilon \text{ slexcy} \,.\, a P b \,.\supset.\, 0 < \text{slinexin}'P$

By use of D20',D21',D23',SP1,SP7,T36,T37, we can prove T38:

T38. $P \varepsilon \text{ slpncy} \,.\, a P b \,.\supset.\, \text{bg}'\text{B}'P \leq \text{bg}'a < \text{nd}'a < \text{bg}'b < \text{nd}'b \leq \text{nd}'\text{B}'\breve{P} \,.\, 0 < \text{slinpnin}'P$

T39. $n \text{ slpnnm } (d, P) \,.\equiv.\, n = \text{Nc}'\vec{P}'a \,.\, P \varepsilon \text{ slpncy} \,.\, a \,\varepsilon\, \text{C}'P$ {L1} {D25'}

T40. $t \text{ cytm } P \,.\equiv.\, P \varepsilon \text{ slpncy} \,.\, t = \text{nd}'\text{B}'\breve{P} + \text{slinpnin}'P$ {L1} {D62'}

From T38, T39, and T40 we can obtain T41:

T41. $n \text{ slpnnm } (d, P) : \supset : \text{nd}'d + \text{slinpnin}'d \leq \text{cytm}'P$

From T39 and Postulate 1A' we can obtain T42:

T42. $n \text{ slpnnm } (d, P) \,.\supset.\, \text{bg}'\text{tr}'a = \text{bg}'a \,.\, \text{nd}'\text{tr}'a = \text{cytm}'P$

From T37 - T42 it is easy to derive successively T43 - T46:

T43. $n \text{ slpnnm } (d, P) : \supset : \text{bg}'d \geq \text{bg}'\text{tr}'d \,.\, \text{nd}'d + \text{slinpnin}'P \leq \text{nd}'\text{tr}'d$

T44. $n \text{ slpnnm } (d, P) : \supset : n = \text{slpnnm}'(d, P) \leq n = \text{slpnnm}'(d, P) \,.\, (\exists x, y) \,.\, x = \text{bg}'d \geq \text{bg}'\text{tr}'d \,.\, y = \text{nd}'d + \text{slinpnin}'P \leq \text{nd}'\text{tr}'d$

T45. $n \text{ slpnnm } (d, P) :. \supset :. (\exists s) : (\exists a, b) : s = \text{slpnnm}'(a, P) \leq n = \text{slpnnm}'(b, P) \,.\, (\exists x, y, c) \,.\, x = \text{bg}'b \geq \text{bg}'c \,.\, y = \text{nd}'b + \text{slinpnin}'P \leq \text{nd}'c \,.\, c = \text{tr}'a$

T46. n slpnnm (d, P) :. ⊃ :. $(\exists Q) : (\exists s) : (\exists a, b)$: s = slpnnm '$(a, P) \leq n$ = slpnnm' (b, P) . $Q = \hat{x}\hat{y}\hat{c}\,[x$ = bg'$b \geq$ bg'c . y = nd'b + slinpnin'$P \leq$ nd'c . c = tr'a] . $\dot{\exists}!\, Q$

By use of D55' we can then prove:

T47. n slpnnm (d, P) : ⊃ : $(\exists Q) : (\exists s)$: $Q\ {}_s\mathrm{tr}_n\ P$

We can then derive T48 and T49:

T48. n slpnnm (d, P) : ⊃ : $(\exists S)$: $S \varepsilon \hat{Q}\,[(\exists s) . Q\ {}_s\mathrm{tr}_n\ P]$

T49. n slpnnm (d, P) : ⊃ : $(\exists S, \lambda)$: $\lambda = \hat{Q}\,[(\exists s) . Q\ {}_s\mathrm{tr}_n\ P]$. $S \varepsilon \lambda$

Finally, by use of D59', we obtain T50 and T51:

T50. n slpnnm (d, P) : ⊃ : $(\exists S, \lambda)$: $\lambda \|_{0 \ldots n}\mathrm{tr}_n \| P$. $S \varepsilon \lambda$

T51. n slpnnm (d, P) : ⊃ : $\exists! \|_{0 \ldots n}\mathrm{tr}_n \|$ 'P

This last theorem is the symbolic logic formulation of Corollary 1 of Postulate 1 (p. 45).

APPENDIX C

INDEX OF LOGICAL SIGNS

(Note: References with an asterisk are to Whitehead and Russell's *Principia Mathematica*. References with the letter W are to pages of Woodger's *The Axiomatic Method in Biology*.

U1 - U16 are on pp. 22 - 26.
D1' - D85' are on pp. 26 - 40.
1A' - 18A' are on pp. 41 - 80.
SP1 - SP7 are on p. 312.
SD1 - SD34 are on pp. 312 - 314.
V1 - V12 are on pp. 311 - 312.

References in parentheses indicate some or, in the case of empirical concepts, all places where the symbol in question is used excepting Appendix B.)

ΔE	(See under Greek letter Δ)
$E_{n,\tau}$	D65.2' (D66', 8A')
$\bar{E}_{n,\tau}$	D79.1' (D79.2', D80', 10A', 14A')
$^{i,j,N'P}E_{n,\tau}$	D65.1' (D65.2')
e	V1 (D19')
e	SD32 (D67')
e^x	SD32
$\mathcal{E}_{n,\tau}$	D63' (D64', D79.1', 2A', 8A', 10A')
E!	*14·02, *30·2, W31 (1A')
(∃)	*9, *10, W25 (D2')
~(∃)	*9·021 (D32')
~(∃ ,)	*9·021, *11·03 (D45')
$\dot{\exists}$!	*25·03, W30 (D30')
$E_{n,\tau}(t)$	D67' (8A', 12A')
ev	D68'
evdspr	D35'
F	D73' (6A')
f	V12 (SD19)
fq	D85' (15A')
fsrn	D45'
ftsc	D48'
$f'(x)$	SD30
g	V12
h	U13' (18A')
H	*304·02 (SD6)
h	V2 (D85')
I	U8' (D69', SP5)
I_n	D77·2' (9A')
$^{i,j,N'P}I_n$	D77·1' (D77·2')
$\Delta\, ^{N'P}{}_{s}I_{n,\tau}$	(See under Greek letter Δ)
i	V2 (D53')
$\mathcal{I}_n$	D70' (D79·1', 9A', 10A')
$_{s}\mathcal{I}_{n,\tau}$	D69' (D70', D71')
$I_n(t)$	D78' (9A', 13A')
incyin	D16'
induct	(See NC induct)
insl	D14'
invl	D66½' (D67', D78', 12A', 13A')
j	V2 (D53')
$^{N'P}J_n$	D76'
k	U14' (18A')
ΔK	(See under Greek letter Δ)
k	V2 (D2')
L	D82' (D85', 16A')
l	V2
$\lim_{x \to y} f(x)$	SD28
$ltfl_m$	D49'
m	V2 (D47')
mernth	D81 (D82', D83')
mspr	D32' (D33')

msprnm	D62½' (D64',D65.1',D71',D77.1')
N	D6.1' (D9',D53',D64',D65.1',D70',D71',D74',D77.1',2A',4A', 8A',9A',11A',14A')
n	V2 (D6')
n!	SD22
N	(See D58' and note after D6)
Nc	*100·01,W36 (D2')
NC induct	*120·01,W36
nd	U16' (D1',D2',D4',D16',D21',D30',D32',D33',D34',D35',D37',D39', D40',D45',D55',D80',1A',SP7)
n/m	*303·01 (SD6)
nx	SD8 (15A', noting $2\sigma^2$ and 2π)
P	V4 (D2')
P^2	*34·02,W34 (D2')
$\vec{P}$	*32·01,W32 (D6')
peer	D44'
pern	D39' (D42',D44')
perndg	D42'
pnfl	D46' (D47',D48',D50',D51',D53')
$P\overline{R}_{sc}Q$	*231·01
prtm	D33' (D62½',D65.1',D66½',D77.1',8A',9A')
ptcd	D66'
Q	V5 (D55')
Q	V4 (D9')
R	D31' (D31',D32',D34',D35',D47',D48',D49',D50',D51',D52',D53', D65.1',D67',D77.1',D78',12A',13A')
r	V2 (D63')
R	V4 (D28')
R_*	*90·01 (SD18)
rcpb	D53'
rdrn	D40'
refl	W34 (D30')
rm	D80'
rn	U4' (D36',SP7)
rnos	D50'
rnth	U9 (D81',D83',D85',14A',18A',SP6)
$R(PQ)$	*233·02
rtln	D29' (D66½',D67',D78',12A',13A')
rtpr	D30' (D31',D34',D48',D50',D52',D53')
rtsr	D11' (D12',D15')
$R(\Theta_g\Theta_g)$	*233·02
s	V2 (D55')
S	V4 (D28')
sb	U2' (D17',D18',D20',D36',SP1)
sdrnth	D83' (D84')
ser	*204·01,W41 (D2')
sin	SD26
sl	D10' (D15')
slcg	U5' (D2',D9',D10',D36',D45',SP2)
slex	U1' (D2',D17',D18',D36',SP1,SP7)
slexcy	D2' (D3',D4',D5',D6',D8',D20')
slexcycg	D9' (D11',D16',D28')
slexin	D5' (D9',D24')
slexnm	D6' (D7',D9',D12')
slexpd	D3' (D9',D20')

slinexin	D4' (D23')
slinpnin	D23' (D21',D30',D32',D33',D34',D37',D38',D39',D40',D45',D55', D62')
slnm	D12' (D13',D14')
slpn	D17' (D61',D82',D84',D85',SP3,SP6,SP7)
slpncg	D19' (D19')
slpncy	D21' (D25',D27',D45',D46',D62',1A',2A',SP4,SP5)
slpncycg	D28' (D29',D30',D62½',D65.1',D71',D77.1')
slpncycn	D20' (D21',D22',D23',D24',D28')
slpnin	D24' (D30',D32',D33',D34',D65.1',D77.1')
slpnnm	D25' (D26',D36',D37',D38',D39',D40',D41',D42',D47',D48',D50', D53',D55',D68',14A',18A')
slpnor	D54'
slpnpd	D22'
slpnsb	D18' (D21',D30',D45',D80')
slrnnm	D36' (D37',D38',D39',D40',D41',D42')
T	D61' (18A')
t	V3 (D1')
t_1	V3 (D80')
t_2	V3 (D80')
T	V6 (D30')
tmsl	D15'
tmslex	D8'
tmslpn	D27'
tr	U6' (D54',D55',D61',1A',SP3)
${}_s\mathrm{tr}_n$	D55' (D56',D58',D59')
${}_s\mathrm{tr}_s$	D55'
${}_s\mathrm{tr}_{s+1\ldots N}$	D58'
$\Vert {}_{0\ldots n}\mathrm{tr}_n \Vert$	D59' (D60')
trans	*201·01 (D30')
trsgnm	D56'
u	V3 (D66½')
U	V6 (D50')
$\bar{U}$	V7 (D66')
v	U12' (18A')
v	W20 (D32')
v	V3 (D67')
w	V3 (D82')
x	V3 (D16')
x^2	SD12 (D83')
x^n	SD23
(ιx)	(See $(\iota\ )$ under Greek letter ι)
xy	SD5 (D67', noting bt)
y	V3 (D55')
z	V3 (D82')
α	V8 (D12')
β	V8
γ	V8
δ	V8
$\Delta^{♀,\ell}$	SD17
$\Delta_*^{♀,\ell}$	(See SD17 and *90·01)
ΔE	D64' (D65.1',D67',D78',3A',11A',12A')
$\Delta^{N'P}I_n$	D75' (D76',D77.1',D78',11A',13A')

$\Delta\ {}^{N'P}I_{s\,\tau}$	D74' (D75')
$\Delta\ {}^{N'P}I_{s\,n,\tau}$	D71' (D72',D73',D74',4A',7A')
ΔK	D72' (D76',5A')
ε	*20·02,W25 (D2')
ζ	V8
Θ_g	*310·03 (For Θ' see *310·011.)
(ι)	*14·01,W31 (D64')
λ	V9 (D12')
λ	V10 (D59')
Λ	*24·02 (SD18)
μ	V9
ν	V9
π	SD27 (15A')
$\Pi_{\varphi a}\ f(a)$	SD20
ρ	V11 (D53')
σ	D84' (15A',17A')
$\Sigma_{\varphi a}\ f(a)$	SD19 (D67')
$\sum_{n=m}^{k} f(n)$	SD21
$(\Sigma_{a\varepsilon\alpha}^{x,♀})\ Q'a$	SD18
φ	(See pp. 14-19 in *Principia Mathematica*)
Φ(x)	SD24
Ψ(x)	SD25
<	SD1 (D2')
>	SD2 (D34')
≤	SD3 (D2')
≥	SD4 (D55')
×	SD6,SD7 (D65.1')
÷	SD9 (D2')
÷	SD10 (D53')
÷	SD11 (D81')
(∃ ,) ( )	SD13 (D35')
'(,)	SD14 (D7')
^^^[ ]	W27,SD15 (D6')
^^^^^{ }	SD15 (D53')
√	SD16 (15A')
$\int_u^v f(x)\,dx$	SD33 (15A')
(,)	*11·01 (D2')
~	W19, *20·06 (D2')
2	*34·02, W34 (D2')
.	W49 (D2')
⊃	*1·01,W21 (D2')
→	*32·01,W32 (D6')
∪̄	*119·01,*119·02,*119·03 (D6.1')
	*31·02,W33 (D8')
:.	W49 (D9')
,	*20·04 (D9')
( )	*9,*10,W25 (D15')
+	*312·02 (D21')
⪽	*23·01,W30 (D30')

ADDENDA

REFERENCES

1. ADRIAN, E.D. *The Basis of Sensation.* New York: W.W.Norton, 1928.
2. BASS, M.J. and HULL, C.L. The Irradiation of a Tactile Conditioned Reflex in Man. *J. comp. Psychol.* 1934, *17*, 47-66.
3. COHEN, M.R. and NAGEL, E. *An Introduction to Logic and Scientific Method.* New York: Harcourt, Brace, 1934.
4. COOLEY, J.C. Outline of Symbolic Logic. Cambridge, Mass. (Distributed in mimeographed form by the Harvard Cooperative Society.) 1939.
5. DEWEY, J. *Logic.* New York: Henry Holt, 1938.
6. DODGE, R. *Human Variability.* New Haven: Yale University Press, 1931.
7. EBBINGHAUS, H. *Grundzüge der Psychologie.* Leipzig: Veit, 1902.
8. EBBINGHAUS, H. *Memory* (Trans.by H.A.Ruger and C.E.Bussenius). New York: Teachers College, Columbia University Press, 1913.
9. EINSTEIN, A. and INFELD, L. *The Evolution of Physics.* New York: Simon and Schuster, 1938.
10. ELLSON, D.G. Quantitative Studies of the Interaction of Simple Habits. I. Recovery from Specific and Generalized Effects of Extinction. *J. exp. Psychol.* 1938, *23*, 339-358.
11. FINKENBINDER, E.O. The Curve of Forgetting. *Amer. J. Psychol.*, 1913, *24*, 8-32.
12. FOUCAULT, M. Les Inhibitions Internes de Fixation. *L'Ann Psychol.*, 1928, *29*, 92-112.
13. GIBSON, E.J. A Systematic Application of the Concepts of Generalization and Differentiation to Verbal Learning. New Haven: Ph.D. Thesis deposited in Yale University Library, 1938.
14. GLAZE, J.A. The Association Value of Nonsense Syllables. *Ped. Sem.*, 1928, *35*, 255-269.
15. HARVEY, W. *Anatomical Studies on the Motion of the Heart* (Trans. by C.D.Leake) (2nd edition). Springfield, Ill.: C.C. Thomas, 1930.
16. HILGARD, E.R. and CAMPBELL, A.A., The Course of Acquisition and Retention of Conditioned Eyelid Reactions in Man. *J. exp. Psychol.*, 1936, *19*, 227-247.
17. HOVLAND, C.I. 'Inhibition of Reinforcement' and Phenomena of Experimental Extinction. *Proc. Nat. Acad. Sci.*, 1936, *22*, 430-433.
18. HOVLAND, C.I. The Generalization of Conditioned Responses. I. The Sensory Generalization of Conditioned Responses with Varying Frequencies of Tone. *J. gen. Psychol.*, 1937, *17*, 125-148.
19. HOVLAND, C.I. The Generalization of Conditioned Responses: II. The Sensory Generalization of Conditioned Responses with Varying Intensities of Tone. *J. genet. Psychol.*, 1937, *51*, 279-291.
20. HOVLAND, C.I. Experimental Studies in Rote-Learning Theory. I. Reminiscence Following Learning by Massed and by Distributed Practice. *J. exp. Psychol.*, 1938, *22*, 201-224.
21. HOVLAND, C.I. Experimental Studies in Rote-Learning Theory. II. Reminiscence with Varying Speeds of Syllable Presentation. *J. exp. Psychol.*, 1938, *22*, 338-353.
22. HOVLAND, C.I. Experimental Studies in Rote-Learning Theory. III. Distribution of Practice with Varying Speeds of Syllable Presentation. *J. exp. Psychol.*, 1938, *23*, 172-190.

23. HOVLAND, C.I. Experimental Studies in Rote-Learning Theory. IV. Comparison of Reminiscence in Serial and Paired-Associate Learning. *J. exp. Psychol.*, 1939, *24*, 466-484.
24. HOVLAND, C.I. Experimental Studies in Rote-Learning Theory. V. Comparison of Distribution of Practice in Serial and Paired-Associate Learning. *J. exp. Psychol.*, 1939, *25*, Dec.
25. HOVLAND, C.I. Experimental Studies in Rote-Learning Theory. VI. Comparison of Retention Following Learning to Same Criterion by Massed and Distributed Practice. *J. exp. Psychol.* (in press).
26. HOVLAND, C.I. Experimental Studies in Rote-Learning Theory. VII. Distribution of Practice with Varying Lengths of Lists. (In prep.)
27. HULL, C.L. The Formation and Retention of Associations Among the Insane. *Amer. J. Psychol.*, 1917, *28*, 419-435.
28. HULL, C.L. The Meaningfulness of 320 Selected Nonsense Syllables. *Amer. J. Psychol.*, 1933, *45*, 730-734.
29. HULL, C.L. The Influence of Caffeine and Other Factors on Certain Phenomena of Rote Learning. *J. gen. Psychol.*, 1935, *13*, 249-274.
30. HULL, C.L. The Conflicting Psychologies of Learning - A Way Out. *Psychol. Rev.*, 1935, *42*, 491-516.
31. HULL, C.L. Mind, Mechanism, and Adaptive Behavior. *Psychol. Rev.*, 1937, *44*, 1-32.
32. HULL, C.L. The Problem of Stimulus Equivalence in Behavior Theory. *Psychol. Rev.*, 1939, *46*, 9-30.
33. JOST, A. Die Assoziationsfestigkeit in ihrer Abhängigkeit von der Verteilung der Wiederholungen. *Z. Psychol.*, 1897, *14*, 436-472.
34. KATTSOFF, L.O. Philosophy, Psychology and Postulational Technique. *Psychol. Rev.*, 1939, *46*, 62-74.
35. LANGER, S.C. *Introduction to Symbolic Logic.* New York: Houghton Mifflin, 1937.
36. LEPLEY, W.M. A Theory of Serial Learning and Forgetting Based upon Conditioned Reflex Principles. *Psychol. Rev.*, 1932, *39*, 279-288.
37. LEPLEY, W.M. Serial Reactions Considered as Conditioned Reactions. *Psychol. Monog.*, 1934, *46*, No. 205.
38. LUH, C.W. The Conditions of Retention. *Psychol. Monog.*, 1922, *31*, No. 142.
39. LYON, D.O. The Relation of Length of Material to Time Taken for Learning, and the Optimum Distribution of Time. *J. educ. Psychol.*, 1914, *V*, 1-9; 85-91; 155-163.
40. MARGENAU, H. Probability, Many-Valued Logics, and Physics. *Philos. Sci.*, 1939, *6*, 65-87.
41. MEUMANN, E. *The Psychology of Learning* (Trans. by J.W. Baird from 3rd German edition). New York: Appleton, 1913.
42. MÜLLER, G.E. and PILZECKER, A. Experimentelle Beiträge zur Lehre vom Gedächtniss. *Zsch. f. Psychol.*, Erg. Bd. 1, 1900.
43. MÜLLER, G.E. and SCHUMANN, F. Experimentelle Beiträge zur Untersuchung des Gedächtnisses. *Zsch. f. Psychol.*, 1894, *6*, 81-190.
44. NAGEL, E. Principles of the Theory of Probability. *International Encyclopedia of Unified Science.* (Vol. I, No. 6) Chicago: University of Chicago Press, 1939.
45. NEWTON, I. *Principia* (Trans. by F. Cajori). Berkeley: University of California Press, 1934.
46. PATTEN, E.F. The Influence of Distribution of Repetitions on Certain Rote-Learning Phenomena. *J. Psychol.*, 1938, *5*, 359-374.

47. PAVLOV, I.P. *Conditioned Reflexes* (Trans. by G.V.Anrep). Oxford: Oxford University Press, 1927.

48. PAVLOV, I.P. *Lectures on Conditioned Reflexes* (Trans. by W.H.Gantt). New York: International Publishers, 1928.

49. RADOSSAWLJEVITCH, P.R. *Die Behalten und Vergessen bei Kindern und Erwachsenen nach experimentellen Untersuchungen*. Leipzig: Nemnich, 1907.

50. RAFFEL, G. Two Determinants of the Effect of Primacy. *Amer.J.Psychol.* 1936, *48*, 654-657

51. REICHE, F. *The Quantum Theory*. New York: E.P. Dutton, 1930.

52. RIETZ, H.L. *Handbook of Mathematical Statistics*. New York: Houghton Mifflin, 1924.

53. ROBINSON, E.S. and BROWN, M.A. Effect of Serial Position upon Memorization. *Amer.J.Psychol.*, 1926, *37*, 538-552.

54. ROBINSON, E.S. and HERON, W.T. Results of Variations in Length of Memorized Material. *J.exp.Psychol.*, 1922, *5*, 428-448.

55. RODNICK, E.H. Characteristics of Delayed and Trace Conditioned Responses. *J.exp.Psychol.* 1937, *20*, 409-425.

56. RODNICK, E.H. Does the Interval of Delay of Conditioned Responses Possess Inhibitory Properties? *J.exp.Psychol.*, 1937, *20*, 507-527.

57. SHERRINGTON, C.S. *The Integrative Action of the Nervous System*. New Haven: Yale University Press, 1906.

58. SHIPLEY, W.C. The Effect of a Short Rest Pause on Retention in Rote Series of Different Lengths. *J.gen.Psychol.*, 1939, *21*, 99-117.

59. SIMLEY, O.H. The Relation of Subliminal to Supraliminal Learning. *Arch.Psychol.*, 1933, No. 146.

60. SWITZER, S.A. Backward Conditioning of the Lid Reflex. *J.exp.Psychol.* 1930, *13*, 76-97.

61. SWITZER, S.A. Disinhibition of the Conditioned Galvanic Skin Response. *J.gen.Psychol.*, 1933, *9*, 77-100.

62. THORNDIKE, E.L. *Mental and Social Measurements*. (2nd edition) New York: Teachers College, Columbia University Press, 1913.

63. WARD, L.B. Reminiscence and Rote Learning. *Psychol.Monog.*, 1937, *49*, No. 4.

64. WARD, L.B. Retention over Short Intervals of Time. New Haven: Ph.D. Thesis deposited in Yale University Library, 1934.

65. WHITEHEAD, A.N. and RUSSELL, B. *Principia Mathematica*. (2nd edition) Cambridge University Press, 1925.

66. WOODGER, J.H. *The Axiomatic Method in Biology*. Cambridge University Press, 1937.

67. YOUTZ, A.C. Functional Equivalence at Comparable Points in Learning and Forgetting: An Experimental Evaluation of Jost's Law. New Haven: Ph.D. Thesis deposited in Yale University Library, 1937.

68. YOUTZ, R.E.P. The Change with Time of a Thorndikian Response in the Rat. *J.exp.Psychol.*, 1938, *23*, 128-140.